Medicine and Biomedical Sciences in Modern History

Series Editors
Carsten Timmermann
University of Manchester
Manchester, UK

Michael Worboys
University of Manchester
Manchester, UK

The aim of this series is to illuminate the development and impact of medicine and the biomedical sciences in the modern era. The series was founded by the late Professor John Pickstone, and its ambitions reflect his commitment to the integrated study of medicine, science and technology in their contexts. He repeatedly commented that it was a pity that the foundation discipline of the field, for which he popularized the acronym 'HSTM' (History of Science, Technology and Medicine) had been the history of science rather than the history of medicine. His point was that historians of science had too often focused just on scientific ideas and institutions, while historians of medicine always had to consider the understanding, management and meanings of diseases in their socio-economic, cultural, technological and political contexts. In the event, most of the books in the series dealt with medicine and the biomedical sciences, and the changed series title reflects this. However, as the new editors we share Professor Pickstone's enthusiasm for the integrated study of medicine, science and technology, encouraging studies on biomedical science, translational medicine, clinical practice, disease histories, medical technologies, medical specialisms and health policies.

The books in this series will present medicine and biomedical science as crucial features of modern culture, analysing their economic, social and political aspects, while not neglecting their expert content and context. Our authors investigate the uses and consequences of technical knowledge, and how it shaped, and was shaped by, particular economic, social and political structures. In re-launching the Series, we hope to build on its strengths but extend its geographical range beyond Western Europe and North America.

Medicine and Biomedical Sciences in Modern History is intended to supply analysis and stimulate debate. All books are based on searching historical study of topics which are important, not least because they cut across conventional academic boundaries. They should appeal not just to historians, nor just to medical practitioners, scientists and engineers, but to all who are interested in the place of medicine and biomedical sciences in modern history.

More information about this series at
http://www.palgrave.com/gp/series/15183

Tinne Claes

Corpses in Belgian Anatomy, 1860–1914

Nobody's Dead

Tinne Claes
Department of History
University of Leuven
Leuven, Belgium

Medicine and Biomedical Sciences in Modern History
ISBN 978-3-030-20114-2   ISBN 978-3-030-20115-9   (eBook)
https://doi.org/10.1007/978-3-030-20115-9

Cover illustration: INTERFOTO / Alamy Stock Photo

This Palgrave Macmillan imprint is published by the registered company Springer Nature
Switzerland AG.
The registered company address is: Gewerbestrasse 11, 6330 Cham, Switzerland

ACKNOWLEDGEMENTS

There is a common theory why writers, and perhaps especially those writers interested in cheerful subjects such as death and the dead body, are often depressed. Writers think a lot and people who think a lot tend to be sad. Add to that long periods of isolation behind a computer screen and the existential crises that come with the pursuit of an academic career, and it seems obvious that writing a book on corpses in late nineteenth-century anatomy is not the best way to ensure your personal happiness.

In the past five years, I have thought about death on a daily basis. I attended a dissecting course voluntarily. I described a behind-the-scenes visit to a crematorium as the highlight of a stay abroad (and this was sincere). I got used to the piercing smell of formaldehyde and alcohol hanging around anatomical museums, and was even allowed to take a few anatomical specimens out of their jars because I was curious to know how preserved flesh felt. I spent months in archives, where the discovery of a complaint on a leaking coffin could make my day.

Believe it or not, through all of this I have been surrounded with people who shared my interests. This book started its life at the University of Leuven, as a PhD dissertation for the research project 'Anatomy, Scientific Authority and the Visualised Body in Medicine and Culture' funded by the University of Leuven and the Research Council of Flanders. I loved being part of this project: the fact that I was able to discuss the history of anatomy in extraordinary detail not only made my research a less lonely experience, but also broadened my views. My first thanks are to my supervisors, Kaat Wils, Raf de Bont and Jo Tollebeek, for their stimulating criticism and relentless support. Without Kaat, this study would not exist: not only because her keen insight and immense knowledge have shaped my own ideas about nineteenth-century science and society, but also because her meticulous feedback on all my academic output has always given me the confidence to carry on while pushing me to become a better scholar. I would like to thank Raf for raising my thinking about the history of science to the next level on many occasions, and Jo for always surprising me with his eye for historical detail.

Veronique Deblon was there when we, two awkward historians, attended a dissecting course for medical students. She was also there in anatomical museums, at conferences and in the archives. I could not have wished for a better doctoral companion and co-author. I am also grateful to Pieter Huistra, Geert Vanpaemel, Sokhieng Au and Jolien Gijbels, who have all strengthened my work in their own way. The members of my research group, Cultural History since 1750, have helped me with insightful feedback and delicious homemade food on lunchbreak seminars. Joris Vandendriessche and Kate Kangaslahti in particular have advanced my research with thought-provoking conversations and comments.

Outside of the safe confinements of the Leuven history department, various academic encounters have helped me to refine and revise my views. I wrote parts of this book at Durham University, during a research stay at two centres with resonant names: the Centre for the History of Medicine and Disease and the Centre of Life and Death Studies. I would like to thank Andreas-Holger Maehle in particular for his warm welcome. Many others have helped me compose my thoughts by showing interest, asking questions, correcting mistakes, giving advice or simply listening. I cannot name all of them, but I would like to mention at least some: Nick Hopwood, Tatjana Buklijas, Hieke Huistra, the participants of the Bodies Beyond Borders conference we organised in Leuven in 2015, the members of the Belgian Network for Medical History, and all archivists and curators who have made this study possible.

The Greek philosopher Epicurus, someone who truly thought a lot, once said that the art of dying well came down to the art of living well. Only living can prepare us to die, and only death can help us to live. The genesis of this book has reminded me of what is important, of what it is that ultimately keeps me from insanity or despair. I could have never finished this research without the support of my friends and family, who make this finite life very much worth living. Margot, Marthe, Sara, Paulien, Stephanie, Valerie: thank you for being my dearest friends. My parents Greet and Jan, brother Bert, and aunt Josfin: thank you for your self-evident support. My grandmother Fien proudly would have added this book to her personal family archives, even though she did not understand any English. My final word of thanks goes to Wederik, who has done literally everything in his power to make my life easier. And he has succeeded. Writing might be known as a depressing experience, but thanks to him, these have been my happiest years so far. I am eager to see what life has in store for us, till death do us part.

CONTENTS

ABBREVIATIONS

ASSB   Archives of the Social Services of Brussels
ASSG   Archives of the Social Services of Ghent
ASSL   Archives of the Social Services of Leuven
BSAP   *Bulletin de la société anatomo-pathologique de Bruxelles*
CAB   City Archives of Brussels
CAL   City Archives of Leuven
SAB   State Archives Brussels
UAB   Archives of the Free University of Brussels
UAG   University Archives of Ghent
UAL   University Archives of Liège

List of Figures

CHAPTER 1

Introduction

Nobody's dead. This phrase, the title of this book, assumes multiple meanings. To start with, it refers to an absence of ownership or belonging with regard to the corpse. The particular matter of the dead body seems to be both indefinable and indeterminate: a conceptual and social no man's land. When death occurs, the body ceases to be a subject, a person, in the strict sense of the word. And yet, the enduring referential relationship to the deceased still prevents the corpse from fully becoming an ownable thing: even dead, the body cannot wholly belong to someone else.

Nineteenth-century anatomists took advantage of this uncertain destiny of the dead body. They argued that corpses could be used in the name of science because they did not belong to anybody. Laws stated that anatomists could use the bodies of those who died in state-funded institutions unless first claimed by their relatives, a course of action that implied the family then took on the costs of burial. The unclaimed dead, mostly meaning the poor, could be used for medical education and research because nobody had declared ownership over them. Since they were nobody's, and often nobodies in the pejorative sense of the word, their bodies ended up on the dissecting table.

The title of this book also points at the continued subjectivity and agency of the dead. *Nobody is dead,* because the deceased live on in the memories and actions of the ones they leave behind. Dying does not fix our identities: through the thoughts and practices of those who survive us,

© The Author(s) 2019
T. Claes, *Corpses in Belgian Anatomy, 1860–1914,* Medicine and Biomedical Sciences in Modern History,
https://doi.org/10.1007/978-3-030-20115-9_1

we can still change after death. Human remains can shift both shapes and meanings as a result of the relations we maintain with them. The corpse is therefore seldom a passive object. Because the deceased continues to be present in the flesh, the dead body is more than just matter.

This ambivalence of the corpse sometimes tied the hands of anatomists. In 1866, for example, a certain Paul drew his last breath in the hospital of St. Pierre in Brussels. He was a painter, 34 years old. The head physician of the hospital Edouard Van den Corput wanted to perform an autopsy because he was unsure about the cause of death. Unfortunately for him, Paul's relatives claimed his body, thereby prohibiting dissection. They wanted to pay their last respects. The presence of the coffin (typically an open coffin) during the funeral service was important to them because they were Roman Catholics. They believed that they could improve the afterlife of Paul by praying over his remains and by letting a priest bless his body. After the funeral service, they would guide his coffin to the cemetery.[1]

Van den Corput was not easily discouraged. He thought that a body buried was a body wasted, especially if interesting scientific facts awaited discovery in its hidden depths. Much like other anatomists in Brussels, he saw the hospital as 'an inexhaustible mine of pathological riches'. It was his job to make sure that these resources 'were scrupulously and largely exploited'.[2] Van den Corput descended the stairs to the morgue, where the body was stored until burial. He opened the lid of the coffin and took off Paul's shirt for examination. He found an unusual swelling at the left side of the body and decided to make an incision. When he put his finger in, he felt that the spleen was abnormally voluminous. He widened the incision and removed the organ. Then he closed the wound, washed the body, put Paul's shirt back on and placed the lid back on the coffin. When Van den Corput presented the spleen at a scientific society a few days later, he apologised because he had not been able to perform a more elaborate examination. He explained to his colleagues that 'the subject was claimed by his family'. 'The autopsy had to happen secretly', while 'the corpse was already lying in the coffin'.[3]

One could argue that Van den Corput treated Paul as a thing. He went to the morgue in order to get his hands on a pathological object, regardless of the wishes of the family. When he later presented the spleen to his colleagues, he did not mention Paul's full name. This interpretation would fit with the views of many historians of anatomy, who have implicitly and explicitly described nineteenth-century anatomical practices as instances of objectification. According to the dominant historical narrative, nineteenth-century anatomists eradicated the humanness of the corpse: they turned social subjects into scientific objects. In the case of Van den Corput, however, such a reading offers only half the story. If he had not taken into consideration that Paul was a man with a loving family, he would have performed a more elaborate autopsy. He would not have cared about stitching up the wound.

The case of Paul's spleen is not unique. As the nineteenth century progressed, so this book argues, anatomists in Brussels increasingly took into account the emotionally charged nature of their research 'material'. This shift was grounded in broader developments happening in the city. The dead began to occupy an important place in the political landscape of Brussels from the 1860s onwards. The Liberal city council wanted to take charge of burials, which had traditionally been under the control of the Roman Catholic Church. On the one hand, churchyards were replaced with communal cemeteries, an evolution that had a profound impact on the culture of death. On the other, civil burials (burials without Catholic ceremony) became a means to express and strengthen anticlericalism as a political force. The Church strongly opposed these changes. As a consequence of these heightened ideological tensions, the 'indecent' burial of hospital patients increasingly came into view. Due to the fact that their burial rituals largely revolved around the presence of an open coffin, Roman Catholics began to use the image of the dissected pauper as a critique of the city's new burial policy.

In the 1880s, other evolutions began to force anatomists to change their habits. Brussels went through an unruly period. The industrialised city, with a large class of working poor, became the heart of social protest in Belgium. A bitter struggle for electoral reform went hand in hand with strikes and demonstrations for better working conditions. In an effort to encourage class consciousness, the emergent Socialist movement drew attention to the unequal treatment of the proletariat in public institutions, including hospitals. Involuntary dissection became a symbol of the unfair treatment of the poor. This critique further gained credibility when the social status of hospital patients changed. Better-off patients started to enter hospital wards, mostly in search of safer surgery. Since these paying patients could choose their own treatment and did not have to participate in medical education or research, the true cost of a hospital stay became tangible and easy to understand for the poor. Protest against involuntary dissection surged. The use of the bodies of the poor, against their wishes, became even harder to justify when consent was introduced as a legal principle in 1889.

The compromise of Van den Corput was exceptional in 1866. At that time, all hospital patients were poor. Relatives were mostly unable to claim the bodies of their deceased loved ones because of a lack of financial means. Since poverty in general was regarded as a sin, there was not a lot of compassion for the dead poor: protests against their involuntary dissection were virtually non-existent. Yet by the end of the century, anatomists had to walk a tightrope to get their hands on dead bodies. Different stakeholders entered the field: families, burial societies, clergy, social reform movements, even the dead themselves. To make matters worse, the discipline of anatomy lost prestige, giving medical scientists from other disciplines room to claim their fair share of cadavers, too. To whom did the body belong? Briefly put, that question became a lot more complex.

This book therefore is as much about anatomists as it is about dead bodies and the social, cultural and ethical sensitivities that surround them. It is a cultural history of anatomy, which systematically points at interconnections between different views of the corpse and the practices conducted in anatomical theatres, more specifically in Brussels from approximately 1860 until 1914. The claims of this book, however, go beyond the specifics of this metropolitan case. As many of the events transforming anatomy in Brussels were part of broader developments, this study offers a starting point for research on the organisation and transformation of anatomy across continental Europe in the lesser-studied second half of the nineteenth century.

Most importantly, this book reassesses and questions previous readings of anatomy as objectification. Apart from a few notable exceptions, to which I return in the next pages, historians have interpreted nineteenth-century anatomical practices as practices of objectification, in which the human body was transformed from a subject to an object. By following the sometimes long and complex journeys of human remains—such as Paul's spleen—from acquisition to disposal, this study sheds light on a conflicting evolution in anatomy, one in which the body could retain or regain its subjectivity and individuality. Taken together, the pages of this book reveal that anatomists in Brussels and beyond increasingly reconciled their interests with the meanings the corpse accrued as the remainder of a person.

1   The Corpse as Object and Subject

Working primarily within Anglo-American fields of study, historians have tended to conceptualise anatomy as an archetype of objectification for various reasons. The research of Ruth Richardson, whose book *Death*,

Dissection and the Destitute (first published in 1987) established the history of anatomy as a fully fledged field of research, was among the earliest to link the practice of anatomy to the commodification of the corpse. Richardson posited that the growing importance of dissection in early nineteenth-century Britain transformed dead bodies, once persons, into objects with a market value. In her words, the body became a 'token of exchange, subject to commercial dealing'.[4] Since then many historians have echoed this argument, drawing attention to the lucrative traffic of both dead bodies (for sale as a whole, in pieces or by the inch) and anatomical specimens.[5] Other studies have shown that, once acquired, anatomical corpses and their parts served various ends, ranging from public enlightenment to medical teaching or research.[6] In such cases, dead persons, much like other objects, functioned as tools to fulfil the needs and goals of the living.

Human remains became things not only in the sense that anatomists exchanged and used them like other objects, but also because they were dehumanised and depersonalised. Ample attention has been given to the words of the famous eighteenth-century anatomist and surgeon William Hunter, who believed that dissection imparted the 'necessary inhumanity' to future physicians.[7] In more contemporary terms, one would say that dissection was key to the development of clinical detachment. In order to become doctors guided by reason instead of emotion, medical students had to be able to distance themselves from the person they were dissecting. By means of the cadaver they were to learn how to disregard the humanness of the many bodies they would later encounter in their careers.

Historians also have pointed at more literal renderings of dehumanisation. Helen MacDonald, for instance, considered dissections as objectifying because persons were disintegrated into a 'mess of fats, fluids, tissues and bones' without form or identity.[8] In the words of Elizabeth Hurren, 'dissection and dismemberment cut away all identifiable dignity'.[9] Historians and museologists studying anatomical preparations have also underlined the removal or transformation of identity. In their view, the fragmentation and conservation of the body went hand in hand with a loss of individual identity, turning persons into tokens of anatomical skill, examples of pathologies or even works of art.[10]

Often these processes of dehumanisation have been seen as an important step in the production of a specific kind of medical knowledge. In line with Nicholas Jewson's influential essay about the 'disappearance of the sick man', objectification in these instances concerned the supersession of bedside medicine by a kind of medicine on the research of tissues and cells, and not on the narrative of the patient.[11] According to historian John Pickstone, the object of medical science, at least in Europe, became 'the body rather than the discursive patient'.[12] From the late eighteenth century onwards, the clinical techniques of pathological medicine subjected the patient to a new form of objectification: physical examinations, autopsy and statistics were all methods that treated the person as an object without identity.[13]

Here, objectification signified the reduction of patients to bodies and their parts, a process quintessential for the construction of 'modern' medical knowledge. Michel Foucault was the first to argue that the scientific vocabularies of anatomy and pathology allowed physicians to turn medicine into a field of knowledge that was both universal and impartial. By classifying tissues and diseases, the subjective patient could become a neutral object of research. According to Foucault, and many scholars after him, anatomy was the foundation of a constructed, analysable body that could be known without the mediation of the person.[14] To gain scientific insight into fertility, for example, anatomists and gynaecologists reduced individual women to their reproductive organs.

In summary, on the one hand historians' analysis of anatomy has dovetailed with aspects of well-known definitions of objectification provided by social philosophers, who coined the term to describe the treatment of persons as things.[15] On the other hand, objectification has stood for the impersonal nature of nineteenth-century medical knowledge. Through mostly implicit associations between objectification as a social and a scientific process, historians have been able to link the scientific agenda of anatomy to its social ramifications, leading to important insights for the social history of medicine. Only in a medical community in which physicians aspired to be scientists rather than philanthropists could the pursuit of knowledge outweigh the preferences of the patient. Only in a social setting in which the doctor was dominant could patients be reduced to their bodies. And only in a context in which the progress of knowledge was regarded as more important than the wishes of the sick could persons end up on the dissecting table against their wishes.

This book, however, argues that these implicit and explicit readings of anatomy as objectification, though valuable, are incomplete and should be carefully historicised. First and foremost, this study contends that anatomy did not necessarily entail the social objectification of the patient. Throughout the chapters of this book, it will become clear that the corpse in anatomy began to transform from a socially disconnected 'thing' to a social subject around the turn of the twentieth century. The wishes of the patient and his or her family grew in significance in Brussels and the rest of continental Europe. In particular when anatomists could no longer claim the corpse unilaterally, they had to reconcile their practices with sensibilities surrounding the body as the remainder of a person. In addition, this study indicates that the patient did not necessarily disappear or lose his or her individuality when becoming an object of research. As a result of the increased importance of pathology and the pursuit of clinically relevant knowledge, individual case histories gained importance in late nineteenth-century anatomy. This book complicates and nuances both social and scientific objectification and the associations between them. It is about placing both the treatment and the meaning of the dead body in their specific historical constellations, and about tracing their evolution over time.

In order to do so, this book builds on reflections from anthropology and material culture studies that stress the ambiguous status of human remains. Drawing on comparative insights into how human material is treated in contemporary biomedicine, scholars such as Helen Lambert and Maryon McDonald have put forward that the meanings of the corpse are always unstable because they are the result of a continuous negotiation between differing interpretations of the body and of death.[16] Human remains can shift from subject to object and vice versa, depending on the relations we have with them: a corpse can simultaneously be 'a deceased subject' for relatives and friends and a 'dead object' for the pathologist.[17] This polysemy of the dead body in practice leads to compromises that attempt to accommodate the goals and sensitivities of several stakeholders. For example, medical scientists, in an effort to allay relatives' objections, might use imaging techniques that allow for post-mortem research without actually cutting into bodies. In the case of Paul's spleen, Van den Corput limited his examination to a part of the body that could be covered up with a shirt in order to circumvent the family's objection.

The corpse is, in other words, constructed by interactions with the living. These interactions in turn are affected by the dead body's material conditions, occurring naturally (for example, by decomposition) or culturally (by cremation, for instance). As early as 1907, the sociologist Robert Hertz linked the social identity of the deceased to the materiality of the corpse. In Hertz' view, death, which he saw as a social process, took time: in certain societies it was only completed when decomposition had ended. Only when the flesh had disappeared did the deceased cease to belong to this world.[18] More recently, Elizabeth Hallam has emphasised that human remains' 'material features, forms and significance emerge through inter-actions with them'.[19] While we process a corpse, for instance in burial ritu-als or in the anatomical theatre, both its form and substance alter: human remains incite different responses dependent on their materiality. For example, teachers often introduce medical students to anatomical studies by working with bones and wet anatomical specimens. By handling these more palatable kinds of human remains, students are able to prepare themselves for dissection.

Building on these insights, the corpse in this book figures as a particular kind of material culture that is both historically and materially fluid, with different meanings and shapes dependent on the context.[20] Both social and material forces determined the trajectory of the corpse through the spaces of anatomy. Social situations—students learning how to dissect, researchers examining tissues under the microscope or families paying their last respects to the deceased—and material realities—the corpse in decomposition, rendered stiff by rigor mortis, or fragmented into pieces—*co-produced* ideas about and practices with the body.[21] In a similar vein, this book considers the dead as active in two ways: socially, because the will of the deceased influenced the post-mortem treatment of his or her body (for example, when family members complied with his or her last wishes) and materially, because the physical conditions of the body deter-mined its treatment (for instance, decomposed corpses are handled differ-ently than embalmed bodies).[22]

Of course, this book is not the first historical study to draw on these kinds of interdisciplinary reflections on the body. In recent research, histo-rians of anatomy have highlighted the 'interpretative flexibility' of ana-tomical objects such as preparations and models.[23] For instance, Rina Knoeff has drawn attention to the various functions of early modern ana-tomical specimens, focusing on the embalmed bodies of Siamese twins prepared by the Dutch anatomist Frederik Ruysch. This preparation at the same time offered scientific knowledge to a medical audience and consola-tion to the parents, who could visit the cabinet. Dependent on the beholder, it was primarily an anatomical object (offering insight into a medical condition) or an emotional subject (a palpable memory of deceased children).[24] Anna Maerker, too, has emphasised the polysemy of anatomical objects when exploring the diverse reactions of medical profes-sionals and lay audiences to wax models of the body in enlightened Europe.[25] In a more general fashion, Samuel Alberti has suggested that anatomical specimens are scientific 'boundary objects' with different meanings in different social settings. In his view, precisely the human nature of anatomical specimens renders a complete and linear scientific objectification difficult, leaving room for alternate, non-scientific, inter-pretations by lay audiences.[26]

Yet the corpses themselves, the raw material necessary for anatomical practices and representations, received less attention in this kind of research, despite their profound influence on the development of anatomy as a science. Dead bodies enable and shape anatomical practices, both as objects and subjects. To name but a few examples, the sex, age, race and pathology of the corpse at the researcher's disposal affect his or her focus and findings, and the availability of bodies may drive the development of medical disciplines. Neurological institutions, for instance, were often connected to mental health hospitals because of the possibility to conduct post-mortems on the brains of psychiatric patients.[27] Furthermore, the material reality and emotional significance of the corpse turned dissections into affective practices of unusual intensity. In the words of Rafael Mandressi, 'the inherent affects of the manipulation of dead bodies fully comprehend the ways in which anatomical learning is produced and thus influence how the latter is shaped'.[28] The disgust evoked by decomposing corpses, for example, allowed anatomists to represent themselves as heroes who strengthened their character in order to gain knowledge, and pre-cisely because their discipline relied on an emotionally charged kind of research material, anatomists had to circumvent or accommodate the wishes of relatives who wanted to pay their last respects.

This book therefore places the dead body at the forefront in all its com-plexity. The novelty of this study is that it aims to show how the interpreta-tive flexibility of the corpse, constantly hovering on the edge between object and subject, influenced anatomy as a discipline *from acquisition to disposal.* Unlike previous research, this book reconstructs the complete journey of the corpse, from the moment the patient drew his or her last breath until the last bits and pieces of his or her body were inhumed or disposed of in other ways. How did the diverse sensitivities surrounding the body and its parts influence the ways in which anatomists held their scalpel? And, the other way around, how did the ways in which anatomists cut and/or divided the body determine its meanings in the following stages of its post-mortem existence?

By relating and comparing diverse post-mortem examinations to each other, this book adds much-needed nuance to the existing historiography.

This study takes into account that a dissected body had different meanings than an autopsied one; that dissection and autopsy implied different incisions in different decades; and that a dissection could still happen after an autopsy, but not the other way around. This book acknowledges that it was possible that certain parts of a body were drained off as waste, whereas other parts of the same body were carefully preserved in an anatomical museum for decades, and still others were buried with (or without) dignity. By studying these diverse anatomical practices together, this book clarifies that the body could be objectified for a limited period of time before it turned into a subject again, and that the objectification of certain organs or tissues did not imply that the rest of the body lost its humanness.

Because the aim of this book is to study an array of practices happening to corpses in anatomy in relation to each other, definitional clarity is important. The terms 'post-mortem practice' or 'post-mortem examination' will be used as umbrella terms for all kinds of autopsies and dissections throughout this book. To gain a more profound insight into the cultural history of anatomy, however, it is mostly important to be more precise. Therefore the table below offers working definitions and essential contextual information on the three main actions around which this book revolves: dissection, autopsy and the preparation of anatomical specimens. Distinguishing these various post-mortem practices from each other is crucial, as they served different ends, incited various emotional responses and were regulated differently. Furthermore, they led to different 'corporeal realities'[29]: dissected, autopsied and/or preserved bodies or body parts did not look, feel or smell the same.

It is important to note that Table 1.1 only provides *working definitions*, meant to frame the rest of this book's discussion. It will become clear that each of these terms was interpreted in complex ways, depending on who was doing the intervention, for what purpose, and in which time period. To name but a few examples, which will be clarified later in this book: dissection invoked different emotional responses before and after the emergence of consent, and being autopsied in 1850 encompassed different incisions than being autopsied in 1900. Ultimately, the aim of this book is to shed light on the unstable and interdependent meanings, scope and functions of various post-mortem examinations: to look at how distinctions between them came about, evolved over time and worked in practice.

Throughout the chapters of this book, it will become clear that the humanness of their research material deeply affected anatomists from acquisition to disposal. This study indicates that dead patients' integrity,

Table 1.1 Key terms and definitions

Term	Definition	Reputation	Legal framework	Corporeal reality
Dissection	Two kinds of dissections were usual in nineteenth-century Belgium. During *systematic dissections*, muscles, vessels, nerves and internal organs were dissected separately according to their bodily system (for example, the skeletal or cardiovascular system). During *topographic dissections*, the body was studied in diverse regions (mostly head, neck, torso, arms and legs) emphasising relations between various structures (such as muscles, nerves and arteries) in that region.	*Social*: Dissection was a social stigma for two reasons: because it was a sign of poverty, and since it was as opposed to regular burial customs. Associations between dissection and criminality existed, but were less strong than in Anglo-American countries. *See* Chaps. 3 and 5. *Scientific*: Dissection was mostly seen as an educational practice in the nineteenth century. *See* Chap. 2.	Unclaimed bodies of persons who died in state-funded institutions could be used for dissection from 1798 onwards until circa 1890, and claiming a body implied that you would pay for the funeral costs. This meant that the poor could not prevent the dissection of a loved one who died in hospital. *See* Chap. 3.	Dissection meant the fragmentation of the body to its extremities, implying extensive cutting and dismemberment. All that was left to bury was a pile of bones denuded of the flesh, which weighed no more than one fourth of the intact body. *See* Chap. 5.

(continued)

Table 1.1 (continued)

Term	Definition	Reputation	Legal framework	Corporeal reality
Autopsy	Whereas autopsy originally meant to look at or observe the surface of a corpse externally (*surface anatomy*), it became increasingly invasive throughout the nineteenth century. After capital punishment, criminals were often autopsied to check if they had actually died (*anatomisation*): a cross-shaped incision was made down the torso and across the chest for examining the major organs in order to check whether someone was medically dead. During *clinical autopsies* in hospitals, corpses were opened up, often through the same cross-shaped incision, in order to search for the cause of death (and hence check the diagnosis). So-called '*complete' autopsies*, which became usual in Belgian hospitals in the late nineteenth century, were more extensive: all body cavities and organs were examined and dissected systematically in order to gain scientific insights on the development of diseases. Note that in all these cases of 'autopsy', organs could be removed and/or preserved for further studies. *See* Chap. 4.	*Social*: Autopsies in general had a better reputation than dissections because they were not opposed to common burial rituals. In fact, the autopsy was still a common procedure for aristocratic and royal families in Belgium in the nineteenth century. *See* Chap. 4. *Scientific*: Autopsies were regarded as scientifically valuable procedures, for both clinical (diagnostic) and scientific purposes. *See* Chap. 4.	Before circa 1890, only unclaimed bodies could be autopsied, although limited autopsies sometimes happened clandestinely. Around 1890, city councils across Belgium gave anatomists the right to autopsy both claimed and unclaimed bodies, as long as the body could be returned 'decently' after the procedure. *See* Chap. 4.	For most of the nineteenth century, autopsies were limited in scope: in most cases, only the organs related to the original clinical diagnosis were checked. Complete autopsies, however, changed the corporeal reality of the autopsied cadaver: all organs could be removed and dissected. Unlike in the case of whole-body dissections, however, bodies were restored after the procedure. Because corpses were sewn back together, and face and hands were left untouched, autopsies were invisible for the untrained eye. *See* Chap. 4.

(continued)

Table 1.1 (continued)

Term	Definition	Reputation	Legal framework	Corporeal reality
Anatomical specimen/ preparation	Anatomical 'specimens' or 'preparations' are preserved body parts. They are used as synonyms in this book, because nineteenth-century anatomists in Belgium seem to have used both terms interchangeably.	*Social:* People were mostly not informed about the preservation of body parts. Still, archival evidence suggests that the preservation of body parts strongly associated with the deceased (such as face, hands, brain) was more sensitive than the preparation of, for instance, tumours. *See* Chap. 5. *Scientific:* The preservation of body parts was regarded as an important scientific endeavour when it considered rare pathological conditions and when clinical information was also kept. Body parts were also kept for educational reasons, mostly for limited periods of time. *See* Chap. 5.	There were no specific laws on the partial preservation of bodies in Belgium. In Brussels, anatomists did have to ask the city council for permission to preserve whole bodies in scientific collections. *See* Chap. 5.	The terms 'specimen' and 'preparation' were used for a diverse range of preserved body parts, including bones and skeletons ('dry' specimens), body parts or foetal bodies kept in preservation fluids ('wet' specimens) and anatomical models made (partially) of artificial materials such as wax and papier-mâché.

autonomy and unicity became more important to anatomists in Brussels in the late nineteenth and early twentieth centuries. Two developments in particular inspired this change: the individualisation of burial rituals and the emergence of medical ethics based on the principle of consent. Even though both topics have been thoroughly studied for nineteenth-century Europe, historians of anatomy have rarely integrated them in their research. It is the ambition of this book to show the importance of these fields for the history of anatomy. Together with the meanings of death and deontology, the significance of the corpse appears to have changed thoroughly—a shift that introduced a turning point in the practice of anatomy.

2   The Individualisation of Death

By the end of the eighteenth century, the idea arose that the dead were a danger for the living. Influenced by miasmatism (the belief that disease was spread by bad air), the problem of overcrowded churchyards entered the political agenda. In bigger cities across Europe, scientists and politicians drew attention to the peril of burying decomposing and stinking bodies in churchyards in the heart of populous districts. They made the case for a more hygienic disposal of corpses: something that needed to happen quickly since the dead kept accumulating at an ever greater rate as cities became more densely populated. These reformers gained the upper hand politically in the second half of the nineteenth century. The result was the replacement of overcrowded churchyards with large communal burial grounds outside of the city. The dead ceased to be buried one atop or alongside the other in soil already saturated with the products of putrefaction, and were put to rest in vast and clean extramural cemeteries instead.[30]

This rise of the cemetery had repercussions far beyond the realm of public health. The foundation of large extramural burial grounds allowed for the replacement of temporal, anonymous burial places by marked graves devoted to an individual or a family, which relatives could visit for several decades. In their pursuit of a swift 'decomposition of the dead without injury of the living', sanitary reformers pursued to bury but one body in each grave.[31] This created new opportunities for commemoration: individual graves came to reflect a culture of death that was more personal.[32] A new attitude towards death emerged, characterised by what Philippe Ariès took to be excessive grief and over-elaborate displays of memory and mourning; it was individualistic rather than communal, quotidian rather than transcendental. Even though the dead were still remembered collectively on special occasions such as holy days, the personal bounds between the dead and their descendants gained importance through increasingly individual commemorative practices. Against this background, the marked grave became a venue for continuing personal relationships with the deceased.[33]

The body played an important role in this new culture of death. Sentimental pre-burial rituals often revolved around the presence of the corpse. In the words of archaeologist Sarah Tarlow, the dead body became 'a place where personal relationships were (re)presented'.[34] Various physical manipulations of the body helped the bereaved to express their relationship with the deceased. In rituals such as the laying out of the dead and the viewing of the body, death became aestheticised. Corpses were rendered suitable for visual display through clothing or make-up, and through the invention of technologies such as embalmment and post-mortem photography. The face of death changed: looking as if they were merely asleep, dead bodies were more readily reminiscent of the deceased. One could argue that the corpse, much like the marked grave, became a means to continue personal bounds with the dead.[35]

Around the same time, the funeral started to express the identity of the deceased. The reason for this was twofold. Firstly, this was part of a process of commercialisation. With the advent of cemetery companies and burial societies, death became a booming business. A decent funeral became a sign of respectability, dividing the successful from the poor. Luxurious coffins, large burial plots and stately tombstones (or precisely the absence of them) reflected the class identity of the deceased.[36] Even for corpses, class and business mattered: embalmment methods, for instance, were patented for funerary contexts and commercialised.[37] Secondly, the burial gained importance as an articulation of ideological denominations. The funeral ritual echoed conflicts between different political stances and religious beliefs, most importantly between Catholics and anticlerical forces.[38] For instance, in his research on French death customs, Thomas Kselman has argued that the funeral service became a way to show respect towards the person the deceased once was. A civic burial, for example, strengthened the identity of the deceased as a freethinker, whereas the bequest of one's body to science could express a belief in the progress of science or a materialist conviction.[39]

One would think that this new culture of death, with its emphasis on the individuality of the deceased, could not have been more opposed to the practices of nineteenth-century anatomists. Various historians have pointed at the disgrace of the anatomy burial as one of the main reasons for popular protest against dissection in the nineteenth century. After all, so they argued, the anatomy burial was the complete opposite of the respectable burial: medical faculties claimed and buried the bodies of the poor regardless of their wishes and without taking their respectability into account.[40] However, this study indicates that anatomists in Brussels took over certain aspects of the new culture of death in the face of growing protest in the late nineteenth century. As they tried to dissociate their practices from the dishonourable burial, their treatment of the corpse changed profoundly, both during and after dissection.

3   MEDICAL ETHICS AND CONSENT

The second half of the nineteenth century was characterised not only by changes in the culture of death, but also by the transformation of medical ethics, which gradually became based on the principle of consent rather than on the authority of the physician. Historians have shown that many European countries introduced consent as a legal principle in the late nineteenth century. Court decisions questioned the traditional paternalism of physicians, obliging them to seek consent for medical interventions and experiments, and to inform the patient adequately before treatment.[41] These requirements gradually found their way to medical advice literature and to the practice of medicine, despite the vagueness of legal instructions concerning the extent of patient information and the widespread acceptance of tacit consent as an authorisation for medical treatments.[42]

As the century progressed, the question of consent further gained importance as a result of developments in medical research. Experiments on living subjects gradually became a methodological requirement for research. Combined with the late nineteenth-century self-confidence of the medical profession, this increased demand for clinical trials led to a growing number of experiments on hospital patients, who were regularly subjected to new methods or treatments.[43] Especially in Germany, known for its early adoption of experimental methods in medicine, this resulted in abuses and scandals. The Neisser case, a trial against a dermatologist who had injected healthy patients with syphilis in the hope of developing a vaccine, incited the first parliamentary legislation on medical experimentation in 1900. The edict, which prohibited nontherapeutic experiments without consent, inspired more stringent legal and political attitudes towards unauthorised medical interventions in continental Europe.[44]

Significantly, the social composition of the hospital changed around the same time. The introduction of health insurance, medical innovations (most importantly the use of antiseptics) and closer contacts with universities started to draw paying patients to hospital wards. As the hospital transformed from a charitable institution for the poor into a prestigious medical centre for all, the subjection of the patient's body to an authoritarian medical regime became problematic. This added to the pressure on physicians to take the consent of the patient seriously. Hospital patients gradually got a say in the way their bodies were treated.[45]

Again, these developments seem at odds with the practices of nineteenth-century anatomists. Historians have shown that dissections mostly happened without the consent of the patient or his or her relatives. Bodies were often transported to the anatomical theatre without informing the family of the deceased. Much like medical experiments, the dissection of the poor for most of the nineteenth century was seen as a 'natural reciprocity', as a justified compensation for state-funded or charitable hospital care.[46] Yet this book argues for the first time that the emergent principle of consent did begin to affect the workings of anatomical theatres in the decades around 1900. The patient's autonomy grew not only during life, but also after death.

4   BRUSSELS AND BEYOND

In short, this book researches if anatomists in Brussels treated the corpse as object or subject from approximately 1860 until 1914. While much of historiographical attention has been given to the first half of the nineteenth century, when the poor were forced to give up their bodies, historians, so this book argues, should start looking at the other half of the century. The decades around 1900 might prove to be a pivotal moment in the social history of anatomy. As the cultural significance of the individual became more important through broader developments in society such as the transformation of the culture of death and the rise of consent, the corpse in anatomy changed from an object into a subject. A shift to altruism began to characterise the practice of anatomy.

The bulk of this history takes place in Brussels, although the other Belgian university cities are integrated in the analysis when they are helpful for highlighting the particularities of the Brussels case. I have chosen Brussels as the main emphasis of this study for various reasons. To begin with, very little is known about the organisation of anatomy in nineteenth-century Belgium. Apart from articles discussing the careers of individual anatomists, the history of anatomy mostly is confined to brief discussions within books about medical faculties or universities. Existing studies focus exclusively on anatomical research and education, paying little or no attention to broader—political, social, religious—contexts.[47] Only a handful of publications have considered anatomy from a different perspective, paying attention to topics such as the architecture of anatomical institutes, the link between anatomy and the arts, or the exhibition of anatomical models on fairgrounds.[48] By offering the first fully fledged study of anatomy in Brussels, this book fills a gap in Belgian historiography.

More importantly, the choice for one in-depth case study follows from Tatjana Buklijas' call to 'take national and regional differences seriously' with regard to the history of death and dissection.[49] Brussels forms a vital case study in a field that is still largely dominated by Anglo-American perspectives. An impressive historiography exists on the context and implementation of the British Anatomy Act of 1832, which was in force not only in Britain, but also—with adaptations—in many of its (former) colonies, including the United States.[50] However, the relatively rare studies on medical centres in continental Europe suggest that key narratives in this Anglo-American historiography, including on grave robbing and on popular protest against dissection, do not apply for contexts with different political systems and cultures of death.[51]

By carefully relating and contrasting original research on Brussels to other national and regional cases, this study first and foremost tries to redress the privileging of Anglo-American perspectives. In addition, this book contributes to the existing literature on anatomy in continental Europe. Until now, historians have mainly looked at important medical centres, such as Paris, Vienna and Leiden.[52] Whereas these cities were undeniably the most influential educational and research centres, crucially shaping modern medicine in continental Europe and beyond, second-rank capitals such as Brussels were far more numerous. As the first encompassing study on such a city, this book aims to offer a point of reference for historians studying anatomy in comparable cities across the continent.

Moreover, Brussels was a particularly interesting second-rank city. It was perhaps not a real metropolis, but was nevertheless a thriving and important place, both locally and within a European context. The city was, after all, the capital of one of the major industrial powers worldwide. By 1900, only Britain had a more impressive gross national product than Belgium, and Brussels was the Belgian city in which wealth had grown the most. Due to its prosperity and central location within Europe, Brussels also was an important cultural centre, attracting artists and intellectuals from near and far.[53]

In other areas, too, Brussels was a city with aspirations. It became the beating heart of a new country in 1830, when Belgium gained independence from the Netherlands. Both the municipal and national authorities had big plans for the city, which had to become a capital worthy of that name. Of particular importance was the Free University of Brussels, which was established with financial support from the Liberal city council in 1834. The municipal authorities wished to turn Brussels into a prestigious university city, which would be able to compete not only with older Belgian universities, but also with important universities abroad. When it came to medical education, their efforts were not without result. In the first fifty years of the university's existence, more than one third of all graduates came from abroad, mostly from Britain and France.

In his recent study on the Brussels medical faculty, Renaud Bardez has suggested that these foreign students were attracted by the abundance of hospital patients—living, dying and dead—on whom they could practise their skills.[54] Close relationships between the hospital and the university characterised medical education in the Belgian capital. Since the end of the eighteenth century, medical education had become practice-based, with an emphasis on internships and clinical courses within public hospitals.[55] In the last third of the nineteenth century, hospitals became part of the city council's 'project of modernity'. The progressive Liberal Charles Buls, who was mayor of Brussels from 1881 until 1899, wanted to turn hospitals into centres of medical research.[56] This central position of the hospital along with the importance of clinic-based education is a welcome contribution to the existing historiography that because of its focus on the Anglo-American case has given much attention to private anatomy schools and the allocation of bodies from the workhouse to anatomists.[57] Conversely, as Fiona Hutton has rightly pointed out, the role of the hospital post-mortem in anatomical teaching and research continues to be poorly studied until today.[58]

The profusion of destitute hospital patients was the result of exponential population growth combined with deep social inequality. Brussels was among the first cities to be hit by the wave of urbanisation that swept over

Europe from the 1830s onwards.[59] The mid-size town rapidly transformed into a jumble of half a million inhabitants. As the nineteenth century progressed, the downsides of urbanisation became increasingly visible. Karl Marx' ideas for the *Communist Manifesto* matured while living in Brussels in the 1840s. He saw that the vibrant city was a paradise for the rich, but a living hell for what he came to call the proletariat. Protests for universal suffrage and social redistribution continued into the late 1890s, and became increasingly violent. Approximately a quarter of Brussels inhabitants were listed as destitute, and hence were at risk of being dissected if they died in hospital.[60]

Most crucially, Brussels forms a well-suited case study to scrutinise the impact of changing burial rituals and medical ethics on anatomy. Because Brussels was a stronghold of Liberal rule, its urban politics were characterised by occasionally fierce ideological disputes. These heightened tensions were manifest in hospitals, where the daily care for patients rested almost entirely in the hands of nuns, and where most patients were Roman Catholics. The outspoken Liberalism of the city council and the university regularly clashed with their beliefs and practices, especially when it came to burials. Through these conflicts, the Brussels case brings to the fore the stance of the Roman Catholic Church, offering an interesting counterpoint to the substantial scholarship on the attitudes of the Anglican (Protestant) Church towards dissection. With regard to medical ethics, Belgium was an early adopter of the principle of consent, which a court decision introduced in 1889.[61] In the following decades, several foreign examples—mostly from Britain, Germany and France—influenced the codification and reinterpretation of the principle of consent, a process that went hand in hand with a reassessment of the legal status of the corpse.

5   Sources and Composition

This book draws on a vast range of sources in order to recover the impact of the death beliefs and practices of different actors on the treatment of the corpse in anatomy. Most importantly, this book is based on extensive research of the rarely used and exceptionally conserved archives of the Social Services of Brussels. Containing all documentation of nineteenth-century hospitals in the city, ranging from meeting reports and regulations to building plans of anatomical theatres and inventories of scientific collections, these archives offer a unique insight into nineteenth-century anatomy. The patient's voice is present in these archives through correspondence and complaints.

In order to clarify the relation between hospitals and universities, or between medical care and medical research, documents preserved in university archives and scientific publications have been analysed. The meeting reports of scientific societies with a special interest in anatomy, most notably the Society for Pathological Anatomy of Brussels (*Société anatomo-pathologique de Bruxelles*), were also important in this respect. Administrative documents preserved in city archives and the minutes of the meetings of the Brussels city council were used in order to shed light on topics such as the transport and burial of dissected and autopsied bodies, the conservation of human remains in scientific collections and the construction of morgues. To get a grasp on public outrage and political conflicts, this study draws on the databases Belgica Press (covering local and national newspapers with different political leanings) and Plenum (containing all parliamentary discussions). Lastly, religious, legislative and deontological works were used for elucidating important contexts for the practice of anatomy.

Apart from an introduction and conclusion, this book consists of four chapters. The introductory chapter builds a picture of the institutional and disciplinary structure of Brussels anatomy and its personnel through the period studied. The other three research chapters are written as an object biography that discusses the life story of anatomical remains with an emphasis on their changing materiality and meaning. These chapters take the reader on a journey through the spaces of anatomy. They reconstruct the sometimes complex trajectories of the corpse: one chapter follows persons from deathbed to dissecting table, considering anatomists' turn from involuntarily acquired to gifted bodies; another looks at the treatment of the corpse in the post-mortem room with a special emphasis on the differences between dissection and autopsy; and the final chapter tracks dissected and autopsied remains until they reach their final destination, whether this was a jar or a coffin. These wide-lens chapters begin with narrative case histories that closely follow the journeys of particular corpses at specific moments in time.

This book reveals that anatomists' treatment of the corpse changed profoundly in the course of the second half of the nineteenth century, under the influence of social policies, ideological tensions, religious sensitivities, cultures of death and broader changes in the field of what we would now call medical ethics. Anatomists increasingly had to reconcile their ways with the diverse meanings that the dead body gathered as the remainder of a person. To a certain extent, so this book argues, they started to treat the corpse as subject rather than object.

NOTES

1. *BSAP* 11 (1866): 3–5.
2. Quotes from *BSAP* 4 (1861): 2.
3. *BSAP* 11 (1866): 5.
4. Ruth Richardson, *Death, Dissection and the Destitute* (London and New York: Routledge and Kegan Paul, 1987), 72.
5. For example: Michael Sappol, *A Traffic of Dead Bodies: Anatomy and Embodied Social Identity in Nineteenth-Century America* (Princeton and Oxford: Princeton University Press, 2002), 98–135; Helen MacDonald, *Possessing the Dead: The Artful Science of Anatomy* (Carlton, VIC: Melbourne University Press, 2010), 74–93; Elizabeth T. Hurren, *Dying for Victorian Medicine: English Anatomy and its Trade in the Dead Poor, c.1834–1929* (Basingstoke and New York: Palgrave Macmillan, 2012); Dániel Margócsy, *Commercial Visions: Science, Trade and Visual Culture in the Dutch Golden Age* (Chicago: University of Chicago Press, 2014).
6. For example: Anna Maerker, "The Anatomical Models of La Specola: Production, Uses and Reception," *Nuncius: Istituto e museo di storia della scienza* 21, no. 2 (2006): 295–321.
7. For example: Ruth Richardson, "A Necessary Inhumanity?," *Medical Humanities* 26, no. 2 (2000): 104–6.
8. Helen MacDonald, "A Body Buried is a Body Wasted: The Spoils of Human Dissection," in *The Body Divided: Human Beings and Human 'Material' in Modern Medical History*, eds. Sarah Ferber and Sally Wilde (Farnham: Ashgate, 2012), 11.
9. Hurren, *Dying for Victorian Medicine*, 30.
10. Samuel J.M.M. Alberti, *Morbid Curiosities: Medical Museums in Nineteenth-Century Britain* (Oxford: Oxford University Press, 2011), 96 and 141–2; Rina Knoeff and Rob Zwijnenberg, "Setting the Stage," in *The Fate of Anatomical Collections*, eds. Rina Knoeff and Rob Zwijnenberg (London: Ashgate, 2015), 3–6; Lisa O' Sullivan and Ross L. Jones, "Two Australian Fetuses: Frederic Wood Jones and the Work of an Anatomical Specimen," *Bulletin of the History of Medicine* 89, no. 2 (2015): 243–66.
11. Nicholas D. Jewson, "The Disappearance of the Sick-Man from Medical Cosmology, 1770–1870," *Sociology* 10, no. 2 (1976): 225–44.
12. John V. Pickstone, *Medicine and Industrial Society: A History of Hospital Development in Manchester and Its Region 1752–1946* (Manchester: Manchester University Press, 1985), 48.
13. Jewson, "The Disappearance of the Sick-Man," 225–44. See also: David Armstrong, "Bodies of Knowledge/Knowledge of Bodies," in *Reassessing Foucault: Power, Medicine and the Body*, eds. Colin Jones and Roy Porter (London and New York: Routledge, 1998), 17–27.

14. Michel Foucault, *The Birth of the Clinic. An Archaeology of Medical Perception*, trans. A.M. Sheridan (New York: Routledge, 1973), 124–48.
15. Most importantly Martha Nussbaum, "Objectification," *Philosophy and Public Affairs* 24, no. 4 (1995): 249–91.
16. Helen Lambert and Maryon McDonald, "Introduction," in *Social Bodies*, eds. Helen Lambert and Maryon McDonald (Oxford and New York: Berghahn, 2011), 1–15.
17. John Troyer, "Embalmed Vision," *Mortality* 12, no. 1 (2007): 22–47; Cara Krmpotich, Joost Fontein, and John Harries, "The Substance of Bones: The Emotive Materiality and Affective Presence of Human Remains," *Journal of Material Culture* 15, no. 4 (2010): 371–84.
18. Robert Hertz, *Death and the Right Hand*, 2nd ed. (Oxford: Routledge, 2004), 35.
19. Elizabeth Hallam, "Articulating Bones: An Epilogue," *Journal of Material Culture* 15, no. 4 (2012): 465.
20. John R. Sofaer, *The Body as Material Culture: A Theoretical Osteoarchaeology* (Cambridge: Cambridge University Press, 2006), xiii-2.
21. On the theory of co-production: Luigi Pellizoni, "Construction, Co-production, and Beyond. Academic Disputes and Public Concerns in the Recent Debate on Nature and Society," *Sociology Compass* 8, no. 6 (2014): 851–64.
22. In other words, corpses have both material (as things) and abducted human agency (as persons) in interactions with the living. See also: Krmpotich, Fontein and Harries, "The Substance of Bones," 373.
23. For more examples, see: Kaat Wils, Raf de Bont, and Sokhieng Au, eds., *Bodies Beyond Borders: Moving Anatomies, 1750–1950* (Leuven: Leuven University Press, 2017).
24. Rina Knoeff, "Touching Anatomy: On the Handling of Preparations in the Anatomical Cabinets of Frederik Ruysch 1638–1731," *Studies in History and Philosophy of Biological and Biomedical Sciences* 49, no. 1 (2015): 32–44.
25. Anna Maerker, *Model Experts. Wax Anatomies and Enlightenment in Florence and Vienna, 1775–1815* (Manchester: Manchester University Press, 2011).
26. Alberti, *Morbid Curiosities*, 164; Idem, "Objects and the Museum," *Isis* 96, no. 4 (2005): 559–71.
27. Elizabeth T. Hurren, "Abnormalities and Deformities. The Dissection and Internment of the Insane Poor, 1832–1929," *History of Psychiatry* 23, no. 1 (2012): 65–77.
28. Rafael Mandressi, "Affected Doctors: Dead Bodies and Affective and Professional Cultures in Early Modern European Anatomy," *Osiris* 31, no. 1 (2016): 119–36, on p. 120.

29. I borrow this phrase from Hurren, *Dying for Victorian Medicine*, 67–8.
30. Thomas W. Laqueur, *The Work of the Dead: A Cultural History of Mortal Remains* (Princeton and Oxford: Princeton University Press, 2015), 388–412. See also: Régis Bertrand, *Mort et mémoire. Provence, XVIIIe-XXe siècles, Une approche d'historien* (Marseille: La Thune, 2001), 21–56. On Belgium: Christophe De Spiegeleer, "Secularisering van stedelijke begraafplaatsen in de tweede helft van de negentiende eeuw in België," in *R.I.P. Aspecten van 200 jaar begrafeniscultuur in Vlaanderen*, ed. Tamara Ingels (Ghent: Liberaal Archief, 2015), 5–21.
31. Julie Rugg, "Constructing the Grave: Competing Burial Ideals in Nineteenth-Century England," *Social History* 38, no. 3 (2013): 328–45, on p. 329.
32. Laqueur, *The Work of the Dead*, 215–38.
33. Philippe Ariès, *Western Attitudes Toward Death From the Middle Ages to the Present*, trans. P.M. Ranum (Baltimore: John Hopkins University Press, 1974), 55–84.
34. Sarah Tarlow, "The Aesthetic Corpse in Nineteenth-Century Britain," in *Thinking through the Body: Archaeologies of Corporeality*, eds. Yannis Hamilakis, Mark Pluciennik, and Sara Tarlow (New York: Kluwer Academic/Plenum Publishers, 2001), 85–95, on p. 94.
35. Julie M. Strange, *Death, Grief and Poverty in Britain 1870–1914* (Cambridge: Cambridge University Press, 2005), 66–97; Troyer, "Embalmed Vision," 22–47.
36. On the United States: Sappol, *A Traffic of Dead Bodies*, 34–8. On France: Thomas A. Kselman, *Death and Afterlife in Modern France* (Princeton: Princeton University Press, 1993), 222–56. On Belgium: Christophe De Spiegeleer, "Sterven, begraven en herdenken van koninklijke en politieke elites in België tussen 1830 en 1940: een culturele en politieke geschiedenis" (PhD diss., Vrije Universiteit Brussel, 2016), 121–7, 296–309, and 316–18.
37. For example: Troyer, "Embalmed Vision," 22–47; Anne Carol, *Embaumement: une passion romantique, France XIXe siècle* (Lyon: Champ Vallon, 2016), 95–127.
38. Christopher Clark and Wolfram Kaiser, eds., *Culture Wars: Secular-Catholic Conflict in Nineteenth-Century Europe* (Cambridge: Cambridge University Press, 2003).
39. Kselman, *Death and the Afterlife*, 125–64.
40. For example: Thomas W. Laqueur, "Bodies, Death, and Pauper Funerals," *Representations* 1, no. 1 (1983): 109–31; Richardson, *Death, Dissection and the Destitute*, 275; Sappol, *A Traffic of Dead Bodies*, 36.
41. On the history of consent in the German lands: Barbara Elkeles, "The German Debate on Human Experimentation between 1880–1914," in

Twentieth Century Ethics of Human Subject Research, eds. Volker Roelcke and Giovanni Maio (Stuttgart: Franz Steiner Verlag, 2004), 18–33; Andreas-Holger Maehle, *Doctors, Honour and the Law. Medical Ethics in Imperial Germany* (Houndmills: Palgrave Macmillan, 2009), 69–94. On France and Britain, see: Emmanuel Betta, "Between Law and Profession: The Origins of Informed Consent (1840–1900)," in *Doctors and Patients. History, Representation, Communication from Antiquity to the Present*, ed. Maria Malatesta (San Francisco: University of California Medical Humanities Press, 2015), 108–33.

42. Maehle, *Doctors, Honour and the Law*, 84–95.
43. On nineteenth-century medical experimentation, see: William Bynum, "Reflections on the History of Human Experimentation," in *The Use of Human Beings in Research. With Special Reference to Clinical Trials*, eds. Stuart F. Spicker et al. (Dordrecht: Springer, 1988), 29–46; Marion Maria Ruisinger, "Geschichte des Humanexperiments. Zur Entwicklung der Forschung am Menschen," in *Standards der Forschung. Historische Entwicklung und ethische Grundlagen klinischer Studien*, eds. Andreas Frewer and Ulrich Schmidt (Frankfurt am Main: Peter Lang Verlag, 2007), 19–35.
44. Elkeles, "The German Debate," 22–8.
45. Alfons Labisch, "From Traditional Individualism to Collective Professionalism: State, Patient, Compulsory Health Insurance and the Panel Doctor Question in Germany 1883–1931," in *Medicine and Modernity: Public Health and Medical Care in Nineteenth- and Twentieth-Century Germany*, eds. Manberg Berg and Geoffrey Cocks (Cambridge: Cambridge University Press, 1997), 18–34; Carl Havelange, "L'hôpital à la croisée des chemins: la question des malades payants," *Annales belges d'histoire des hôpitaux et de la santé publique* 25, no. 1 (1987): 83–94; Keir Waddington, "Unsuitable Cases: The Debate over Outpatient Admissions, the Medical Profession and Late-Victorian London Hospitals," *Medical History* 42, no. 1 (1998): 26–46.
46. See Chap. 3.
47. For example: Stéphane Louryan, "Un Portrait des Enseignants d'Anatomie humaine à l'Université Libre de Bruxelles," *Revue Médicale de Bruxelles* 29, no. 1 (2008): 63–9; Carl Havelange, "Rupture ou Continuité? La Création de la Faculté de Médecine de l'Université de Liège en 1817," in *Regards sur 175 ans de Science à l'Université de Liège, 1817–1992*, ed. Anne-Catherine Bernes (Liège: Université de Liège, 1992), 42–52.
48. Guido Everaert et al., "Het Anatomisch Instituut in het Bijlokehospitaal te Gent," *Stadsarcheologie* 11, no. 2 (1997): 4–21; Chaké Matossian, ed., *Art, Anatomie: trois siècles d'évolution des réprésentations du corps* (Brussels: Part de l'œil, 2007); Chloé Pirson, *Corps à corps: les modèles anatomiques entre art et médecine* (Paris: Mare et Martin, 2009).

49. Tatjana Buklijas, "Cultures of Death and Politics of Corpse Supply: Anatomy in Vienna, 1848–1914," *Bulletin of the History of Medicine* 82, no. 3 (2008): 570–607, on p. 573.

50. For example: Richardson, *Death, Dissection and the Destitute*; Hurren, *Dying for Victorian Medicine*; Sappol, *A Traffic of Dead Bodies*; MacDonald, *Human Remains: Dissection and Its Histories* (New Haven and London: Yale University Press, 2006); Idem, *Possessing the Dead*; Fiona Hutton, *The Study of Anatomy in Britain, 1700–1900* (London: Pickering and Chatto, 2013).

51. Buklijas, "Cultures of Death," 570–607.

52. Emmanuelle Godeau, *"L'esprit de corps": sexe et mort dans la formation des internes en médecine* (Paris: Editions de la Maison des sciences de l'homme, 2007); Buklijas, "Cultures of Death," 570–607; Idem, "Public Anatomies in Fin-de-Siècle Vienna," *Medicine Studies* 2, no. 1 (2010): 71–92; Hieke Huistra, *The Afterlife of the Leiden Anatomical Collections: Hands On, Hands Off* (London and New York: Routledge, 2019).

53. Gita Deneckere, *1900: België op het breukvlak van twee eeuwen* (Tielt: Lannoo, 2006), 19; Jane Block, ed., *Belgium, the Golden Decades (1880–1914)* (New York: Peter Lang, 1997).

54. Renaud Bardez, "La Faculté de médecine de l'Université Libre de Bruxelles: entre création, circulation et enseignement des savoirs, 1795–1914," (PhD diss., Université Libre de Bruxelles, 2016), 172–3.

55. Ibid., 190–254.

56. Charles Buls and Marcel Bots, *Het dagboek van C. Buls* (Ghent: Liberaal Archief, 1987), 103–4.

57. For example: Helen MacDonald, "Procuring Corpses: The English Anatomy Inspectorate, 1842 to 1858," *Medical History* 53, no. 3 (2009): 379–96; Hurren, *Dying for Victorian Medicine*.

58. Hutton, *The Study of Anatomy*, 12.

59. Anneleen Arnout, *Streets of Splendour: Shopping Culture and Spaces in a European Capital City (Brussels, 1830–1914)* (London and New York: Routledge, 2019), 9.

60. Eric Min, *De eeuw van Brussel: biografie van een wereldstad, 1850–1914* (Amsterdam and Antwerpen: De Bezige Bij, 2015), 12–13.

61. Julie de Ganck, "De verzorging van het vrouwelijk geslacht: een maatschappelijke kwestie?," *Historica* 16, no. 2 (2014): 9–16.

CHAPTER 2

Anatomy Is Done?

In 1865, the Royal Academy of Medicine of Belgium (*Académie royale de médecine de Belgique*) held a competition in which participants were asked to describe the recent history of the medical sciences in Belgium. In the winning essay, the Brussels physician Léon Marcq paid much attention to developments in surgery and obstetrics, as well as to recent research in the fields of ophthalmology (a branch of medicine that deals with the structure, functions and diseases of the eye), epidemiology and parasitology. In Marcq's opinion, the importance of these medical disciplines could not be overestimated, as they would 'turn medicine into a science rather than an art'.[1] Conversely, anatomy, which had been the catalyst for medical progress in the previous century, was not as promising anymore. 'As the necessary foundation for the others', it remained an important part of medical education, but its scientific potential was exhausted. 'Descriptive anatomy is not a science in which you can discover much anymore. *It is done.*'[2]

Marcq's assessment of anatomy fits with the views of medical historians, who have shown that anatomy ceased to be 'the star of the sciences of life' in the nineteenth century.[3] In the long eighteenth century—from the days of Andreas Vesalius in the mid-sixteenth century until circa 1800—anatomy had been quintessential not only to the progress of medicine but also for understanding Creation as a whole. Historian Andrew Cunningham has convincingly argued that 'old anatomy' was not confined to dissection but had many facets, including the study of bodily functions and patho-

© The Author(s) 2019
T. Claes, *Corpses in Belgian Anatomy, 1860–1914*, Medicine and Biomedical Sciences in Modern History,
https://doi.org/10.1007/978-3-030-20115-9_2

logical alterations (of dead human and animal bodies, and of live animal bodies).[4] Around 1800, however, anatomy's position and scope as a discipline altered as a result of the demise of natural philosophy (meaning the study of nature as God's Creation) and the rise of specialisation. These changes heralded an end to the 'old anatomy'. The discipline was not the major science of life anymore, pushed to the background by disciplines such as physics, chemistry and experimental physiology.[5]

However, anatomy emerged in new forms. No longer a study of Creation and the divine, it gradually became a medico-scientific pursuit of death, dearth and disease. In the crowded hospitals of late eighteenth-century Paris, physicians began to follow patients from the sickbed to the autopsy table. By systematically relating symptoms (observed at the bedside) with pathological alterations in corpses (seen during autopsies), they tried to locate diseases in particular places within the body that could be empirically scrutinised. Since an intimate knowledge of the healthy body was a prerequisite for recognising what was abnormal or diseased, both dissections and autopsies were quintessential.[6]

In the course of creating 'Paris hospital medicine', however, pathology based on the seats of diseases transformed from an *anatomical* into a *clinical* subject. Eighteenth-century pathological anatomists' approach had been absolutely anatomical, primarily concerned with the inspection of the dead body, working back from morbid lesions to the patient's complaints during life. The new Parisian approach worked the other way around: it began from signs observed at the bedside and worked forwards, aiming for confirmation in the post-mortem room. In the eighteenth century, a lesion discovered post-mortem pointed back to a symptom; in the nineteenth, a sign pointed forwards to an internal lesion. Visible facts continued to be fetishised, but anatomy was no longer the end point of the life sciences.[7]

Anatomy continued to transform in the second half of the nineteenth century, when the laboratory replaced the post-mortem room as the centre of medical research.[8] This shift went hand in hand with the emergence of what John Pickstone has called a new 'way of knowing'. Around 1850, an analytic-comparative type of knowledge, in which scientists described and classified objects and phenomena through their division into deeper structures and functions (the basic principle of dissection), was surpassed by the ideal of controlling life phenomena through experimentation. While the dead body and its components were a favoured object of analytic-comparative research, the new experimental way of knowing mainly relied on animal experimentation.[9] Around the same time, the discipline of anatomy was reclassified as a dead and static science. Precisely because the shift in what counted as scientific medicine—from empirical observation to experimentation—went hand in hand with a reduction of the scope of anatomy, anatomists were left scrambling to salvage the prestige of their discipline. To come back to Marcq's words, their scientific endeavour was increasingly considered 'done'.

Curiously enough, anatomy strengthened its position within academic medical curricula in the same period. In Belgium, the law on higher education of 1876 turned anatomical exercises into an obligatory part of the medical training.[10] The reasons for this were twofold. On the one hand, dissections were meant to train students' hands and characters so that they would become skilled and cool-headed physicians. On the other hand, microscopic anatomy echoed the increasingly scientific agenda of universities. Microscopic exercises in anatomy introduced students to laboratory research, and therefore reflected the wish to educate medical scientists rather than practitioners. As the basis of both clinical and research practices, anatomy cut across late nineteenth-century debates on the purpose of universities, which gradually evolved from 'professional schools' to research centres.[11]

This introductory chapter discusses the changing status of anatomy, with a focus on the intellectual landscape of Brussels and the period after 1850. What role did anatomy play in the clinic, in medical research and in university education, and how did this evolve under the influence of emergent scientific ideals and medical disciplines? First, this chapter sheds light on anatomy's role in the clinic, focusing on the introduction and evolution of the 'Parisian model' in Brussels hospitals (Sect. 1). Clinicians continued to participate in the field of anatomy and left their mark on the methods through which it was studied and taught. The second section of this chapter pays attention to the scientific status of anatomy (Sect. 2). When the idea that anatomy was an auxiliary discipline became more and more prevalent in the second half of the nineteenth century, various anatomists tried to maintain scientific status by integrating new methods and theories into their work, and became key figures in the establishment of new medical disciplines. Lastly, this chapter shows how these new clinical and research approaches to anatomy were translated into medical education (Sect. 3).

1   A LOYAL SERVANT: ANATOMY IN THE CLINIC

Historians have situated the roots of modern medicine in Parisian hospitals, where, as aforementioned, a clinical approach to anatomy developed in the late eighteenth century. Three simple yet effective innovations improved clinicians' diagnoses: palpation (feeling the body), auscultation (hearing the body) and autopsy (looking inside the body). By following patients from diagnosis to deathbed to autopsy table, physicians started to relate abnormalities, determined during life, with pathological alterations in corpses.[12] Anatomy quickly became the bedrock of the clinic: not only did physicians test and correct diagnoses after death, they also relied on the knowledge they acquired during post-mortems to diagnose living patients by comparing their symptoms with results from previous autopsies. In Michel Foucault's much-cited words, this centrality of anatomy led physicians to gradually replace the question 'what is the matter with you?' with the question 'where does it hurt?'[13]

The idea of localising diseases was not entirely new. In his magnum opus *On the Seats and Causes of Disease* (1761), Giovanni Battista Morgagni, today regarded as the founding father of pathological anatomy, had already anchored pathologies in organs. The novelty of early nineteenth-century pathology was the scale: diseases were increasingly situated in tissues rather than organs. Tissue pathology was, in essence, the privileging of a taxonomy of layers within the body as the primary seats of disease, and initially was an eighteenth-century product developed by well-known medical scientists such as John Hunter and Matthew Baillie.[14] Tissue pathology only found its ultimate institutional setting in the crowded hospitals of post-Revolutionary Paris. The abundance of patients—living, dying and dead—allowed Xavier Bichat to develop a more clinical approach to anatomy, bridging the gap between the external signs of disease and their internal manifestations on a large scale. The most important innovation of 'Paris hospital medicine' was that it merged an anatomical approach with a clinical one: the dead body was used to diagnose and understand the living. The Parisian tradition came to be marked by attempts to situate diseases in tissues, together with a systematic clinical observation of large numbers of patients, sometimes by means of statistical methods.[15]

Although these innovations were rather disappointing from the patient's point of view as they did not have any direct therapeutic effect, they did permit physicians to sketch more authoritative portraits of diseases. Whereas in humoral medicine disease had been viewed as a disruption of balances, affecting the body as a whole, pathologies were now located in particular places within the body, allowing for empirical observation.[16] This possibility to relate signs to lesions gave medicine an objective foundation that bolstered physicians' scientific authority. Their clinical sight, meaning their capacity to link visible pathologies with medical symptoms, allowed them to represent themselves as scientists.[17] The power of anatomy, in part, rested on its illumination of previously opaque spaces in the body; additionally, relating someone's pain to a specific space inside the body was a way to construct meaning in the face of suffering.[18]

Parisian medicine also signified a specific kind of medical training that revolved around clinical courses and dissection (which ceased to be a public spectacle, becoming a closed event for medical students instead). By alternating bedside teaching with dissecting courses, clinical diagnosing was linked to pathological anatomy. In the dissecting room, students became acquainted with the structure of the healthy body, after which they formed an image of pathological alterations by means of specimens and autopsies. Generally speaking, education in Parisian hospitals, often referred to as the *Ecole pratique*, marked a move towards a more practical education based on hands-on experience rather than theoretical knowledge.[19]

Foreign students, who flocked to Paris in great numbers to gain their own experience with hospital patients and cadavers, soon exported this model to all major cities of Europe.[20] As France occupied the Southern Netherlands (now Belgium) from 1795 to 1815, it is perhaps no wonder that the Parisian approach to anatomy already made its entry into large cities such as Brussels, Antwerp, Ghent, Leuven and Liège at the very beginning of the nineteenth century. Leading figures of emergent medical schools (*écoles de médecine*) had studied in Paris[21] or decided to go there in order to be able to copy the system at their return. One of them was the Brussels physician Jean-Baptiste Terrade, who had offered private dissecting courses in his own home since 1799, leading to complaints from neighbours on 'the disgusting view of cadaveric remains and their pestilent emanations'.[22] Terrade returned home with a Parisian degree in 1804; one year later, he founded a medical school together with Jean-Baptiste Van Mons and Antoine Curtet. With the public hospital of St. Pierre as their main operating base, they established a practice-based education after the French example.[23]

Public hospitals in other large cities of the Southern Netherlands, too, became the heart of medical education in this period. Often building on existing relations with private teachers such as Terrade, medical schools and public hospitals developed a strong, symbiotic relationship. The best students, those who distinguished themselves during a specially organised exam, received internships, allowing them to gain the hands-on experience they sought after. They were divided into two categories: 'externs', who assisted with medical treatments during the day, and 'interns'—the most prestigious position—who stayed in hospital permanently. Interns followed up the treatments of patients on a day-to-day basis, often assisted their teachers with the preparation of their lessons and were regularly on duty during the night. In return for this much-needed help from students, public hospitals provided medical schools with 'clinical cases' (patients on whom students could practise their skills) and with corpses for anatomical courses. As a result of this cooperation, city hospitals evolved from care institutions to hubs for medical education and research in the Parisian tradition.[24]

Anatomy was part of a broader, practice-based education within these hospitals. It was taught by clinicians and surgeons, who highlighted the clinical application of anatomical knowledge. The diagnostic value of anatomy was most important. For instance, Curtet, who replaced Terrade as the main teacher of anatomy in Brussels in 1806, encouraged his students to base their diagnoses on tissue pathology, more specifically on the nosography of Philippe Pinel, the chief physician of the Salpêtière hospital.[25] Much like in Paris, where Bichat and his followers taught their students 'general anatomy' (meaning that tissues were units of analysis for anatomy, physiology and pathology), anatomical demonstrations were always linked to the study of function and disease. In the words of the statutes of the Brussels *école de médecine*, the course of anatomy was 'not limited to enumerating and describing the different parts of the human body, but introduces [students] to their uses and functions; simultaneously, it touches upon the different diseases all of them may contract'.[26]

After the defeat of Napoleon in 1815, the Southern Netherlands became part of the Kingdom of the Netherlands. Under this new regime, medical training gradually became centralised in universities, which were (re-)established in Liège, Ghent and Leuven. In these cities, the new medical faculties quickly absorbed the existing medical schools, while maintaining their relationships with public hospitals. The system of interns and externs stayed in place; the number of practical clinical courses augmented.[27]

And nonetheless, criticism on medical education grew. Philippe-Antoine Marcq, a former Parisian student and the father of the physician mentioned in the introduction of this chapter, complained as late as 1821 that medical students had 'to go abroad in search of insights that the current mode of instruction denies them'.[28] The main problem was that the hospitals attached to universities, especially in Leuven, were too small, offering students too few clinical cases and dissecting opportunities. As a result, many physicians thought that the medical training remained overly theoretical.[29]

However, the situation was different in Brussels. As the city did not receive a university due to its proximity to Leuven, the old medical school continued to exist (in part because of its strong relationship with the hospital, which had become dependent on students' help).[30] Because students with scientific aspirations were more likely to opt for university studies, even more emphasis was placed on practice-based education in the clinic.[31] This emphasis on practical knowledge was also reflected in dissecting courses, which had a more prominent place within the curriculum than at universities, both in hours and in intensity. Anatomy remained the basis for clinical diagnosing. As I will show in the next pages, anatomy continued to be grounded in clinical practice in Brussels throughout the entire nineteenth century.[32]

After the outbreak of the Belgian Revolution in 1830, medical education changed again. The law on higher education of 27 September 1835 centralised medical training in universities and dissolved other medical schools. Two state universities, those located in Liège and Ghent, remained in place. In addition, two private universities were established. In line with its historical roots, the University of Leuven, which Pope Martinus V had originally founded in 1425, became a Catholic institution under the oversight of the Belgian bishops. In an effort to transform the city into a capital worthy of that name, Brussels received a university for the first time. In 1834, the Free University of Brussels was founded as a private university of Liberal signature, with financial support from the city council.[33]

As the Belgian government did not have the authority to inspect education in the Catholic University of Leuven and the Free University of Brussels because the freedom of education was an important principle of the new constitution, the organisation of a central exam had to guarantee that students who graduated from different universities had similar competencies and knowledge.[34] Anatomy was an important part of this exam. To become a licensed physician, students had to pass exams on human anatomy and comparative anatomy. Attesting to the Parisian tradition, knowledge of pathological anatomy was part of the exam on 'general therapeutics' due to its diagnostic potential.[35] Even though hands-on dissections were not compulsory until 1876, all universities offered dissecting courses to their students, sometimes in exchange for a supplementary fee per cadaver or body part.[36]

The transition from *école de médecine* to medical faculty went smoothly in Brussels. The relationship with the St. Pierre hospital was maintained and connections with other public hospitals within the city (most notably with the hospital of St. Jean) were established in the following decades. The teaching staff and the scientific collections remained largely the same. The curriculum, too, hardly changed throughout the first half of the nineteenth century: medical training continued to attest to the Parisian example, revolving around the disciplines of anatomy, physiology and pathology.[37] Anatomy and the clinic remained closely knit: anatomy continued to be practised and taught by clinicians and surgeons who only spent part of their time in dissecting rooms. For example, the first Brussels professor of anatomy, Pierre-Joseph Graux, combined his chair with the teaching of clinical courses in the hospital of St. Pierre and continued his work as a surgeon.[38] His successor Jean-Joseph Crocq also was a man of many interests. In the words of one of his successors, Crocq not only aspired to be 'professor of this and that' but also was a private practitioner and a clinician.[39] In Crocq's view, anatomy, first and foremost, 'elucidate[d] clinical studies'.[40]

However, from around 1850, anatomy increasingly came under siege, supplanted by newer, experimental disciplines. In German research centres and universities, fields such as experimental physiology, comparative anatomy, microscopic pathological anatomy and embryology flourished while macroscopic anatomy lost prestige. A new conception of science became grounded in experimentation rather than post-mortem examination; its conceptual stratum was the cell rather than bodily tissues. Tellingly, when the German physiologist Joseph-Antoine Spring, in his capacity as rector of the University of Liège, deplored Belgian students' lack of scientific aspiration in his influential opening speech of the academic year of 1860, he meant that students were not interested in experimentation and only focused on their practice-based education in the clinic. In his view, universities produced medical practitioners, but not scientists. In order to stimulate students' *esprit scientifique*, Spring urged medical professors to pay more attention to 'developing and transmitting good methods of observation and experimentation'.[41] In Spring's speech, the clinic was not presented as the centre of medical science; instead the clinic had become a centre of medical practice.

Historians have argued that anatomists' responses to these changes in the scientific landscape can be divided into roughly two categories. A first and largest group of anatomy teachers, who often were only part-time instructors, viewed anatomy as a stepping stone to more prestigious posts in clinical medicine or surgery. A second group of anatomists, who will take centre stage in the next section of this chapter, primarily identified themselves not as clinicians but as scientists and insisted that anatomy was a science in its own right. They minimised the role of macroscopic anatomy, instead emphasising research in fields such as embryology, histology, comparative anatomy and anthropology. In part, this distinction corresponded with a generation gap, with the second response being more characteristic of younger anatomists.[42]

In Brussels, where connections with the hospital were exceptionally strong, the clinical approach to anatomy persisted. Anatomists defensively asserted that dissection provided students with indispensable knowledge about the structure of the body. In their view, anatomy remained the most certain kind of medical knowledge due to its empirical nature and therefore was 'the method par excellence [...] to escape erroneous judgements and false speculations, causes of so many miscalculations in science and so many dangers in practice'.[43] This does not mean, however, that nothing changed. Laurens de Rooy has shown that anatomists in the Netherlands, a neighbouring country of Belgium, struggled to maintain their status in the second half of the nineteenth century. In the face of criticism on exaggerated specialisation and classification, they underlined the relevance of anatomy for the practice of medicine. Precisely in their efforts to prove their continued importance, Dutch anatomists renewed their discipline.[44] A similar development took place in Brussels: in their quest to prove the practical value of anatomy, clinicians resorted to new methods and approaches.

Most importantly, a new conception of pathology began to rock the foundations of Parisian medicine, which had been rooted in a firm relationship between bedside medicine and tissue pathology. Following the work of the German pathologist Rudolf Virchow, diseases were increasingly situated in cells rather than in tissues or organs. Larger seats of disease were not completely disregarded but were no longer viewed as sufficient. In the words of historian Russell Maulitz, Virchow added a 'new layer of seeing' to pathology, as the gross appearance of diseased organs or disturbed tissues was now overwritten, but not obliterated, by the study of pathological cells.[45] For Virchow and his followers, the microscope was as important as the dissecting knife: both microscopic research and macroscopic analyses were necessary for gaining an understanding of diseases.[46]

At the same time, the adoption of experimental methods turned pathology in a different direction, more closely related to physiology, (micro) biology and chemistry. As early as 1847, Virchow posited that 'pathological anatomy only deals with the products of disease, not with the disease itself'.[47] In his opinion, pathologists had to focus on the physiology of the diseased body, as mere anatomical studies did not elucidate the causal relations between symptoms. Experiments on living organisms were to replace post-mortems because 'only experiments can show a certain occurrence in its dependence on a certain condition'.[48] Yet Virchow's views were still localist, as he searched for the seats of disease within cells. In his research on cancer, for example, the surgical extirpation of pathological body parts had to confirm the presence of cell-based 'clean margins'. Although Virchow understood cancer as a physiological process, the clinical application of his ideas came down to an anatomical localisation of diseased cells.[49]

These Virchovian conceptions of diseases quickly found their way to Belgian universities and hospitals. For the discipline of pathological anatomy, which became a separate chair around 1850, the microscope began to equal the dissecting knife in importance. Professor Gottlieb Gluge, born in Germany and trained at the University of Berlin, introduced microscopic exercises on pathological tissues in Brussels in the late 1840s. He believed that autopsies could only be instruments for pathological research if they integrated microscopic analyses.[50] His successor Guillaume Rommelaere shared this opinion and established a course on microscopic anatomy (*anatomie de texture*) concerning both the healthy and the diseased body. In 1866, after a study trip to England, Rommelaere asked the university board for the foundation of a specialised autopsy service aimed at pathological-anatomical research, on both a macroscopic and a microscopic scale.[51]

This autopsy service, the first pathological laboratory in Belgium, began operations in 1878. The university service was located within the hospital, indicating its strong ties with clinical diagnosing. In the following decade, other Belgian teaching hospitals established similar services, modelled after the Brussels example.[52] These autopsy services had a twofold function: for students, they were an extension of the anatomical theatre; for clinicians, they served as a kind of histological laboratory. Students could encounter a wide variety of pathologies and practise their surgical techniques during autopsies, whereas clinicians could ask specialised pathologists to conduct post-mortem research, also on a cellular level, in order to confirm or refute diagnoses made during patients' lives.[53]

Yet there was almost no room for experimentation. Significantly, when microbiological experiments became more important towards the end of the century, the Free University of Brussels founded new specialised hospital services (such as the Laboratory for Clinical Pathology in 1889) and research institutions (for instance, the Institute for Hygiene, Bacteriology and Therapeutics in 1894).[54] In the Laboratory for Clinical Pathology in particular, researchers were less interested in Virchovian localisation (linking certain diseases to specific malignant cells or bacteria) and occupied themselves with clinical chemistry instead, for example, through the analysis of blood and urine samples. As a result, the autopsy service remained closely linked to the Parisian tradition of 'clinical' pathological anatomy. Microscopic research grew in significance but newer conceptions of pathology found ground elsewhere. In this regard, the Brussels situation was similar to the development of pathology at universities in Austria-Hungary, where pathological anatomy remained an independent discipline as newer approaches were institutionalised at separate chairs.[55]

One of the main research interests in the field of pathological anatomy was the study of cancer and the classification of tumours. In 1860, over half (74 out of 133) of the preparations added to the pathological collection of Brussels were tumours, and tumours continued to be particularly well-represented in consecutive years.[56] Assistants to dissecting courses—such as prosectors (who prepared corpses and specimens for anatomical demonstrations) and lecturers in charge of anatomical demonstrations—often published in the field of cancer research.[57] Rommelaere, too, combined macroscopic and microscopic analyses in his research on diverse types of tumours.[58] Much like in the Parisian tradition, anatomy in these studies essentially remained an exercise in clinical diagnosing: the post-mortem study of tumours determined possible treatments during life, such as surgical techniques that could be used to remove them.

The Society for Pathological Anatomy of Brussels (*Société anatomo-pathologique de Bruxelles*) endured as a stimulus for research based on anatomical localisation throughout the second half of the nineteenth century. This scientific society, which Gluge had founded in 1857, was attached to both the university and the teaching hospitals. At least until 1914, members during bi-weekly meetings discussed anatomical specimens in relation to bedside observations. The importance of pathological anatomy for clinical diagnoses was repeatedly stressed in the society's journal.[59] As late as 1889, in his opening speech of the academic year, the vice president of the society and former professor of anatomy Jean-Joseph Crocq portrayed the discipline as the foundation of empirical medicine:

> It is pathological anatomy that renders medicine positivistic: take it away, and the interpretation of symptoms is groundless. [...] Without the security and the control of pathological anatomy, clinical practices become abstract, incomprehensible and utterly unfruitful.[60]

Crocq's views were widely shared. In general, clinicians in late nineteenth-century Brussels continued to value localist approaches to pathology, on both a macroscopic and a microscopic level. They viewed post-mortem research as the basis of, and as a control mechanism for, newer strands of pathology. Pathological anatomy was seen as a certain foundation on which newer approaches could build, and pathologist-anatomists maintained their position as diagnostic arbiters. As a result, anatomy and pathology partly overlapped, even after the establishment of separate chairs.

Links between anatomy and surgery, too, remained remarkably strong in Brussels hospitals. Unlike in other European countries, dual appointments— that is, professors being employed not only by the university, but also by the hospital—existed in Brussels and other Belgian university cities until the late nineteenth century.[61] Many teachers primarily identified themselves as surgeons and taught anatomy as the servant of surgery. For instance, Louis Deroubaix (professor of anatomy from 1841 until 1894) and his colleague Joseph-Guillaume Sacré combined their chair with a position as a surgeon in the hospital of St. Jean. In the words of Deroubaix, anatomy was 'the uncontested basis of surgical practices', for it trained 'the senses and eyes to perceive [...] the products of abnormal processes'. Additionally, dissections provided an opportunity to train one's hands. In Deroubaix' opinion, surgeries could only be performed with a sure touch after having gained experience with the scalpel in the anatomical theatre. Before handling living bodies, surgeons had to improve their manual skills and cutting techniques on corpses.[62]

This emphasis on the value of dissecting experience for surgery had a long history. Surgeons used to be a separate professional group whose medical training and social standing differed from those of physicians.

Contrary to regular physicians, they traditionally occupied themselves with 'manual work'—everything that involved blood, ranging from amputations to pulling teeth—for which anatomical knowledge was indispensable. Anatomical exercises had been part of the curriculum to become a surgeon in the Southern Netherlands since the early modern period; in fact, physicians only started to compete with them for corpses when anatomy gained prestige in the eighteenth century.[63] Late eighteenth- and early nineteenth-century private teachers, such as Terrade in Brussels, mostly included 'operations of surgery' in the programme of their courses on anatomy.[64]

Anatomy in earlier centuries mainly served as a means to enhance the reputation of surgery, but by the time Deroubaix made his plea for dissection the tables had turned. In the second half of the nineteenth century, anatomists in various countries sought to enhance their reputation by linking their discipline to the work of surgeons. After all, the results and public image of surgeons had improved because of the use of antiseptics and anaesthetics. In his research on Britain, Salim Al-Gailani has shown that anatomists stressed the clinical value of their work to safeguard their positions at medical faculties, for example, by claiming that anatomical representations would do 'the clinician a real service' and allow for safer surgery.[65] In a similar vein, Deroubaix tried to defend the importance of his chair by relating it to surgical practices.

New ways of dissecting and representing the body, such as topographic and sectional anatomy capitalised on surgeons' success. Topographic anatomists studied the body by means of diverse regions (head, neck, arms, legs, etc.) emphasising relations between various structures (muscles, nerves and arteries, etc.) in that region. They argued that this approach allowed for a better prediction of the impact of surgical interventions. A similar argument was made for anatomical cross-sections, a representational technique that gained popularity in the 1870s and 1880s, partly due to the rise of the refrigerator, as it was based on transverse sections of frozen bodies.[66]

In Belgium, teachers who either had started their career as, or still worked as surgeons introduced topographic anatomy at universities. Topographic anatomy became a compulsory part of the curriculum in the last third of the nineteenth century.[67] In this period, dissections following a regional method gained ground because it was believed that they taught students a more practical kind of knowledge, while systematic dissections, in which muscles, vessels, nerves and internal organs were dissected separately according to their bodily system, 'led students to neglect the complete overview and focus on details'.[68] In addition, topographic dissections allowed students to dissect in 'the most economical way' (meaning that fewer corpses were needed), for all regions of the body could be dissected in the course of two semesters.[69]

Manuals of topographic anatomy often included cross-sectional representations of the body because these images made it possible to show mutual relations between different tissues and organs 'truthfully'.[70] By representing the body as if it was sliced into sections, anatomists translated the three-dimensional complexity of the body into flat images that fitted on a book page without the need for drawing in perspective or other technical interventions.[71] Anatomical cross-sections (preserved 'slices' of the human body) also became part of Belgian university collections in the last third of the nineteenth century. For example, Edouard Bouqué, who was responsible for both anatomical and surgical demonstrations at the University of Ghent, proudly reported that the collection of descriptive anatomy contained approximately 130 anatomical cross-sections in 1888.[72] Transverse representations of the body, both preparations and models, also found their way to university collections in Brussels and Liège, yet, in most cases, it is unclear when and by whom.

In spite of their efforts and innovations, however, anatomists lost prestige. The idea that anatomy was in a scientific impasse prevailed. Although nobody denied that anatomy remained quintessential for medical practitioners, the discipline became of secondary importance for medical scientists. This was a result of a more general transformation of the scientific landscape: as a consequence of medical specialisation and the rise of the laboratory, the dissecting room began to exude an atmosphere of tradition rather than progress. By the 1870s, a younger generation of research-minded anatomists described Parisian medicine as 'the old system'. Despite his efforts to link anatomy to surgery, Deroubaix would ultimately be remembered as 'always being a quarter century behind'.[73]

2   A 'Science Accessoire': Anatomy in Medical Research

The mid-nineteenth century was a period of transition for the medical sciences in general and for anatomy in particular. As Western societies rapidly moved in the direction of specialisation of labour, knowledge and expertise, also the field of medicine fell apart into a growing number of disciplines. Medical specialisations developed, focusing on specific populations (such as women, children or the insane), state needs (e.g. public health or forensic medicine), parts of the body (ophthalmology or dentistry, for instance) or methods (like histology and radiology).[74] In the last third of the nineteenth century, these disciplines entered a process of institutionalisation, establishing their own journals, societies and professorial chairs. This troubled anatomists: while medical specialties claimed a growing number of body parts—such as the eye (ophthalmology), the brain (neurology) or the female reproductive system (gynaecology)—anatomy seemed on the verge of being pulled apart by newer disciplines.

At the same time, medicine became grounded in experimental sciences. An experimental way of knowing began to shape the medical sciences.[75] In the second half of the nineteenth century, laboratory medicine based on (experimental) chemistry, physiology and histology surpassed the clinical-anatomical method of science. German universities and research centres, rather than Parisian hospitals, were the most important motors of this evolution.[76] This so-called scientification of medicine, too, worsened the position of anatomy. Whereas macroscopic anatomy was closely linked to tissue pathology, late nineteenth-century pathological research increasingly relied on microbiological analyses.[77] On top of that, the living body continued to gain importance in medical research as a result of the emergence of new techniques of representation and examination, such as the introduction of medical photography in the mid-nineteenth century or the invention of X-rays in 1895.[78]

Against this background, anatomy threatened to be reduced to an auxiliary discipline. In the essay with which this chapter opened, Marcq indeed defined macroscopic anatomy as a *science accessoire* without its own scientific agenda. He thought that anatomical publications focused on 'form rather than content', offering medical students and practitioners helpful representations of the body without adding new insights. This lack of scientific aspiration was related to the belief that the study of anatomy was 'done'. Marcq thought that macroscopic anatomy, one of very few sciences that is regarded as having a finite scope, had ceased to generate new theories and empirical results because anatomists had discovered and described the 'last fibres' in the previous decades. In his view, this explained 'why these professors have not been able to publish anything'.[79]

Marcq was partly right, even though anatomists affiliated to Belgian universities had published and continued to publish around the mid-nineteenth century. Following patriotic aspirations fuelled by competition with the rich anatomical tradition of the Netherlands (from which Belgium had recently gained independence), several Belgian anatomists published treatises in the 1840s and 1850s.[80] Their works were popular and highly valued. In 1862, the Leuven professor of anatomy Etienne Van Kempen even won the award for the best Belgian medical monograph, a prestigious prize that the Ministry of Internal Affairs awarded every five years. According to the jury report, however, the novelty of his prize-winning manual lay less in scientific findings than in its clarity, composition and correctness.[81] Veronique Deblon has shown that the popular works of the Brussels anatomist Constant Crommelinck, too, were appreciated for achieving a clear and useful synthesis rather than for adding new knowledge.[82] This enhanced the view of anatomy as an auxiliary discipline, serving foremost educational purposes.

However, Marcq's idea of medical science did not necessarily correspond with anatomists' views. His judgement followed from an ideal of science that was only beginning to take shape and which was not uncontested. Numerous studies from the last two decades have stressed the heterogeneity of scientific practices and knowledge. Although the idea of the 'scientification' of medicine through processes of experimentation and quantification retains its force, newer scholarship has diversified the plot, illuminating a variety of meanings of science in nineteenth-century medicine.[83] In the words of Michael Hagner, 'science meant different things to different actors in different contexts'.[84] For those anatomists who continued to view the clinic as the heart of medicine—and they were, as we have seen in the previous pages, quite numerous in Brussels—anatomical localisation remained a pillar of what they thought was 'scientific' medicine.

Moreover, by reducing anatomy to dissection and autopsy, Marcq did not recognise that anatomists were a diverse group. In fact, a younger generation of anatomists probably would have agreed with Marcq's ideas about macroscopic anatomy, but they no longer viewed the scalpel as the essence of their discipline. In line with the reconfiguration of the scientific landscape, these scientists integrated new approaches into their field and became (often without knowing it and while still calling themselves anatomists) key figures in the establishment of seceded, specialised branches of research. To put it briefly, new disciplinary boundaries developed only for anatomists to cross, or rather ignore, them.

Similar developments took place elsewhere. In his study on the Netherlands, De Rooy linked the mid-nineteenth-century idea of an anatomical crisis to the drifting apart of physiology (the study of bodily functions) from anatomy (the study of bodily structures). Physiology—as a new, redefined discipline—became a front-runner in the development of experimental medicine, whereas the scientific relevance of anatomy came under question. Yet a few Dutch anatomists reclaimed their status as modern scientists through the incorporation of new approaches—experimental, embryological and evolutionary—in their work.[85] For the United States, Michael Sappol asserted that while gross anatomy ceased to generate new theories or major discoveries, the discipline's progeny—meaning newer disciplines such as anthropology or microbiology—were highly productive.[86] In her masterful study on the disciplinary formation of biology in Germany, Lynn Nyhart argued that the meanings of anatomy in the middle decades of the nineteenth century 'can only be understood by means of its relations to the competing intellectual currents and expanding institutional possibilities of the time'.[87]

The analysis put forward in the next pages builds on these insights. The goal is to do justice to the complexity of anatomy and its relationship with other disciplines in the second half of the nineteenth century. In Brussels and other Belgian university cities, a relatively small group of career academics continued to identify themselves as anatomists while they shifted between emergent disciplines. Disciplinary boundaries between anatomy, physiology, comparative anatomy, embryology and anthropology were permeable, allowing for mutual interactions between scientists, theories and methods. Yet at the same time its seceded disciplines took shape, anatomy was redefined: the establishment of new fields of research went hand in hand with shifts in what the discipline of anatomy was believed to entail.

Anatomy and Physiology

Historians have mostly discussed the development of anatomy and physiology from the point of view of medical specialisation, drawing attention to their gradual growing apart from the eighteenth century onwards. In the second half of the nineteenth century, the foundation of university laboratories devoted to experimental physiology was said to reinforce this process of separation.[88] Physiology came to symbolise the 'scientification' of medicine, as its dependence on quantitative diagnostic methods, chemical analyses, microscopes and other instruments reflected the emergent ideals of quantification, precision and mechanical objectivity.[89] As such, physiology became a model for the medical sciences, even though its methods were never fully applied or accepted in all its aspects, since many practitioners continued to interpret medicine at least in part as an art based on intuition and experience.[90]

While the physiological laboratory began to ensure the scientific character of medicine, the dissecting room became a wholly different world, invested with an aura of faded glory and tradition. In disciplinary histories of physiology, this dichotomy was the starting point of a story of 'emancipation', in which physiologists fought their way to freedom by replacing post-mortems by experiments on living organisms. For instance, the historian Gerald Geison has put forward that 'the liberation of physiology from anatomy' in the 1870s marked the beginning of a new golden age for British physiology. Only when physiology both institutionally and methodologically distanced itself from anatomy, the discipline gained importance for medical research.[91] Historians working on France and Germany, too, related the late nineteenth-century prime of physiology to its separation from anatomy and to its alliance with experimental sciences.[92]

More recently, however, continued associations between anatomy and physiology have come into view. Historians have shown that anatomists sometimes incorporated new physiological methods. The corpse could be interpreted in physiological terms: post-mortem research could be related to processes in the living body rather than to patients' symptoms and signs. The Viennese clinician Josef Skoda, for instance, used experiments on cadavers and patients in order to construct a classification of chest sounds.[93] This kind of post-mortem experimentation led to collaborations between anatomists and physiologists, and occasionally, like in Skoda's research, resulted in clinical applications.[94]

In Belgium, too, the discipline of physiology continued to be linked to anatomy, even though it was institutionalised relatively soon. Already during the French period, in 1806, courses in physiology were separated from anatomy in the medical school of Brussels.[95] In the second half of the nineteenth century, physiologists further distanced themselves from anatomy. Whereas before they believed that the functions of body parts were determined by their form (their anatomical structures), they now started to interpret the body in biochemical terms. They gradually replaced the idea of 'animated anatomy' by a concept of bodily functions that did not depend on anatomical structures. The well-known Liège physiologist Léon Frédericq, for example, argued that physiology had to become 'applied chemistry, physics, mechanics and mathematics' in order to understand bodily functions.[96] Accordingly, physiologists started to rely on new research methods based on the natural sciences. German scientists strongly influenced this shift. Experimental methods were stimulated by German professors appointed in Belgium (such as Théodor Schwann in Liège and Gottlieb Gluge in Brussels), as well as by Belgian physiologists who had stayed at foreign research institutes (such as Richard Boddaert in Ghent and Gustave Verriest in Leuven). In the 1880s and 1890s, the universities of Brussels and Liège founded specialised physiological institutes, modelled after German research centres.[97]

Physiology gradually became a discipline independent from anatomy, with its own research agenda and methods. This did not mean that anatomical studies were disregarded. In 1889, the members of the Society for Pathological Anatomy of Brussels, for instance, asserted that dissection and normal anatomy 'laid the foundation for physiology'.[98] As a kind of next step in anatomical studies, physiological experiments became complementary to anatomical demonstrations. The idea was that anatomy taught students how to make observations (to 'see'), whereas physiology taught them how to conduct experiments themselves (to 'do').[99] In other words: anatomy remained important, but only as a stepping stone to physiology.

The careers and publications of physiologists continued to show close connections with anatomy. Many professors of physiology started their academic lives as prosectors or had been professors of anatomy. A few of them, including Gluge in Brussels, at times, combined both professorial chairs.[100] Scientific contributions to the fields of anatomy and physiology were often written by the same authors.[101] Also spatially, physiological institutes remained closely connected to anatomical theatres. In Brussels and Liège, the construction of a specialised physiological laboratory happened simultaneously with the building of a new anatomical institute, located at the same scientific site.[102]

Furthermore, the corpse continued to play an important role in the research of physiologists, who often conducted experiments on both living and dead bodies. The well-known Brussels physiologist and biologist Paul Héger, for example, studied the circulation of blood (a traditional topic of research for physiologists) in dissected organs to get a grasp on the physiological functions 'situated in the organ itself', detached from the nervous system.[103] In Liège, Fredericq used both vivisections and demonstrations on the corpse to show students the workings of the body. To measure the elasticity of the lungs, for instance, students had to 'insert a glass tube in the windpipe of a human cadaver'.[104] Dead bodies were especially important for the study of the human organism. Because it was not morally acceptable to cut open a living person, researchers often drew conclusions on human physiology from comparisons between vivisections on animals and dissections on humans.[105]

In fact, as Susan Lawrence has eloquently pointed out, 'studying the internal parts of living things often required researchers to make dynamic systems into static objects, to stop change in order to grasp it'.[106] The apparatus of physics and quantitative measurements were often used on preparations of (dead) isolated tissues or organs. The research of Héger is an example of such an approach, as dead tissue was used to study the mechanical aspects of blood circulation. Vivisectional approaches, too, were often aimed at the determination of both the causal conditions of, and the anatomical locations for, physiological functions. In his studies on intoxication, for example, the French founding father of physiology François Magendie discovered the role of the cerebellum in maintaining animal equilibrium by vivisection of the cranial nerves.[107] Van Kempen used a similar approach in his research on the spinal marrow, in which he localised specific functions in specific anatomical structures.[108] Although understanding function became the main goal of physiology, anatomical localisation remained important.

Similar to nineteenth-century developments in the field of pathological anatomy, the main difference between these studies and those from earlier periods was the sequence of scientific deduction. Whereas eighteenth-century anatomists investigated bodily structures in order to gain an insight into function, nineteenth-century physiologists worked the other way around. They altered and mutilated organs in order to see which functions were consequently lost. In this way, they deduced the functions of the removed part. As Claude Bernard, one of Magendie's students, explained: 'instead of proceeding from the organ to the function' as anatomists did before, experimental physiologists 'start from the physiological phenomenon and seek its explanation in the organism'.[109]

To conclude, one could say that the disciplinary divisions between anatomy and physiology were easily crossed in practice, both by individual scientists and in collaborations between anatomists and physiologists. Despite its institutionalisation, the discipline of physiology remained related to anatomy in the careers of physiologists, as well as in institutions, publications and research. However, despite the continued association between both disciplines, the rise of experimental physiology did change the image of anatomy profoundly. Early practitioners of 'new' physiology redefined anatomy as a dead and static science, which was merely preliminary to physiology.

Comparative Anatomy and Embryology

The disciplinary boundaries between human and comparative anatomy were equally permeable. Much like in other European countries, comparative anatomy (the comparative study of bodily structures of different species) remained an important professorial chair at all Belgian medical faculties, even after the institutionalisation of disciplines such as zoology, veterinary medicine and biology.[110] Because studies in comparative anatomy mostly had an underlying interest in comparing animal structures with human ones, anatomists easily shifted between both disciplines. To take but one telling example, in 1883, Jean Wehenkel quit his job as head of the Brussels autopsy service in order to become the principal of the newly founded school for veterinary medicine.[111]

Comparative anatomy traditionally, had been interwoven with a wide range of subjects in natural history, philosophy and theology. Its main goal was the hierarchical arrangement of living forms, from simple plants to humans, into groups that reflected a unifying plan for natural diversity. As part of the more general demise of what Cunningham has called 'old anatomy', this started to change in the first decades of the nineteenth century.[112] The French naturalist Georges Cuvier abandoned the idea of a single hierarchy of animal life by distinguishing between different body forms, such as vertebrates and invertebrates. He also included extinct creatures in his taxonomies.[113] At the Catholic University of Leuven, the comparative anatomist Pierre-Joseph Van Beneden adopted, extended and debated Cuvier's work, for example, in his studies on molluscs and fossils.[114]

Around 1850, two developments further pushed comparative anatomy into different directions. Under the influence of Darwinian evolution and embryology, comparative anatomists shifted their attention from grasping the overall design of nature to the processes underlying development, both of species and of individuals.[115] On the one hand, embryology, which had already played an important role in early nineteenth-century comparative anatomy, received a new layer of meaning through evolutionary theories. On the other hand, anatomists began to map the development of individuals from fertilised ovum until birth.

Evolutionary morphologists studied the changing shapes of embryos in order to shed light on the process of evolution. They did so largely under the influence of the recapitulation theory, which stated that the development of an individual reflected the evolution of his or her species. For Germany and the Netherlands, respectively, Nyhart and De Rooy have demonstrated that anatomists claimed this field, even though evolution had long been considered as a physiological category. Evolutionary theories were a welcome response to criticism on the descriptive nature of anatomy. They promised to turn anatomy into an explanatory science again: the comparative study of anatomy and embryology would not only serve scientific classification but would also verify, refute or refine theories on descent.[116]

In Belgium, evolutionary morphology developed as a discipline perched between comparative anatomy and the emergent biological sciences. In the words of historian Raf de Bont, it was 'an intellectual rather than an institutional success'. The subject was divided between different courses (zoology, comparative anatomy, physiology and anatomy) and between the faculties of medicine and sciences.[117] For example, even though the most important evolutionary morphologist in Belgium, Edouard Van Beneden (the son of Pierre-Joseph Van Beneden) worked as a zoologist at the faculty of sciences in Liège, he often hired students of medicine to conduct research on embryological development. Among his pupils were Charles Van Bambeke, who would become the first professor of embryology at the University of Ghent, as well as Albert Brachet. Remarkably, the first professorial chairs in embryology were established at faculties of medicine in the 1870s, often on the initiative of anatomists and physiologists, such as Héger in Brussels or Van Bambeke in Ghent.[118] Although biologists gradually claimed the study of animals and their embryos, the topic of evolutionary morphology figured as a scientific refuge for anatomists affiliated to medical faculties until at least the turn of the twentieth century.

Albert Brachet became the most important embryologist in Brussels. After his time in the laboratory of Van Beneden, he further specialised in embryology in the laboratories of the German embryologist Gustav Jacob Born and the Scottish anatomist John Cunningham. Brachet became professor of anatomy at the Free University of Brussels after his return in 1904. Two years later, he also received the chair in embryology. The appointment of Brachet was a turning point: for the first time, the chair of anatomy in Brussels was not in the hands of a clinician, but of a research-minded embryologist.[119] Throughout his career, Brachet crossed many disciplinary boundaries. He combined comparative with human anatomy, descriptive with experimental embryology. As a researcher, Brachet tried to elucidate the underlying causes of development by conducting experiments on (animal) embryos, a discipline he came to call 'causal embryology'. As a teacher, he sought the total integration of anatomy and embryology because he considered the study of organogenesis indispensable for an understanding of anatomical complexes.[120]

Apart from evolutionary aspects, embryology offered anatomists new empirical possibilities because the changing forms through which minute specks developed into adult shapes were still unknown. Although the German anatomist Samuel Thomas von Soemmerring had already published the first developmental series of human embryos in 1799, studies on human development only really took off in the 1880s, following the work of the Swiss anatomist Wilhelm His.[121] His' serial representations of human development were based on a comparative study of embryological specimens, which he had preserved by means of his own invention: the microtome. Using this instrument, fixed embryos (embedded in paraffin blocks) could be cut into very thin slices that could then be stained. Large-scale models of these sliced embryos enabled anatomists to envision and study their anatomical details.[122] These kinds of models quickly found their way to anatomical collections abroad, including in Brussels.[123] Anatomists also invented their own methods of preserving and representing embryos. In Brussels, for example, the embryologist Polydore Francotte was important for the spread of methods to fix, stain and photograph embryological specimens. He even invented and patented his own microtome.[124]

Before embryos could be preserved and studied, they had to be found. In the second half of the nineteenth century, anatomists and gynaecologists with growing ambition sought to collect human embryos, for example, by researching blood loss from pregnant women and examining uteruses during autopsies. In 1871, for instance, a medical student presented a foetus of the first month at a society meeting of the Society for Pathological Anatomy of Brussels. He had found this specimen by chance through the microscopic analysis of blood loss from a woman who had not known she was pregnant. In the student's own words, his discovery was 'one of the rare good fortunes that happen to an observant practitioner'.[125] On other occasions, the society's members discussed embryological specimens in order to shed light on the development of the human foetus in utero or on the evolution of humans as a species.[126] As a result of these collecting efforts, the anatomical museum of Brussels contained 63 embryonic and foetal specimens in 1872; a number that continued to grow in the following decades.[127] In the same year, the Ghent collection of descriptive human anatomy already housed an impressive 292 embryonic and foetal specimens.[128] By 1888, this number had risen to approximately 500 macroscopic specimens (including models) and, in Van Bambeke's words, 'thousands' of microscopic slides.[129]

This interest in embryos and foetuses can be explained by what Nyhart has called an 'explorer's motivation'. Nyhart has argued that human embryology gave anatomists an opportunity to re-establish their discipline as a science of empirical discovery. Embryology was not yet 'done'; on the contrary, it offered seemingly endless possibilities since structures needed identification in every stage of development. The mission of mapping developing structures fitted with a long anatomical tradition, in which lending one's name to a newly discovered structure signified eternal fame.[130] Moreover, the morphological study of embryos did not require a paradigmatic leap because the primary aim was still the visualisation and classification of bodily structures. By exchanging the dissecting knife for the microscope and the adult corpse for the embryo, late nineteenth-century anatomists could revive their discipline without the need to replace an analytic-comparative way of knowing with an experimental one.

Anatomy and Anthropology

The emergent discipline of anthropology also allowed anatomists to reclaim scientific status without having to spend hours and days in the laboratory. Many members of the Society for Anthropology of Brussels (*Société d'anthropologie de Bruxelles*), which was founded in 1882, had a background in anatomy. The society, in essence, was a copy of a similar society in Paris (*Société d'anthropologie de Paris*), where the French anatomist and anthropologist Paul Broca since 1859 propagated a new kind of anthropology. Broca heavily relied on anthropometry, the measurement of human physical qualities. The most important value was the cephalic index, which measured the relation between the maximum width and length of the head.[131]

The Brussels professor of history Léon Vanderkindere had called for the establishment of an anthropological society in Belgium since 1872. He believed that measuring skulls was the most certain method to gain an insight into the racial origins of cultures, something he deemed quintessential in order to understand their political and social history. For this reason, he researched the origins of Belgians by means of military statistics on body length in the 1870s. However, as a historian, Vanderkindere lacked the anatomical knowhow necessary for the increasingly complex anthropometrics of the 1880s.

Even though Vanderkindere was the initiator and the first president of the Society for Anthropology of Brussels, the physicians Victor Jacques and Emile Houzé soon took the lead. Jacques and Houzé published their first anthropometric studies on ethnology independent from one another in the early 1880s. Jacques studied old skulls dug up at an obsolete churchyard in Brussels; Houzé compared the cephalic index of Flemish and Walloon skulls. In later years, Jacques became more interested in archaeology, leaving the fields of physical anthropology and ethnology to Houzé.[132]

By the 1890s, Houzé had become the leading man of the society. For most of his career, he combined his passion for physical anthropology with a position as a physician at the Brussels hospital of St. Jean, where he measured the skulls of his patients. He regularly added preparations (mostly skeletal remains) to the anatomical collection of the university and frequently attended the meetings of the Society for Pathological Anatomy of Brussels. Houzé finally left the hospital in 1904, when he attained the chair of anthropology at the Free University of Brussels. Other members of the Brussels anthropological society also had a background in anatomy, including the criminal anthropologist Léon Warnots and Edouard Willems, who succeeded Houzé as professor of anthropology in 1921.[133]

The studies presented at the Brussels Society for Anthropology initially were very diverse, ranging from archaeology to evolution theory, history or ethnology. As historian Leen Beyers has shown, physicians dominated physical anthropology and its fields of application within the society until the First World War. Those contributions with an emphasis on anatomical methods concerned various branches of research, such as palaeoanthropology (the study of skeletal remains in order to elucidate the early development of humans), criminal anthropology (studying the 'typical' appearance of criminals) or teratology (a term that mostly signified the study of congenital anomalies). Members of the society also discussed anatomical specimens.

As the century progressed, however, anthropologists increasingly distanced themselves from the medical sciences. They started to pay more attention to environmental influences, such as social class or religion. Anatomists could easily contribute to anthropological studies as long as they revolved around the physical differences between races or around the anatomical history of (certain groups of) humans. Only when the importance of non-physical factors increased, anthropology gradually ceased to be a scientific refuge for anatomists.[134]

Similar evolutions took place in other disciplines. Adherents of new disciplines and research methods by the end of the nineteenth century increasingly represented themselves as *different* from traditional, macroscopic anatomy. As a result, anatomy was gradually reduced to dissection and the mere description of form. As the process of specialisation progressed, newer approaches in anatomy received their own names, professorial chairs and institutions. Due to this readjustment of the disciplinary landscape, anatomists became nearly invisible: whenever they did something else than dissecting or describing macroscopic structures, they were no longer called anatomists. Yet precisely this redefinition of anatomy enabled it to become the core of medical education.

3   THE MOTHER SCIENCE: ANATOMY IN MEDICAL EDUCATION

In 1849, Gluge wrote an essay on recent developments in anatomy and physiology in Belgium. In the spirit of the times, he especially stressed the importance of experimental approaches for the advancement of medical research. In order to design and conduct these experiments, however, Gluge deemed a notion of anatomical structures necessary. In his view, anatomical knowledge was a prerequisite to gain an insight into pathology and physiology. Much like one could not recognise pathologies without knowledge of the healthy body, one could not research bodily functions without a notion of anatomical structures.

> Anatomy and the description of the organs of men [...] is the scientific basis for physiology and the art of medicine. The progress of these last sciences is always dependent on those of the first.[135]

Gluge's reasoning was not unique. While anatomy was reduced to an auxiliary discipline, it was increasingly regarded and represented as the 'mother science'. Perceived as a closed body of descriptive knowledge, anatomy became the basis on which other, explanatory sciences could build.

As such, anatomy was the bedrock of medical education. Whereas Marcq posited that anatomy was 'done' as a science, he regarded the discipline as 'indispensable' for the training of physicians. In his opinion, students had to acquire a basic anatomical knowledge before they could practise other branches of medicine. Anatomy was 'an intermediary', 'a stepping stone' and 'a driving force', without which medical students could not move forward.[136] Almost half a century later, in 1907, the Ghent anatomist Hector Leboucq still used the same reasoning to argue that 'anatomy had to stay the basis of medical education'. In his words, anatomy was 'never to be abandoned', because it was the 'foundation' for, and the 'connecting thread' between specialised disciplines.[137]

Educational reformers appear to have shared this opinion, as they strengthened the position of anatomy in the last quarter of the nineteenth century. The Belgian law on higher education of 1876 turned anatomical exercises into an obligatory part of the medical curriculum. From that moment onwards, students learned the basis upon which the rest of their education would rest in the dissecting room: every one of them was required to dissect himself (or, in rare instances, herself). By compelling future physicians to dissect during their training, the law of 1876 confirmed an existing practice. As aforementioned, dissection had become an essential part of medical education around 1800. Following the Parisian model, students learned the normal structures of the body during anatomical demonstrations and hands-on dissections, before they were instructed about pathological alterations by means of anatomical specimens and autopsies. The knowledge acquired in post-mortem rooms foremost served clinical diagnosis. Students were taught to diagnose living patients by thinking ahead: by thinking about what their cadavers would look like.

Anatomy teachers preferred hands-on dissections to demonstrations because students were supposed to not only see but also feel the difference between healthy and diseased body parts. Apart from colours and sizes, dissection manuals described how organs felt. Students had to know 'the tactile sensation' of the healthy body in order to recognise pathologies like sclerosis (hardening) or oedema (fluid accumulation).[138] Because the shape and touch of organs were different after putrefaction or preservation, it was important that they performed dissections on *pieces fraîches*.[139] In addition, a familiarity with the feeling of body parts allowed them to diagnose patients by means of palpation, and to perform surgeries blindly through small incisions. As aforementioned, many teachers—such as

Deroubaix in Brussels—regarded dissection as the basis not only of clinical diagnosis but also of surgical practices, mostly because it gave students an opportunity to train their manual skills.

Apart from skilful hands, students also acquired the nerves of steel needed for surgeries in the dissecting room. Throughout the nineteenth century, anatomists argued that dissections would harden them to stand the challenges of their future career, for it took an 'imperturbable soul' to overcome the stench of decomposing cadavers.[140] The prolonged conservation of corpses in ill-aired dissecting rooms turned dissections into physically challenging experiences. In Brussels, for example, corpses were often on the table for over a week. When taking into account that hardly any form of preliminary embalming was done until the 1880s, it becomes easy to imagine that dissections were—literally and figuratively—stomach turners.[141] To reconcile these physical hardships, students smoked pipes and made pleasantries. Historians have shown that dissecting rooms were rife with bawdy conversations, sexual undertones and black humour, coping strategies to suppress fear and inconvenience. The shared burden of dissection bonded groups of young men together in a medical community based on male comradeship.[142]

Dissection also forced students to emotionally distance themselves from the persons they had their hands on (or in). Through the cutting and fragmenting of bodies, students had to learn how to keep a cool head during medical interventions. On top of that, the transgressive act of fragmenting bodies, an infraction of usual burial rituals, was thought to raise students above common 'superstitions'. As such, dissections symbolised the physician's mastery of the body, as well as the supremacy of rationality over the emotions. For all these reasons, dissection was a central constituent of medical identity. In anatomical manuals and in memoires, students' entrance into the dissecting room was described as a rite of passage that marked the start of their medical career.[143] On photographs, too, dissections symbolised students' transformation into physicians (Illustration 2.1).[144]

Because of its social and symbolic implications, the requirement of dissection was one of the reasons why medicine was not considered as an appropriate career for women in the late nineteenth century.[145] In her research on Britain, Laura Kelly has shown that handling the dead, naked body—sometimes of men, and in the company of male students and teachers—allegedly conflicted with female modesty. Additionally, it was feared that fragmenting dead bodies would distress women since their personalities

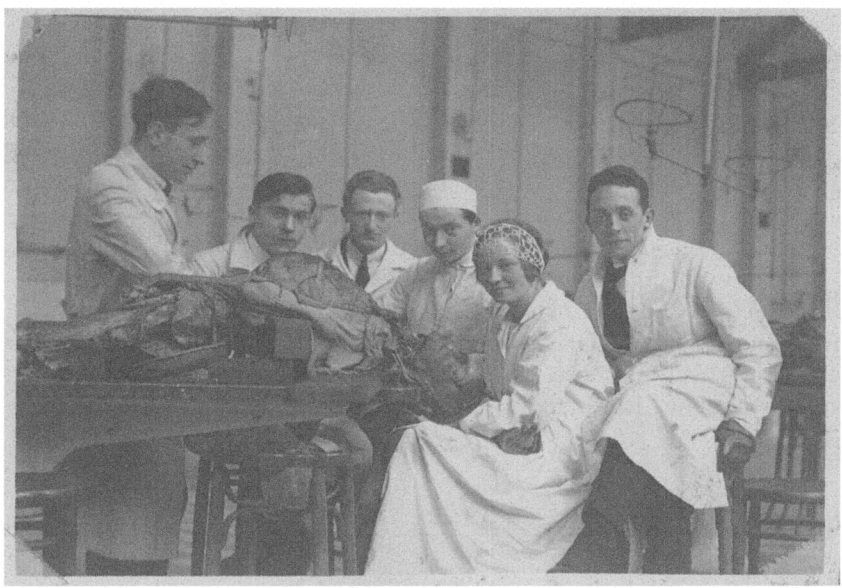

Illustration 2.1   Staged dissection at Parc Léopold, early twentieth century. This portrait is particularly interesting because a woman is dissecting alongside her male colleagues. © Archives of the Free University of Brussels, 2Y2/45.7

and emotions were more 'delicate' than those of their male counterparts.[146] In Brussels, too, the grim nature of the dissecting room presented problems with regard to female students. In 1881, for instance, Arthur Bonmariage, who taught a hygiene course to future teachers—many of them women—caused a public scandal by using *pieces véritables* in his lessons on anatomy. According to newspaper articles, teaching 'young girls' anatomy on the cadaver instead of by means of wax models and images was not only 'a truly terrifying invasion of naturalism into education', but also 'painful to watch, and unhealthy from a moral point of view'.[147] On other occasions, female dissectors were described as 'persons from an intermediary sex' that symbolised the downfall of traditional norms and values, and posing a threat to the ideal of marriage:

Women who dissect [...] maybe these are the norms of the future. But we are not there yet. Until today, the piano, rather than science, is the favourite passion of our sisters and companions.[148]

It is not surprising then, that despite the gradual acceptance of women in medical training from the 1880s onwards, very few of them specialised in anatomy until well into the twentieth century. One notable exception was Bertha De Vriese, who received her medical degree from the University of Ghent in 1900 and whose research on the blood vessels of the brain stem is a standard work until today.[149] In a group photo of a conference on anatomy in Jena from 1904, she is—quite tellingly—registered as the only woman on film, surrounded by bearded men in suits (Illustration 2.2).

The law on higher education of 1876 not only rendered dissections obligatory. Exercises in microscopic anatomy, too, obtained a central place in the curriculum. It is important to note, however, that these exercises served a different purpose. As the basis of clinical diagnosis and as a means to train students' hands and minds for surgery, dissections were foremost

Illustration 2.2   Group portrait of the participants of the anatomical conference of Jena, 1904. Berta De Vriese is sitting between all her male colleagues on the first row, fourth position to the right. © University Archives of Ghent

of clinical importance. Although microscopic anatomy was also said to 'render great service to the clinic', it was primarily seen as a vehicle of scientific medicine because it introduced students to research practices.[150]

In this regard, the inclusion of microscopic exercises in the medical curriculum echoed changes in universities' ambitions. In the second half of the nineteenth century, and especially from the 1870s onwards, medical faculties gradually supplemented their traditional task of educating physicians with an orientation towards research. Increasingly, the curriculum was aimed at not only the formation of medical practitioners by means of clinical courses, but also the training of medical scientists by means of laboratory research. German universities constructed laboratories for both pedagogical and scientific reasons, instructing students in routine scientific manipulations and offering opportunities to conduct more advanced research.[151] In their quest to reform medical education, adherents of—what they called—'scientific medicine' repeatedly referred to this example.[152] As Geert Vanpaemel has shown, the precise meanings of this German model varied largely among European countries and universities. 'German' science was not a template ready to be copied: it was not a fixed and well-defined system, but a malleable concept. As an umbrella term for a multitude of values associated with modern science, the German model was used and appropriated in various countries to address everything in domestic science systems that needed reform.[153]

At Belgian medical faculties, comparisons with Germany were mainly used to strengthen the role of the natural sciences in medical training and to introduce laboratory research and experimentation in the curriculum.[154] As early as 1844, Van Kempen argued that the progress of the German medical sciences followed from close collaborations between the faculties of medicine and the natural sciences.[155] In the following three decades, this argument repeatedly came to the fore in discussions on the number and the level of natural sciences taught to medical students in the preparatory candidature. Before starting their actual medical training, all students had to follow courses on physics, chemistry and biology. These natural sciences were increasingly represented as the pure sciences on which the applied science of medicine depended.[156] Significantly, the natural sciences were also seen as the foundation of medical experimentation. The histologist Jean-Baptiste Carnoy, for instance, called them 'the indispensable basis' for medical research. In his opinion, cellular biology specifically was 'the first principle of every scientific method'.[157]

Because of this importance attached to cellular research, the introduction of microscopic anatomy in the medical curriculum became an important goal for advocates of scientific medicine. Even before microscopic exercises became compulsory in 1876, anatomists and pathologists at Belgian universities—such as Gluge and Rommelaere in Brussels—had established courses to introduce their students to microscopic work.[158] In their view, microscopic exercises were 'an indispensable complement to theoretical courses'. To raise the level of medical training in Belgium to the bar set by Germany, students had to be trained to 'handle the microscope themselves in order to gain an exact insight into the structures of our organs'.[159] Microscopic exercises had to stimulate students' *esprit scientifique* by introducing them to the context of the laboratory. The idea was that it would inspire talented students to pursue their scientific aspirations during doctorates or by participating in scientific competitions.[160]

In the 1870s and 1880s, similar ambitions led to the establishment of a multitude of specialised courses. In general, the pursuit of a more scientific medical education went hand in hand with an expansion of the curriculum. The idea was that specialised courses allowed for more room for in-depth analyses and experimentation.[161] This had a profound impact on the teaching of anatomy. For example, in 1840 Théodor Schwann still held the all-encompassing chair of *Anatomie générale, descriptive, pathologique, l'organogénèse et les monstruosités* (general anatomy, descriptive anatomy, pathological anatomy, the embryological development of organs and teratology) at the Catholic University of Leuven. Forty years later, in 1880, this chair was divided among three professors. Charles Ledresseur was responsible for topographic anatomy and obstetric anatomy; Gustave Verriest taught pathological anatomy, histology and exercises in microscopic anatomy; and Emile Venneman occupied himself with the anatomy of the eye, descriptive anatomy and dissections.[162] In Brussels, too, the subject of anatomy was gradually divided into different courses, such as descriptive anatomy, anatomical demonstrations (dissections and microscopic exercises), topographic anatomy, pathological anatomy, comparative anatomy and embryology.[163]

The establishment of these specialised courses, however, did not happen without resistance. Especially in Brussels, where clinically oriented professors remained dominant, fear existed that an emphasis on science and an exaggerated specialisation would impede students from their primary goal: namely to become physicians. The Brussels professor Crocq, for example, opposed the generalisation of laboratory courses:

Think of a young man who studies medicine. You put him in a laboratory where he, for three months, works on a study on pus globules or fibrin fibres in blood, and for what purpose? To become a general practitioner in Wolverthem or Steenockerzele [small villages in Belgium]. In my opinion, he could have spent his time a lot better.[164]

As early as 1872, Louis Deroubaix proposed to create a separate scientific institute for those interested in disciplines such as physiology and histology. He, like other professors of the Free University of Brussels, including Crocq and Rommelaere, wanted to organise scientific education outside of the university. In their opinion, the majority of students benefitted most from a practice-based education in the clinic. For the relatively small number of students aspiring to a career as a scientist rather than as a physician, so they argued, a separate institute for higher education (*Institut central des hautes études*) could be created. This idea was partly inspired by financial reasons, as the Free University of Brussels, unlike the state universities of Ghent and Liège, did not receive state funding to establish laboratories and research institutes and therefore struggled to keep up.[165] The rector of the Catholic University of Leuven, Alexandre Namêche, also opposed the adoption of specialised courses based on experiments and laboratory work. He worried that an emphasis on research would alienate students from the practice of medicine, turning them into scientists instead of doctors.[166] Despite this resistance, however, specialised courses were gradually introduced at all Belgian universities, including in Brussels.

Yet disciplines which could be taught in the hospital, and which were regarded as clinically relevant, became part of the official curriculum more easily. It is telling in this respect that the law on higher education of 1876 rendered many clinical courses obligatory. Laboratory exercises, conversely, remained elective, with microscopic anatomy as the only exception.[167] For instance, while exercises in pathological and topographic anatomy became compulsory parts of the curriculum, experimental embryology did not.[168] The reason was the scientific status of embryology. While autopsies were closely related to clinical diagnoses and topographic anatomy to surgical practices, these exercises were deemed useful for the formation of practitioners. Embryology, by contrast, was associated with medical research rather than with clinical practice, and therefore was not considered as an essential part of physicians' general training.

As a result of their absence from the official curriculum, only a handful of students with scientific aspirations attended specialised research courses, which mostly had the reputation to be very difficult. Adherents of scientific medicine deplored this status quo. In their opinion, the exclusion of newer disciplines from the curriculum led to a growing discrepancy between medical education and research. As one of them lamented in 1876: 'It goes without saying that the old framework is not adapted to modern science.'[169] The laws on higher education of 1890 and 1891 only partially incorporated this grievance. While embryology became part of the candidature, the programme of the doctorate hardly changed. Clinical exercises continued to have priority over research.[170] It appears that, although research gradually gained importance in the medical training, it stayed only of secondary consideration. Despite the orientation towards Germany, the emphasis of medical education continued to be on observation rather than experimentation, medical practice rather than medical science, and the dissecting room and the clinic rather than the laboratory. While anatomy was in decline as a scientific discipline, it stayed at the core of medical education as a primarily clinical practice.

4   Conclusion

The mid-nineteenth-century transformation of anatomical practices was part of larger developments within the medical field. In an era of medical specialisation, anatomy fell apart into a growing number of subdisciplines, focusing on certain types of bodies, body parts or methods. At the same time, a new conception of physiology, with an emphasis on experimentation, helped to rearrange the disciplinary order among the medical sciences associated with anatomy. Whereas physiology came to embody the new ideal of scientific medicine, anatomy was increasingly associated with the mere description of form. A dichotomy between observational and experimental sciences emerged, differentiating anatomy from physiology or anatomical pathology from other forms of pathology. It was through this adjustment and expansion of the disciplinary landscape that anatomy was redefined as a descriptive science, confined to observation and taxonomy, without a larger scientific agenda.

Anatomy began to lose prestige when experimentation replaced empirical observation as the main criterion for 'scientific' medicine. This turn in what counted as science, however, was gradual and subject to discussion. In the second half of the nineteenth century, the meanings of science and anatomy were heterogeneous, as they were influenced by the ongoing transformation of the medical landscape. In Brussels, for example, most anatomists primarily identified themselves as clinicians and continued to view the clinic as the heart of medicine. For them, anatomical localisation remained quintessential.

Even anatomists who agreed with 'modern' scientific standards did not consider their discipline 'done'. For them, dissection was not the essence of their identity as anatomists anymore. They incorporated new techniques and theories in their research. In hindsight, it is easy to say that they were pioneers of new, seceded disciplines. Yet at the time, medical specialisation was a gradual process. Most seceded anatomical disciplines were not housed in specialised research institutions until the 1880s and 1890s and were integrated into the medical curriculum even slower. As Nyhart has convincingly shown for the German context, anatomy during this ongoing restructuring developed as a cluster of research orientations and methodologies rather than as a well-defined discipline.[171] In a similar vein, the careers and publications of Belgian anatomists show that anatomy was reshaped in connection with its competing subdisciplines. In response to criticism on the mere descriptive nature of their work, research-minded anatomists looked for ways to re-establish their status as modern scientists. This led them to incorporate methods into their work that are not traditionally viewed as being part of the anatomical enterprise.

First and foremost, the microscope created new possibilities. As the traditional goal of identifying the macroscopic structures of the human body seemed 'done', microscopic anatomy offered a vast new area of research. By integrating microscopic analyses, anatomists hoped to turn their discipline into a science of empirical discovery again. This in part also explains their interest in (evolutionary) embryology, for anatomical structures needed identification at every stage of their development.[172] Secondly, whereas anatomy is usually associated with what Pickstone has called a 'comparative-analytic way of knowing', anatomists in reality incorporated experimental methods in their work. Experiments were often conducted on both living and dead organisms, thus blurring the lines between anatomy and physiology. Thirdly, anatomists reframed their work within evolutionary or ethnological theories in order to give the description of anatomical structures a larger, explanatory goal.

The image of anatomy as a *science accessoire* was powerful nonetheless. Although the division between anatomy and its seceded disciplines was not immediately reflected in the practices of scientists, it did change the perception of the discipline of anatomy profoundly. Increasingly, when anatomists occupied themselves with something other than the description of macroscopic structures, it was no longer called anatomy or was qualified as a specific kind of anatomy. In the pursuit of their own professorial chairs and institutions, adherents of new disciplines represented their work as *true* science, in opposition to anatomy. In this process, termed 'disciplinary boundary work' by Thomas Gieryn, anatomy was gradually pushed out of the scientific realm.[173] However, precisely this redefinition of anatomy enabled it to become, to come back to Marcq's words, 'the foundation for the others'.[174] Perceived as a closed body of descriptive knowledge, anatomy became the basis on which newer medical sciences were built.

Accordingly, anatomy's position in the medical curriculum was strengthened. The Belgian law on higher education of 1876 compelled students to participate in macroscopic and microscopic anatomical exercises. On the one hand, dissections were important tools to train medical practitioners. They were seen as the foundation of clinical diagnosis and surgery, and as a means to prepare students' hands and minds for a medical career. Microscopic exercises, on the other hand, were above all meant to rouse students' interest in medical research, as they introduced them to the context of the laboratory. Anatomy, at least at a cellular level, was seen not only as an important part of medical training but also as an important research skill. As the basis of clinical and research practices, anatomy fulfilled a dual purpose. In so doing, the discipline befitted debates on the task of universities, which were increasingly expected to train both medical practitioners and scientists. On top of that, specialised anatomical disciplines—embryology, microscopic pathological anatomy, obstetric or ocular anatomy, and so on—became part of the medical curriculum from the 1870s onwards, mostly as elective courses.

In the following chapters, it will become clear that this ambiguous status of anatomy (simultaneously auxiliary and foundational), as well as the expansion of the disciplinary landscape, affected the distribution and treatment of anatomical material. As Tatjana Buklijas has pointed out, new disciplines 'all requested their share of corpses for research and instruction'.[175] The competition between emergent subdisciplines had an impact on the distribution of the crucial resource of human cadavers. Chapter 4 will show that an analysis of the body supply to several disciplines allows historians to study universities' and hospitals' priorities. Were corpses primarily used for medical research or education? Did the distribution of cadavers follow the scientific hierarchy of medical disciplines, or were they divided equally between research and teaching courses, thus reflecting the twofold mission of late nineteenth-century universities? Chapter 5 addresses similar questions with regard to anatomical collections. Were anatomical specimens, for example, sometimes destroyed during medical experiments for research purposes, or was their preservation as teaching tools deemed more important? Did anatomists reinterpret old collections within the framework of new disciplines?

First, however, the next chapter takes a step back and discusses how patients moved from the deathbed to the dissecting table. The following pages describe the processes through which patients *became* research material before turning to patients' actual use as research material. This chapter has provided some important background information to help understand why anatomists' claims to dead bodies became increasingly problematic as the nineteenth century progressed: the scientific prestige of their discipline diminished and other medical sciences' need for corpses increased. Yet the next chapter argues that medical faculties' access to corpses in the late nineteenth century was, above all, impeded by ideological conflicts about the treatment and the burial of the poor.

NOTES

1. Léon Marcq, *Essai sur l'histoire de la médecine belge contemporaine* (Brussels: Manceaux, 1866), 155.
2. Ibid., 130. Own emphasis.
3. Andrew Cunningham, *The Anatomist Anatomis'd: An Experimental Discipline in Enlightenment Europe* (Farnham: Ashgate, 2010), 368.
4. Ibid., 19.
5. Ibid., 361–89.
6. Russell C. Maulitz, *Morbid Appearances: The Anatomy of Pathology in the Early Nineteenth Century* (Cambridge: Cambridge University Press, 2002), 36–104; Idem, "Pathology," in *The Cambridge History of Science: The Modern Biological and Earth Sciences*, vol. 6, eds. Peter J. Bowler and John V. Pickstone (Cambridge: Cambridge University Press, 2009), 367–81.
7. Cunningham, *The Anatomist Anatomis'd*, 380–2.
8. For example: William F. Bynum, *Science and the Practice of Medicine in the Nineteenth Century* (Cambridge: Cambridge University Press, 1994), 92–117; Andrew Cunningham and Perry Williams, eds., *The Laboratory Revolution in Medicine* (Cambridge: Cambridge University Press, 1992).
9. John V. Pickstone, *Ways of Knowing: A New History of Science, Technology and Medicine* (Manchester: University of Chicago Press, 2001), 106–61.

10. Renaud Bardez, "La Faculté de médecine de l'Université Libre de Bruxelles: entre création, circulation et enseignement des savoirs, 1795–1914" (PhD diss., Université Libre de Bruxelles, 2016), 256–65.

11. Pieter Dhondt, *Un double compromis: enjeux et débats relatifs à l'enseignement universitaire en Belgique au XIXe siècle* (Ghent: Academia Press, 2011), 129–32.

12. On Paris hospital medicine: Maulitz, *Morbid Appearances*, 36–104; Cunningham, *The Anatomist Anatomis'd*, 380–2.

13. Michel Foucault, *The Birth of the Clinic. An Archeology of Medical Perception*, trans. A.M. Sheridan (New York: Routledge, 1973), xvi.

14. Maulitz, "Pathology," 367–81.

15. On Bichat: Maulitz, *Morbid Appearances*, 1–35. On the importance of the urban context of Paris for the development of clinical medicine: Dora Weiner and Michael J. Sauter, "The City of Paris and the Rise of Clinical Medicine," *Osiris* 18 (2003): 23–42.

16. On the impact of the localisation of diseases on the doctor-patient relationship: Nicholas Jewson, "The Disappearance of the Sick-Man from Medical Cosmology, 1770–1870," *International Journal of Epidemiology* 38, no. 3 (2009): 622–33.

17. Cf. Foucault, *The Birth of the Clinic*, 124–48.

18. For a similar argument on X-Rays: Daniel S. Goldberg, "Suffering and Death among Early American Roentgenologists: The Power of Remotely Anatomizing the Living Body in Fin de Siècle America," *Bulletin of the History of Medicine* 85, no. 1 (2011): 1–28. On the link between dissection and the construction of a scientific self, see: Michael Sappol, *A Traffic of Dead Bodies: Anatomy and Embodied Social Identity in Nineteenth-Century America* (Princeton and Oxford: Princeton University Press, 2002), 2–3.

19. Laurence Brockliss, "The New Paris Medical School and the Invention of the Clinic," in *The French Experience from Republic to Monarchy, 1792–1824: New Dawns in Politics, Knowledge and Culture*, eds. Mairé F. Cross and David Williams (Basingstoke: Palgrave Macmillan, 2000), 120–39; Thomas Neville Bonner, *Becoming a Physician: Medical Education in Great Britain, France, Germany and the United States, 1750–1945* (New York and Oxford: Oxford University Press, 1995), 103–41; Othmar Keel, *L'avènement de la médecine Clinique moderne en Europe, 1750–1815: politiques, institutions et savoirs* (Montréal: Géorg éditeur et presses de l'université de Montréal, 2002).

20. Maulitz, *Morbid Appearances*, 109–60; John H. Warner, *Against the Spirit of the System: The French Impulse in Nineteenth-Century American Medicine* (Baltimore and London: Princeton University Press, 1998), 32–65.

21. For example Nicolas Ansiaux and Jean-Nicholas Comhaire in Liège.
22. J.B. de Tramasure, police officer, to the Brussels City Council, 12 April 1799, folder 5, Fonds de l'instruction publique, CAB. Quoted in: Pieter Dhondt, "La situation précaire de l'enseignement supérieur dans les départements belges entre 1797 et 1815," *Revue belge de philologie et d'histoire/Belgisch tijdschrift voor philologie en geschiedenis* 82, no. 4 (2004): 939.
23. Bardez, "La Faculté de médecine," 37–42.
24. Ibid., 47–51.
25. Veronique Deblon, "Anatomie gereanimeerd" (PhD diss., KU Leuven, 2020).
26. Louis-Joseph Urban, *Établissement d'une école de médecine à Bruxelles, avec l'approbation du ministre de l'intérieur* (Brussels: Imprimerie Urban, 1805), 5–6.
27. Dhondt, *Un double compromis*, 63–8.
28. Philippe-Antoine Marcq, *De l'état actuel de l'enseignement médical en Belgique* (Brussels: Demanet, 1821), 13; J.R. Marinus, "Nécrologie de Philippe Antoine Marcq, mort en 1837, à l'age de 40 ans," *Bulletin médical belge* 19 (1837): 153–6.
29. Dhondt, *Un double compromis*, 67.
30. Deblon, "Anatomie gereanimeerd."
31. This was due to the fact that students could only obtain a medical degree after university studies. Students at the medical school of Brussels graduated as *officiers de santé*, a degree that was more limited although it allowed for starting a medical practice in rural areas. In comparison to the training offered at universities, the education offered at medical schools (in Brussels, but also in Antwerp) was more practice-based, less time-consuming and cheaper.
32. Bardez, "La Faculté de médecine," 233–9.
33. Dhondt, *Un double compromis*, 89–102.
34. Ibid., 133–50.
35. *Loi organique de l'enseignement supérieur du 27 septembre 1835, modifiée par la loi du 15 juillet 1849* (Brussels: G. Stapleaux, 1849): 16.
36. See, for example: "Règlement pour l'amphithéâtre d'anatomie et les salles de dissection," *Annuaire de l'Université catholique de Louvain* 1 (1837): 37–41.
37. Bardez, "La Faculté de médecine," 89–98.
38. Stéphane Louryan, "Un portrait des enseignants d'anatomie humaine à l'Université Libre de Bruxelles entre 1834 et 1905," *Revue médicale de Bruxelles* 29 (2008): 64–5.
39. Quoted in: Louryan, "Un portrait des enseignants," 65.
40. *BSAP* 32 (1883): 1.

41. Joseph-Antoine Spring, "De l'esprit scientifique à notre époque et dans nos universités," *Annales des Universités de Belgique* 2 (1860–1863): 386–96, on p. 392. See also: Dhondt, *Un double compromis*, 181–96.

42. Salim Al-Gailani, "The "Ice Age" of Anatomy and Obstetrics: Hand and Eye in the Promotion of Frozen Sections around 1900," *Bulletin of the History of Medicine* 90, no. 4 (2016): 622–5; John H. Warner and Lawrence J. Rizzolo, "Anatomical Instruction and Training for Professionalism from the 19th to the 21st Centuries," *Clinical Anatomy* 19 (2006): 408.

43. *BSAP* 3 (1860): 3.

44. Clinically-oriented anatomists also renewed their discipline in other countries, see: Laurens de Rooy, *Snijburcht: Lodewijk Bolk en de bloei van de Nederlandse anatomie* (Amsterdam: Amsterdam University Press, 2011), 92–118; Al-Gailani, "The Ice Age of Anatomy," 620.

45. Maulitz, "Pathology," 369.

46. See also: Cay-Rüdiger Prüll, "Pathology and Surgery in London and Berlin, 1800–1930: Pathological Theory and Clinical Practice," in *Pathology in the 19th and 20th Centuries: The Relationship between Theory and Practice*, ed. Cay-Rüdiger Prüll (Sheffield: EAHMH Publications, 1998), 71–99, on p. 73.

47. Rudolf Virchow, "Ueber die Standpunkte in der wissenschaftlichen Medicin," *Archiv für pathologische Anatomie und Physiologie und für klinische Medicin* 1 (1847): 1–19, on p. 10.

48. Ibid., 17.

49. Maulitz, "Pathology," 375. See also: Wilson I.B. Onuigbo, "The Paradox of Virchow's Views on Cancer Metastasis," *Bulletin of the History of Medicine* 36, no. 5 (1962): 444–9.

50. Bardez, "La Faculté de médecine," 180.

51. Guillaume Rommelaere, *Des institutions médicales et hospitalières en Angleterre* (Brussels: Bols-Wittouck, 1866), 33.

52. See Chap. 4.

53. Bardez, "La Faculté de médecine," 326–30.

54. On these specialised services : Claire Dickstein-Bernard, "Naissance des services spéciaux dans les hôpitaux belges au XIXe siècle: Réflexions sur le cas bruxellois," *Annales de la Société Belge d'Histoire des Hôpitaux et de la Santé publique* 23–24 (1985–1986) : 49–66.

55. Hans-Heinz Eulner, *Die Entwicklung der medizinischen Spezialfächer an den Universitäten des deutschen Sprachgebietes* (Stuttgart: Ferdinand Enke Verlag, 1970), 506 and 641–9.

56. Léon Marcq, "Coup d'oeil sur l'histoire de l'anatomie pathologique," *BSAP* 7 (1862).

57. For example Jean Bougard, *Etudes sur le cancer* (Brussels: Hayez, 1882). See also: Stéphane Louryan, "L'encadrement des travaux pratiques d'anatomie entre 1834 et 1940," *Revue médicale de Bruxelles* 33 (2012): 117–24.
58. Louryan, "Un portrait," 65.
59. For example: *BSAP* 18 (1870): 6; *BSAP* 28 (1879): 9–10.
60. *BSAP* 38 (1889): 5.
61. Other examples of professors with dual appointments are Adolphe Burggraeve, Léon Warnots, Remy-Joseph Lavise and Auguste Brunin.
62. Louis Deroubaix, "Clinique chirurgicale de l'Hôpital St. Jean," *Annales de l'Université de Bruxelles. Faculté de médecine de Bruxelles* 4 (1883): 2.
63. Deblon, "Anatomie gereanimeerd."
64. See also: Toby Gelfand, "The "Paris Manner" of Dissection: Student Anatomical Dissection in Early Eighteenth Century Paris," *Bulletin of the History of Medicine* 46, no. 2 (1972): 99–130.
65. Al-Gailani, "The Ice Age of Anatomy," 620.
66. On segmental anatomy: De Rooy, *Snijburcht*, 92–118. On anatomical cross-sections: Al-Gailani, "The "Ice Age" of Anatomy," 611–42.
67. Charles Ledresseur, *Résumé du cours d'anatomie des regions professé à l'université catholique* (Louvain: Peeters-Ruelens, 1882) and Deroubaix, "Clinique chirurgicale," 2. Note that in Belgium topographic anatomy was often termed 'anatomie des régions'.
68. Félix Putzeys, "Préface," in Daniel John Cunningham, *Guide de dissection, résumé d'anatomie topographique*, trans. Pierre Kuborn (Liège:Nierstrasz, 1890), v.
69. Ibidem.
70. For example, in the translation of Cunningham's manual (note 68) and in the much-used manual Paul Jules Tillaux, *Traité d'anatomie topographique avec applications à la chirurgie* (Paris: Asselin, 1879).
71. Al-Gailani, "The "Ice Age" of Anatomy," 611–42.
72. Edouard Bouqué, Collection d'anatomie humaine descriptive, 1888, folder 44 (102), UAG.
73. Louryan, "Un portrait," 65.
74. Weisz, *Divide and Conquer: A Comparative History of Medical Specialisation* (Oxford: Oxford University Press, 2006).
75. Pickstone, *Ways of Knowing*, 135–61.
76. On German laboratory medicine: Bynum, *Science and the Practice of Medicine*, 95–103. On this evolution in Belgium: Dhondt, *Un double compromis*, 361–84.
77. Bynum, *Science and the Practice of Medicine*, 92–117.
78. Cay-Rüdiger Prüll, *Medizin am Toten oder am Lebenden? Pathologie in Berlin und in London, 1900–1945* (Basel: Schwabe Verlag, 2004), 423–44.

79. Marcq, *Essai sur l'histoire*, 130.

80. For example: Etienne Van Kempen, *Traité d'anatomie descriptive et d'histologie spéciale* (Leuven: Vanlinthout, 1854); Idem, *Manuel d'anatomie générale* (Leuven: Vanlinthout, 1860); Adolphe Burggraeve, *Cours théorique et pratique de l'anatomie* (Ghent: Impens, 1840). On the patriotic aspirations of Belgian anatomists: Veronique Deblon and Pieter Huistra, "Het geheim van de anatoom: de ontwikkeling van de Belgische anatomie in de negentiende eeuw," *Studium: Tijdschrift voor Wetenschaps-en Universiteitsgeschiedenis/Revue de l'Histoire des Sciences et des Universités* 9 (2016): 202–16.

81. "Prix quinquennaux décernés à MM les professeurs Van Kempen et Van Beneden," *Annuaire de l'Université Catholique de Louvain* 27 (1862): 405–6.

82. Veronique Deblon, "Imitating Anatomy: Recycling Anatomical Illustrations in Nineteenth-Century Atlases," in *Bodies Beyond Borders: Moving Anatomies, 1750–1950*, eds. Kaat Wils, Raf De Bont, and Sokhieng Au (Leuven: Leuven University Press, 2017), 115–37.

83. See, for example, various essays in: Frank Huisman and John H. Warner, eds., *Locating Medical History: The Stories and their Meanings* (London and Baltimore: John Hopkins University Press, 2004); David Cahan, "Looking at Nineteenth-Century Science: An Introduction," in *From Natural Philosophy to the Sciences: Writing the History of Nineteenth-Century Science*, ed. David Cahan (Chicago: University of Chicago Press, 2003), 3–15.

84. Michael Hagner, "Scientific Medicine," in *From Natural Philosophy to the Sciences: Writing the History of Nineteenth-Century Science*, ed. David Cahan (Chicago: University of Chicago Press, 2003), 74.

85. De Rooy, *Snijburcht*, 18–19.

86. Sappol, *A Traffic of Dead Bodies*, 314.

87. Lynn K. Nyhart, *Biology takes Form: Animal Morphology and the German Universities, 1800–1900* (Chicago: University of Chicago Press, 1995), 67.

88. De Rooy, *Snijburcht*, 26–34; Nyhart, *Biology takes Form*, 65–90.

89. Hagner, "Scientific Medicine," 66.

90. Christopher Lawrence, "Incommunicable Knowledge: Science, Technology and the Clinical Art in Britain 1850–1914," *Journal of Contemporary History* 20 (1985): 503–20.

91. Gerald L. Geison, *Michael Foster and the Cambridge School of Physiology: The Scientific Enterprise in Late Victorian Society* (Princeton: Princeton University Press, 1978), 13–47, on p. 27.

92. John E. Lesch, *Science and Medicine in France: The Emergence of Experimental Physiology, 1790–1855* (Cambridge, MA: Harvard University

Press, 1984); Kathryn M. Olesko, "Commentary: On Institutes, Investigations and Scientific Training," in *The Investigative Enterprise: Experimental Physiology in Nineteenth-Century Medicine*, eds. William Coleman and Frederic L. Holmes (Berkeley: University of California Press, 1988), 295–332; Michael Hagner and Hans-Jorg Rheinberger, *Die Experimentalisierung des Lebens: Experimentalsysteme in den biologischen Wissenschaften 1850–1930* (Berlin: Akademie, 1993), 7–27; Eulner, *Die Entwicklung der medizinischen Spezialfächer*, 46–65.

93. Tatjana Buklijas, "Cultures of Death and Politics of Corpse Supply: Anatomy in Vienna, 1848–1914," *Bulletin of the History of Medicine* 82, no. 3 (2008): 570–607, on p. 589.

94. Prüll, *Medizin am Toten*, 253–63.

95. Bardez, "La Faculté de médecine," 40–2.

96. Léon Frédericq, "L'enseignement de physiologie à l'Université de Berlin," *Revue de Belgique* 3 (1881): 118–37, on p. 136–7.

97. Robert Halleux, "Naar de kern van het leven: de biologie," in *Geschiedenis van de wetenschappen in België, 1815–2000*, vol. 1, eds. AndréeDespy-Meyer, Robert Halleux, Jan Vandersmissen, and Geert Vanpaemel (Brussels: Dexia/La Renaissance du livre, 2001), 289–304; IsidoorLeusen, *125 jaar fysiologie in de faculteit geneeskunde van de universiteit te Gent, 1817–1942* (Ghent: Archief RUG, 2000), 31–46; Smaller physiological laboratories were already established at Belgian universities in the 1870s: Bardez, "La Faculté de médecine," 313–17.

98. *BSAP* 38 (1889): 5.

99. For example Medical faculty of the university of Leuven to rector Alexandre Namèche, 5 March 1874, folder 22, Anatomisch Amfitheater, ASSL.

100. Other examples are Charles Poelman and Richard Boddaert in Ghent, or Théodor Schwann in Liège.

101. For example: Louis Deroubaix, *Traité des fistules uro-génitales de la femme: comprenant les fistules vésico-vaginales, vésicales cervico-vaginales, uréthro-vaginales, vésicales cervico-utérines, vesico-utérines, urétéro-vaginales et urétérales cervico-utérines* (Brussels: Manceaux, 1870).

102. Liliane Viré, "La "Cité scientifique" du Parc Léopold à Bruxelles, 1890–1920," *Cahiers Bruxellois: Revue de l'histoire urbaine* 19 (1974): 86–179.

103. Paul Héger, *Expériences sur la circulation du sang dans des organes isolés* (Brussels: Manceaux, 1873), 3–16.

104. Léon Fredericq, *Manipulations de physiologie: guide de l'étudiant au labouratoire pour les travaux pratiques et les démonstrations de physiologie* (Paris: J.B. Baillière, 1892), 127.

105. Cunningham, *The Anatomist Anatomis'd*, 295–9.

106. Susan C. Lawrence, "Anatomy, Histology and Cytology," in *The Cambridge History of Science*, 265.

107. On these different types of physiological experimentation, see: Richard L. Kremer, "Physiology," in *The Cambridge History of Science*, 342–66, on pp. 346–7.

108. Etienne Van Kempen, *Expériences physiologiques sur la transmission de la sensibilité et du mouvement dans la moelle épinière* (Brussels: Mortier, 1859).

109. Cunningham, *The Anatomist Anatomis'd*, 374–5.

110. De Rooy, *Snijburcht*, 39–42 and 202–7; Nyhart, *Biology takes Form*, 35–102.

111. J. Wehenkel to the commission of hospitals, 3 May 1883, folder 106, Autopsies, Affaires générales, ASSB.

112. Cunningham, *The Anatomist Anatomis'd*, 375–9.

113. Lawrence, "Anatomy, Histology and Cytology," 271–2.

114. On Pierre-Joseph Van Beneden: Raf de Bont, *Darwins kleinkinderen: de evolutietheorie in België 1865–1945* (Nijmegen: Uitgeverij Vantilt, 2008), 45–6, 59–60 and 81–2.

115. Lawrence, "Anatomy, Histology and Cytology," 272.

116. On Germany: Nyhart, *Biology takes Form*, 65–102. On the Netherlands: De Rooy, *Snijburcht*, 52–91.

117. De Bont, *Darwins kleinkinderen*, 102.

118. Ibid., 80–102.

119. Louryan, "Un portrait," 68–9.

120. Jacques G. Mulnard, "The Brussels School of Embryology," *International Journal of Developmental Biology* 36 (1992): 17–24.

121. Nick Hopwood, "Producing Development: The Anatomy of Human Embryos and the Norms of Wilhelm His," *Bulletin of the History of Medicine* 74, no. 1 (2000): 29–79.

122. Nick Hopwood, "Giving Body to Embryos: Modeling, Mechanism and the Microtome in Late Nineteenth-Century Anatomy," *Isis* 90, no. 3 (1999): 462–96.

123. These models are still part of the university collection today.

124. Polydore Francotte, *Résumé d'une conférence sur la microphotographie appliquée à l'histologie, l'anatomie comparée et l'embryologie* (Brussels: Manceaux, 1887); Idem, *Manuel de technique microscopique: applicable à l'histologie, l'anatomie comparée, l'embryologie et la botanique* (Brussels: Lebègue, 1885). On Francotte's microtome: Brian Bracegirdle, *A History of Microtechnique: The Evolution of the Microtome and the Development of Tissue Preparation* (London: Heinemann, 1978), 224.

125. "Hydropisie du cordon sur un fœtus de 6 semaines," *BSAP* 20 (1871): 51–4.

126. For example: "Un embryon enveloppé de toutes ses membranes," *BSAP* 2 (1859): 29; "L'importance de l'anatomie pathologique pour

l'élucidation des problèmes de l'anatomie comparée," *BSAP* 32 (1883): 145–52.

127. E. Carpentier, Catalogue des pièces contenues dans la collection de société anatomo-pathologique de Bruxelles séant à l'université libre, 24 November 1872, folder 69, Collections scientifiques, Affaires générales, ASSB.

128. Procès-verbaux de récolement, 1874, folder 4A 25 (82), UAG.

129. Charles Van Bambeke, Accroissements des collections d'histologie et d'embryologie pendant les années 1886–1887–1888, 27 September 1888, folder 4A44 (102), UAG.

130. Nyhart, *Biology takes Form*, 86.

131. Jean-Claude Wartelle, "La Société d'Anthropologie de Paris de 1859 à 1920," *Revue d'histoire des sciences humaines* 10 (2004): 125–71.

132. Leen Beyers, "Rasdenken tussen geneeskunde en natuurwetenschap: Emile Houzé en de Société d'Anthropologie de Bruxelles 1882–1921," in *Degeneratie in België 1860–1940: een geschiedenis van ideeën en praktijken*, eds. Jo Tollebeek, Geert Vanpaemel en Kaat Wils (Leuven: Leuven University Press, 2003), 43–78.

133. Louryan, "L'encadrement," 117–24.

134. Beyers, "Rasdenken," 43–78.

135. Gottlieb Gluge, "Sur les progrès que l'anatomie et la physiologie humaine ont faits dans les derniers temps en Belgique," *Bulletin et mémoires de l'Académie royale de médecine de Belgique* 8 (1849): 684–99, on p. 686.

136. Marcq, *Essai sur l'histoire*, 130–3.

137. Hector Leboucq, *L'anatomie humaine et les tendances modernes de la morphologie* (Ghent: Annoot-Braeckman, 1907), 18–19.

138. For example Thomas Harris, *Manuel d'autopsies ou méthode de pratiquer les examens cadavériques au point de vue clinique et médico-légal*, trans. H. Surmont (Brussels: Manceaux, 1888), 37.

139. For example Adolphe Swaen to Louis Trasenter, 15 June 1882, folder 254, Faculté de médecine 1882–1912, UAL.

140. 'An imperturbable soul' is a phrase from the Belgian anatomist Jozef François Kluyskens, quoted in: Veronique Deblon, "Een nieuw beroep en een onverstoorbare ziel: de rol van de anatomie in de gezondheidszorg rond 1800," in *Vesalius: het lichaam in beeld*, ed. Geert Vanpaemel (Leuven: Davidsfonds, 2014), 86–93, on p. 91.

141. Dossier concernant les mesures prises pour la désinfection des amphithéâtres, 1883, folder 107, Autopsies, Affaires générales, ASSB.

142. On France and the United States: Emmanuelle Godeau, *L'esprit de corps: Sexe et mort dans la formation des internes en médecine* (Paris: Maison des sciences de l'homme, 2007), 27–39; Sappol, *A Traffic of Dead Bodies*, 74–97.

143. For example Armand Colard, *Souvenirs du vieux Saint-Pierre* (Brussels: Arcsia, 1952).

144. On dissections (and photographs of dissections) as rites of passage: John Harley Warner and James M. Edmonson, *Dissection: Photographs of a Rite of Passage in American Medicine, 1880–1930* (New York: Blast Books, 2009).

145. On women and medical education in Belgium: Denise Keymolen, "Feminisme in België. De eerste vrouwelijke artsen (1873–1914)," *BMGN Low Countries Historical Review* 90 (1975): 38–58.

146. Laura Kelly, "Anatomical Dissections and Student Experience at Irish Universities c.1900s–1960s," *Studies in History and Philosophy of Biological and Biomedical Sciences* 42, no. 4 (2011): 467–74.

147. Quotes from "Encore M. Bonmariage," *Journal de Bruxelles*, 27 February 1881 and "Kadaver-onderwijs," *Het Handelsblad*, 22 February 1881.

148. Quotes from: "Lettres parisiennes," *Journal de Bruxelles*, 28 November 1881 and "Lettres parisiennes," *Journal de Bruxelles*, 7 February 1887. Another telling example is: "La vie anglaise," *Indépendance belge*, 2 March 1892.

149. Petra Gunst and Fien Danniau, "De Vriese, Bertha (1877–1958)," UGentMemorie, Accessed 10 August 2015. http://www.ugentmemorie.be/personen/de-vriese-bertha-1877–1958.

150. Richard Boddaert, *De l'importance des études pratiques en médecine* (Ghent: Hebbelynck, 1870), 10.

151. Dhondt, *Un double compromis*, 261–4 and 285–306.

152. For example: Boddaert, *De l'importance*, 18; Gustave Verriest and Jean-Baptiste Carnoy, "De l'organisation des études médicales," *Revue médicale* 2 (1883): 44–53.

153. Geert Vanpaemel, "The German Model of Laboratory Science and the European Periphery (1860–1914)," in *Sciences in the Universities of Europe, Nineteenth and Twentieth century: Academic Landscapes*, eds. Ana Simões, Maria Paula Diogo, and Kostas Gavrogly (Dordrecht: Springer, 2015), 211–25.

154. Dhondt, *Un double compromis*, 361–84.

155. Etienne Van Kempen, "Rapport adressé à M. le Ministre de l'Intérieur, sur l'état de physiologie en Allemagne," *Annales des Universités de Belgique* 3 (1844): 1054–7.

156. Jean-Baptiste Carnoy, *Révision de la loi de 1876. Les programmes des examens de sciences naturelles et de médecine* (Leuven: Fonteyn, 1889), 27–9.

157. Ibid., 25.

158. Dhondt, *Un double compromis*, 363–5.

159. Charles Delcour, *Situation de l'enseignement supérieur donné aux frai de l'état. Rapport triennal 1871–1872 et 1873* (Brussels: Gobbaerts, 1876), viii.

160. Boddaert, *De l'importance*, 12–15.

161. Dhondt, *Un double compromis*, 369–74.

162. Based on a comparison of the course schedules in *Annuaire de l'Université catholique de Louvain* (Leuven: Vanlinthout et Vandenzande), 1840 and 1880.

163. Louryan, "L'encadrement," 117–24.

164. Jean-Joseph Crocq, [no title], *Bulletin de l'Académie royale de médecine de Belgique* 11 (1877): 857, quoted in: Dhondt, *Un double compromis*, 368.

165. Dhondt, *un double compromis*, 368. See also: Bardez, "La Faculté de médecine," 260–4.

166. Dhondt, *Un double compromis*, 238.

167. Ibid., 367–74.

168. *Recueil de documents concernant la révision de la loi du 20 Mai 1876 sur la collation des grades académiques et les programmes des examens universitaires* (Brussels: Adolphe Mertens, 1883), 5.

169. Gustave Verriest, "De l'organisation des études médicales," in *Révision de la loi de 1876*, 2.

170. *Règlement pour la collation des grades institués par les lois du 10 avril 1890 et du 3 juillet 1891* (Leuven: Van Linthout, 1911), 13–15. Note that in 1890, topographic anatomy became part of the medical candidature (before, it had only been a compulsory course in the doctorate).

171. Nyhart, *Biology takes Form*, 67.

172. Ibid., 86.

173. Thomas Gieryn, *Cultural Boundaries of Science. Credibility on the Line* (Chicago: University of Chicago Press, 1999), 1–35.

174. Marcq, *Essai sur l'histoire*, 130.

175. Buklijas, "Cultures of Death," 586.

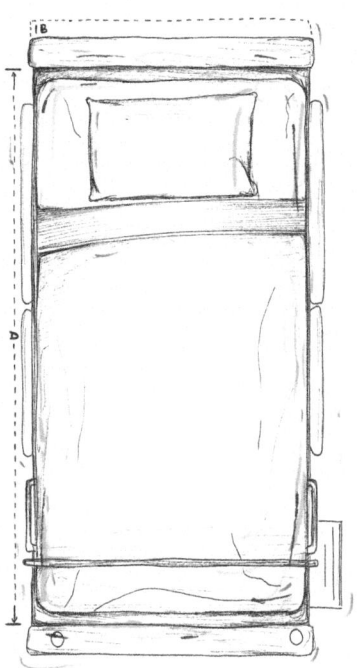

CHAPTER 3

From Deathbed to Dissecting Table: Acquiring Anatomical Material

1   PREAMBLE: THE COCO AFFAIRE, 1887

March 2, 1887. A man leans on his shovel. Around him are freshly dug graves, some with wreaths made from wild flowers, some with self-made crosses, others with nothing at all. It is the part of the cemetery where the poor are buried, far away from the main entrance, far away from the shrines and memorial stones of the rich. Almost every day a simple hearse towed by two horses finishes its route here, stacked with plain wooden coffins. It stops at all public hospitals in Brussels, collects their dead and brings them to their final resting place. It is a chilly Wednesday morning.

The man lifts his shovel and continues digging. Soil. Stones. Debris. More stones. Whereas the wealthy are able to buy their dead an eternal resting place, the poor are supposed to decompose within five years so their graves can be reused. Yet decomposition is slow in the stony ground of the cemetery. Too little air, too little bacteria. The city council's efforts to fill the graves with sabulous clay, transported from the other part of town, are only partly successful. Graves are refilled while the previous owners are still lying there, putrefied but not vanished.

Finally the shovel hits a coffin. With the help of two policemen, the gravedigger hoists the rectangular box up. It is heavy, unusually heavy for

© The Author(s) 2019
T. Claes, *Corpses in Belgian Anatomy, 1860–1914*, Medicine and
Biomedical Sciences in Modern History,
https://doi.org/10.1007/978-3-030-20115-9_3

a coffin that should contain the remains of a young woman. The inspector of the cemetery, Mr. Vandevorst, walks up. 'Open it', he commands impatiently. 'Pull the nails out'.

The day before, Vandevorst had received a phone call from alderman De Mot. If he could exhume the coffin of a certain Cécile Amadou, who had been buried last Friday? De Mot had told him about the mysterious letter in his mailbox. How he had opened the envelope and found a piece of paper with cut-out letters from newspapers. 'The Negress who died in the hospital of St. Jean was not buried', it said.

The gravedigger feels Vandevorst's eyes boring into his back while he is tries to pry the nails loose with his cold hands. Six nails, one in each corner, two in the middle. The lid of the coffin comes off slowly, revealing a stovepipe, stones and firewood. The gravedigger turns his head to face Vandevorst, who is looking at the policemen with a meaningful gaze. They all wonder the same thing, even though nobody voices the question out loud. Where is the body of Cécile Amadou?

*   *   *

Cécile Amadou had been coughing for weeks. A persistent cough that irritated her throat and made it painful to breath. At first she thought that she had caught a regular cold, as she had done a million times before, wandering through the rainy streets of Brussels. For weeks she coughed up phlegm and mucus, felt feverish and listless, began to lose weight.

She refused to go to the hospital. She knew that most people left that awful place on their back. Feet first in a coffin carried out by those drunk and rude boys, *garçons d'amphithéâtre* they were called, through the backdoor that looked out the Rué Pachéco. But then she started to cough up blood.

Four months before her empty coffin was exhumed, Cécile Amadou reluctantly entered the St. Jean hospital. Her body felt weak. She had woken up in a pool of her own sweat. She was not well-rested as dreams had disturbed her sleep. She had to do something. It was 9 in the morning on the 5th of November when she signed the hospital papers and was brought to a bed in a ward for the poor.

A few hours later she overheard a crowd of students surrounding her sickbed. They were discussing her clinical diagnosis with their teacher. Taking into account her background as a 'woman of poor morals and ill repute', they decided she probably had tuberculosis. Much like syphilis and alcoholism, this dreaded pulmonary disease often affected prostitutes and 'other degenerates'. One student, observing her lifeless complexion, sunken cheeks and emaciated body, suggested that the bacteria might already be in her blood.

Cécile Amadou's hopes of walking out alive started to melt. She should never have come here. It was 7 in the evening on the 23rd of February when she died. However, she would not leave the hospital feet first in a coffin.

*　*　*

When Emile Houzé heard that the Negress had finally died, his hands shook in excitement. Even though he spent most of his days wandering the hospital wards of St. Jean as a physician, his true passion was physical anthropology. Five years ago he had published his first anthropometrical study on the anatomical differences between Flemish and Walloon skulls. In the same year he became a founding member of the Society of Anthropology of Brussels. He started to measure the skulls of hospital patients who died in his ward, and occasionally prepared one of those skulls for the anatomical museum. He also published studies on several African skulls, which were brought to him from the Belgian Congo.

He rushed to the office of the hospital director, Mr. Mascart. Houzé needed his permission to dissect and conserve the skeleton of Cécile Amadou. If her family and friends had not claimed her remains, there would be no problem. If they had, he would have to be creative. One thing was for sure: he would not let this opportunity slip away. A skull of the 'pure Negro type' was too valuable for science to be wasted to the earth.

Her body had been claimed, said Mr. Mascart. The funeral service would take place in the chapel of the hospital on the 25th of February. The only way to conserve the body of Cécile Amadou for the scientific collection would be to ask her family, but it was unlikely that they would grant their permission. Or, perhaps, he could make sure that he only took the parts of the body that nobody would miss?

While Cécile Amadou's family and friends were praying for her salvation, Houzé sneaked into the mortuary, where her coffin remained unattended. Together with a handful of students, he forced the nails out of the closed coffin, took the body out, and left a stovepipe, stones and firewood in its place. Half an hour later, the mourners left the chapel and followed a coffin filled with rubbish to the cemetery, all the while under the impression that they were in the presence of their deceased friend.

* * *

The gravedigger is standing next to the opened grave, meanwhile chilled to the bone. Newspaper reporters are pacing the cemetery, which has been transformed into a crime scene. Two representatives of the public prosecutor's office, Mr. Wellens and Mr. De Roo, are inspecting the empty coffin he dug up this morning. The stovepipe has already been removed in order to serve as evidence in their case. Wellens and De Roo walk away to talk to Vandevorst. They are taking the coffin with them as well. Finally, Vandevorst orders him to close the grave. He may go home.

Over the next few days he follows the story in the newspapers. The theft of the corpse causes anxiety in poor districts not too far from his home. For the poor, the empty coffin is a reminder of their sad fate. They cannot afford the care of a private physician. If they become ill, they have the choice between death and the hospital. If they opt for the latter, their bodies are used for medical experiments while alive, and for dissection when dead. It is unfair. Nobody wants to end up like Cécile Amadou.

He also reads about the judicial investigation. Mascart, Houzé and the *garçons d'amphithéâtre* have been interrogated. Mascart is suspended from his job for at least one month. Quickly after he had left the cemetery on the 2nd of March, the police invaded Houzé's home. They found a few of Cécile Amadou's bones drying in an oven, whereas others were still boiling in a large kettle. Apparently Houzé had put the rest of her body in the coffin of another hospital patient, who had been buried at his cemetery on the 28th. He wondered if he would have some more digging to do in the following days.

Even though most newspapers thought the theft of Amadou's corpse was scandalous, Houzé never denied the facts. He firmly believed that science legitimised his actions. A few journalists agreed with him. One of them reported that 'the little affair received an exaggerated amount of attention it did not deserve'.

The gravedigger did not agree. Times are changing, he thought before he closed the newspaper.[1]

2   INTRODUCTION

In 1888, professor of anatomy Louis Deroubaix presented a plan of action with the objective of acquiring more corpses for education. He wanted to take a few measures to improve the 'deplorable situation' he was in. The number of bodies had become insufficient as a result of the greater emphasis on hands-on dissection in the curriculum and of the growing student population. Due to a lack of teaching 'material', he had been forced to suspend his lessons, causing 'a certain disarray among the students'.[2]

In his plan, Deroubaix insisted that the existing hospital regulations had to be strictly followed, which meant that the bodies of deceased patients could only be buried without prior dissection if the funeral had been paid in full. He also proposed to make the regulations more stringent. He thought it was a good idea to ask families for financial compensation for both the funeral and the hospital treatment, or at least to raise the funeral costs. Finally, he proposed that the university 'could pay a premium for each body left to science'.[3] Deroubaix' plan reflected a way of thinking in which death turned patients into negotiable commodities: as material objects, corpses could be used as a means of compensation, and could be compensated for.

The proposal of Deroubaix marks a transitory period in the history of the acquisition of corpses for dissection. On the one hand, its reasoning builds on a reciprocal logic: the use of the bodies of poor hospital patients is seen as a fair compensation. The idea is that the hospital is entitled to the corpse because of the care provided. On the other hand, the suggestion to 'pay a premium' attests to changing attitudes towards the ownership of patients' corpses. By his willingness to pay for their acquisition, Deroubaix implicitly admitted that bodies do not *naturally* belong to the hospital. They had to be obtained from the patients themselves (during life) or their families (after death).

This shift is at the heart of this chapter, which links discussions concerning the procurement of corpses for dissection in Brussels to larger changes in attitudes towards death and the destitute.[4] Stirred by social unrest, the forced dissection of the poor increasingly came under question from the 1860s onwards. By looking at this debate from multiple perspectives, this chapter sheds new light on the acquisition of bodies. Special attention will be paid to ideological tensions and the codification of the principle of consent. While historians have mainly studied the issue of consent in relation to medical experimentation and surgery, the case of Brussels suggests that it had a profound impact on the procurement of corpses for dissection, inspiring the emergence of anatomical donation in the twentieth century.[5]

Because the procurement of corpses for dissection only changed gradually, the developments discussed in this chapter occasionally reach back to the early nineteenth century and go on until the 1960s. My focus, however, is on the decades around the turn of the twentieth century, from the 1880s until the 1910s. As will become clear throughout this chapter, this was a pivotal period in the history of the acquisition of anatomical material. Different political, ethical and legal developments came together, obliging anatomists to change their ways. Dead bodies ceased to be regarded as physicians' property and came to be seen as an extension of the living patient instead.

This chapter deals with the acquisition of corpses for dissections (complete divisions of the body into anatomical structures) as opposed to autopsies (partial post-mortem examinations, mostly—but not always—aimed at determining the cause of death, see also Table 1.1). I use the terms 'post-mortem' and 'post-mortem examination' for both dissections and autopsies. These distinctions are crucial because the social purposes and cultural sensitivities surrounding these two operations differed greatly. While autopsies were often performed on members of the higher social classes to serve family interests, dissections were mostly imposed on the poor for the benefit of medical teaching and research.[6] Whereas the outward appearance of the corpse was, at least nominally, respected during autopsies, dissections transformed the corpse into an unrecognisable mess of tissues and bone, hence prohibiting common burial rituals.[7] Most importantly, for the British context, Helen MacDonald has drawn attention to the different legal frameworks for dissection and autopsy. While there were laws on the distribution of corpses for dissection, the rules concerning autopsies were unclear, and therefore more lenient.[8] In the next chapter, in which I look more closely at the differences between dissections and autopsies, I show that in Brussels, too, not all post-mortems were regarded or regulated in the same way.

This chapter explains why donated bodies began to replace unclaimed bodies in Brussels. In the first half of the nineteenth century, the use of the bodies of the poor, without their consent, was described as a 'natural reciprocity': as a righteous compensation for their care or burial (Sect. 3). This began to change as a result of ideological debates on the treatment of the poor (Sect. 4). In the 1860s, the dissection of indigent patients acquired symbolic importance in Catholic-Liberal conflicts over burials, and in the 1880s, the Socialist Party represented the dissected pauper as the epitome of class injustice. Then, I argue that this resistance against the medical use of the poor was further intensified by the codification of the principle of consent in 1889, and by a legal reassessment of the corpse in 1899 (Sect. 5). Together, these interconnected developments led to a shift in the procurement of corpses for dissection from the turn of the twentieth century onwards. As indigent patients gained ownership of their own bodies, anatomists increasingly had to comply with the standard of consent. By studying the early days of anatomical donation (Sect. 5), this chapter sheds light on a neglected topic in the social history of anatomy, as well as on the changing meanings of burial rituals, which began to reflect the identity and wishes of the deceased.

3   A 'NATURAL RECIPROCITY'

In 1948, the Brussels professor of forensic medicine Maurice De Laet wrote an essay on the legal framework for autopsies and dissections in Belgium. To his regret, he found that Belgian legislation was 'relatively silent on the authorisation or interdiction of post-mortems'.[9] There was only one national law on the subject, which dated back to *3 Vendémiaire, an VII*—or 24 September 1798. On that day, the French republican government gave anatomists in the Southern Netherlands the right to use the corpses of paupers who died in state-funded hospitals.[10] In Paris, poor hospital patients had been used for post-mortem research and education from the early eighteenth century onwards.[11] The great number of bodies made available in this way had turned the French capital into a medical centre par excellence, attracting foreign students and researchers in search of hands-on experience.[12] In 1798, the republican government established legislation on this (existing) practice of dissecting hospital patients by naming the *Hôtel-Dieu* and *Charité* hospitals official 'suppliers' of medical schools. The system was also exported to newly acquired territories.[13] The procurement of corpses for dissection in the Southern Netherlands was founded on the basis of French legislation that, as De Laet noticed 150 years later, stayed in place despite two regime changes. The law remained operative when the Southern Netherlands became part of the Kingdom of the Netherlands in 1815, and after Belgian independence in 1830.

In the Southern Netherlands, the law of 1798 confined the right to dissect to qualified teachers of anatomy, who had to attend to a few formalities before they could take up the dissecting knife 'for strictly didactic and scientific reasons'. Most importantly, they had to register with the police and arrange the distribution of dead bodies with the local authorities. The procurement of corpses became a primarily municipal matter. The distribution of anatomical material could differ between cities, as it was arranged by mutual agreement between medical faculties, city councils and *commissions administratives des hospices civils* (administrative commissions for hospitals and other public healthcare institutions within a municipality).[14] In principle, infractions against these arrangements did not fall under criminal law, but under local law.[15] The Belgian situation in this regard resembled the regulation of post-mortems in Germany, where the decision as to how to deal with the corpses of hospital patients depended on the regulations of the individual hospital, and therefore differed per locality and region.[16] On a more general level, the limited legal framework on dissection in Belgium fits with the conclusions of historians of crime, who have emphasised that the nineteenth century still knew high levels of discretionary justice. Law was made as often on the ground, by local practice, as it was by central authorities.[17]

Apart from small differences, regulations in Belgian university cities were comparable with each other in the nineteenth century. Only *corps abandonnés* ('abandoned corpses') or *corps non-réclamés* ('unclaimed corpses') were used for dissection. In Brussels, for example, relatives could claim the bodies of deceased hospital patients, and hence prevent their dissection, within 24 hours after the establishment of death. In Ghent, Leuven and Liège, too, family members could claim bodies within a limited period of time. Unclaimed corpses ended up in the anatomical theatre after 36 or 48 hours. The right to claim bodies also implied financial responsibilities. In all university cities, paying for the burial was a requirement to claim the body of a deceased hospital patient. Families sometimes also had to reimburse the cost of treatment, mostly when corpses were in short supply.[18]

Both conditions—time and money—frequently prevented relatives from claiming the bodies of their loved ones. Especially in Brussels, where dissection could happen after no more than 24 hours, families were often late. Sometimes their relatives' corpses were fragmented before they even knew about their passing. This was probably an intended result of hospitals' system of notification. When a patient died, hospitals informed his or her family by mail. Instead of a *lettre express*, they sent a registered letter. The result was to be expected:

> The addressee, who is often absent when the mailman arrives, [...] only receives this letter two or three days after his or her relative's death. Therefore it happens that decent citizens arrive at the hospital when the body is already at the anatomical institute.[19]

This led to heart breaking situations. On 14 January 1887, for example, Martin Cuypers died at the Brussels St. Pierre hospital at seven o'clock in the morning. A form letter that confirmed his death was sent to his father, Pierre, who received the sad news around three in the afternoon. Because in the letter, he was asked to communicate 'all useful information with regard to the deceased' in order to organise the funeral, he first visited the burial society of which his son was a member.[20] He only made it to the hospital the next morning. When he asked to pay his son a last visit, he was told 'he was already buried'.[21] In reality, the body of Martin Cuypers had been brought to the dissecting room one hour before, a mere 25 hours after death. His father was one hour late. In response to a complaint, the hospital employee stated that he had 'judged it necessary to tell the father that the burial had already taken place' because 'the cadaver was in no state to be exhibited anymore'.[22]

This was not an isolated incident. Two years earlier, a certain Mme. Pouillon arrived at the Brussels hospital of St. Jean to visit her husband. She simultaneously found out that she was widowed—she had not yet received the news of the death of her husband—and that she had 'missed the burial'.[23] These examples show that secrecy surrounded the procedure to claim bodies. In general, anatomists believed that being honest and open with 'little enlightened people', loaded with all sorts of 'regrettable prejudices', would result in a sharp decrease in the number of bodies available for dissection.[24] This explains both why they lied to families about the fate of their deceased relatives (who were consigned to the scalpel rather than to the grave) and why the form letter for families was formulated vaguely (stating they should inform the hospital about their funeral preferences, rather than making clear they had to claim the body in order to avoid dissection).

The limited timeframe to claim bodies in part followed from practical reasons. Bodies had to be moved to the dissecting room quickly because putrefaction hindered their use for education and research. Corpses in decomposition were more difficult to examine, as tissues changed in size and shape, and because they carried health risks. Despite the gradual introduction of embalming and other preservation methods, anatomists throughout the nineteenth century complained that a significant number of bodies, in summer up to one fourth of all corpses, arrived in the anatomical theatre 'in an advanced state of putrefaction, which prohibited their use for dissection'.[25] Putrefaction was particularly problematic for those interested in neurology, 'since the nervous system, the bone marrow tissue especially, is altered by microbes of decay in a great number of cases'.[26]

When families did arrive on time to claim the body, they faced another challenge: paying for the burial of their deceased relative. Members of the lower classes of society regularly ended up in the anatomical theatre against their or their families' wishes because of insufficient financial means. Again, this was not an unfortunate side effect of regulations, but a desired outcome. Burial costs were increased when corpses were in short supply, often by only offering middle-range coffins. In 1891, for example, the medical faculty of Liège complained about the provision of caskets 'at a low price', for it endangered practical anatomy lessons.[27] In Brussels, too, Deroubaix thought 'an increase in the price of burial' would help to solve the shortage of bodies for dissection.[28]

Mostly this kind of reasoning was rooted in a notion of indebtedness. The idea was that by facilitating the advancement of medicine with their bodies, indigent patients settled their debts with society.[29] In the words of a much-read French medical dictionary, the dissection of the poor was 'a natural reciprocity'.[30] In the view of physicians, the use of the bodies of hospital patients was a fair compensation for the care or the burial provided. According to the medical faculty of Brussels, for example, 'the right to be buried without an autopsy' was inseparable from the duty to bear expenses. To give the poor the opportunity to bury their dead without compensation for the services they had received was considered to be 'an exaggerated emphasis on individual liberties'.[31]

The burden of debt not only determined the post-mortem use of the poor, but also influenced their treatment during life. In his groundbreaking work *The Birth of the Clinic*, Michel Foucault described the provision of hospital care as a tacit contract between the rich and the poor. He argued that care financed by rich citizens and politicians was repaid by generally poor hospital patients, whose bodies became subject to the medical gaze.[32] More recently, Grégoire Chamayou has posited that indigent patients reimbursed society with their 'cognitive surplus value', with which he meant the knowledge contained in their bodies.[33] Similar analyses were already put forward in the nineteenth century. For example, the French physician Séverine Icard, whose publications were also popular in Belgium, noted that there was 'a tacit contract between science and poverty'. In his opinion, poor patients paid 'the highest price for care': namely *le don complet qu'il fait de sa personne à la science*, the complete donation of 'their selves' to science.[34]

It was indeed on the bodies of the poor that future physicians learned how to examine, diagnose and research. Nicolas Ansiaux, professor of surgery at the University of Liège, for instance, taught his students that hospital patients 'were destined to teach them the medical practice'.[35] Students refined their medical techniques during clinical lessons, so that they could use them on a rich clientele once they got their degree. Drugs, treatments and surgical procedures were also improved by tests on hospital patients, mostly without their consent. Dissection was the apogee of this healthcare system, in which assistance was only granted in exchange for the subjection of one's body to science.[36]

For most of the nineteenth century, little opposition seems to have existed against this practice in Belgium. Physicians and politicians argued that the provision of medical care (or a burial) made up for the strict conditions. The left-wing progressive Liberal politician Paul Janson, for example, asserted that 'when entering the hospital, where one is surrounded with dedicated and intelligent care, it is natural to accept the consequences of the hospital regime'.[37] Patients followed this reciprocal logic in written complaints. They stated, for instance, that their cooperation in clinical lessons or in the collection of human material for research had not been rewarded with the treatment they sought: 'despite everything, I have not had surgery yet'.[38]

The same went for corpses. Dissections were described as justified if patients had received proper care. If there had been no treatment, for instance if the patient had died in the ambulance, hospitals were not legally entitled to perform an autopsy or dissection on the body.[39] Similarly, families sometimes complained that their relatives had been dissected 'even though' they had only stayed in the hospital for a short time.[40] It is important to note that historical evidence is insufficient to argue that the mostly illiterate poor were routinely informed about, and agreed to, the use of their bodies for medical purposes. On the contrary, sources suggest that medical faculties tried to keep the regulation on dissection a secret. Yet the few remaining complaints nevertheless suggest that the logic of exchange underpinning the working of hospitals was not questioned fundamentally before the late nineteenth century.

This tied in with more general attitudes towards poverty. Even though an elaborate network of private charitable organisations, often related to the Roman Catholic Church, existed in Belgium, hospitals and other institutions for the 'infirm' or 'recalcitrant' poor, such as psychiatric institutions, homes for the elderly, orphanages or workhouses, were often at least in part state-funded. For the poor staying in these state institutions, care was rarely unconditional. Following a utilitarian philosophy, they mostly had to submit themselves to a regime of underpaid labour and restricted freedom before they received assistance. By obliging the poor to perform hard labour, governments wanted to procure some kind of 'utility' from their institutionalisation, while urging them to return to a life of wage work as soon as possible. Although most studies on the utilitarian character of nineteenth-century Belgian poor relief have focused on workhouses and houses of correction—institutions for beggars and vagrants rather than for the (working) poor at large—hospitals reflected the same logic.[41] In addition, it appears that hospitals were part of the same network of institutionalised care. When corpses were in short supply, for example, medical faculties often acquired bodies from other state-funded institutions, including psychiatric institutions and prisons.[42]

Severe responses to poverty stemmed from an optimistic belief in self-improvement and individual responsibility. The Belgian constitution of 1831, as well as the political culture of the first half of the nineteenth century more generally, conveyed a liberal range of thought. Politicians mostly interpreted human behaviour in terms of possibility and responsibility. The idea was that the individual, whose freedom was guaranteed by the constitution, was able to choose his or her profession and living conditions. Politicians believed that the 'worthy' poor would be able to provide for themselves through hard work, morality and precaution. Only a few indigent people were believed to 'deserve' assistance, for example, those incapable to work because of invalidity. Poverty was mostly seen as a condition for which one was individually accountable. It was a result of laziness or immorality. In houses of correction, for instance, the 'undeserving' poor were to be 'disciplined' and 're-educated'. In accordance with this way of thinking, the few social measures taken before the 1880s and 1890s aimed for individual empowerment instead of structural change.[43] The use of the bodies of the destitute within state-funded hospitals fitted with this larger system of poor relief, in which poverty was considered a personal mistake one had to pay for.

Yet not everyone shared these harsh attitudes towards the poor. As aforementioned, there existed a large network of private charitable organisations in Belgium, which mostly provided assistance on the basis of Christian ideals. Their idea of care did not require material compensation, although acts of altruism were often financed or performed to assure one's own salvation.[44] Significantly, clergymen and women were also important within public healthcare institutions. The daily care for patients rested almost entirely in the hands of nuns for most of the nineteenth century, as the first training programmes for lay nurses were only established in the 1880s. As late as 1900, 18.30 per cent of all nuns (5738 women) in Belgium were involved in hospitals. In comparison with lay staff, this number is impressive: in 1910, only 1644 lay women were active as care-takers, nurses and helpers in all medical establishments, hospices and poor relief services.[45] From these figures, it is clear that Christian charitable ideals influenced the workings of hospitals, even if they were state-owned.

In fact, as a result of different ideas about the 'utility' of charity, conflicts between medical faculties and hospitals were not uncommon. Commissions of hospitals, in which representatives of the Church had a seat, sometimes made it easier for families to claim bodies. In Liège, for instance, the price of coffins was decreased several times in the 1880s. This led to shortages for anatomists.[46] In Brussels, too, anatomists thought that the commission of hospitals had made it too easy to claim corpses. In their opinion, low burial fees resulted in 'an excessive scarcity of bodies'.[47] In a meeting of a scientific society in 1874, the anatomist Jean Joseph Crocq argued that the hospital's proposal to soften regulations, more specifically the proposal to give families more time to claim bodies, attested to 'an exaggerated respect for cadavers'.[48]

Another reason why involuntary dissection was sometimes regarded as too severe was the continued association with the post-mortem punishment of criminals. Even though under Belgian penal law, post-mortem punishment did not exist—unlike in Britain, where the Murder Act of 1752 stipulated that 'some further terror and peculiar mark of infamy be added to the punishment'—links between dissection and criminality existed.[49] Before corpses of hospital patients became available for dissection in 1798, the bodies of convicts had been the main source of anatomical material. The unclaimed bodies of prisoners continued to end up in anatomical theatres until the mid-twentieth century in Brussels and Ghent.[50] Anatomical displays further strengthened this association. Medical faculties collected inmates' skulls throughout the nineteenth century: first for phrenological research, later within the framework of criminal anthropology.[51] In popular anatomical museums, which toured Belgian fairgrounds from the early nineteenth century until the 1930s, skeletal remains and wax models of famous criminals were a common attraction.[52] As a result, dissection came to signify both poverty and crime in the popular imagination. To mobilise criticism against involuntary dissection, for example, politicians argued that treating the bodies of the poor in the same way as (or even worse than) those of convicts turned dissection into a punishment for a poverty-stricken life at public expense:

> Today, all bodies of persons, who died in hospital, are delivered to the faculty of medicine in the interest of science. [...] You would not say, that even the penal code, in article 10, prescribes that the body of an executed criminal is to be brought to his or her family when claimed.[53]

Belgium was not unique in this respect. Across Europe, anatomists regularly dissected the bodies of the poor without their approval in the eighteenth and nineteenth centuries. In this period, bodies of prisoners, the only legal source of anatomical material in early modernity, were gradually complemented or replaced by bodies of indigent hospital patients.[54] Much like in Belgium, both the growing importance of dissection for medical education and changing attitudes towards poverty informed this shift.

In her pioneering research on Britain, Ruth Richardson drew attention to the relationship between dissection and poverty politics. She convincingly argued that the Anatomy Act of 1832 paved the way for a political system 'which antagonised the poor as a class'.[55] The Anatomy Act tried to prevent clandestine exhumation and body trading by allocating the corpses of those dying in public institutions, most importantly the workhouse, to medical faculties. The sacrifice of the corpses of the destitute had to protect the remains of the propertied classes. This reflected the redistribution of power in the reformed electorate, benefitting the middle classes but suppressing the voteless poor. Significantly, Richardson described the Anatomy Act as an 'appendage' of the New Poor Law (1834), which established a harsher poverty relief system.[56]

In other European countries, too, the misfortune of poverty could qualify a person for dissection. In the German *Länder*, the distribution of corpses to medical faculties went hand in hand with debates on the deserving and undeserving poor.[57] In the city of Haina, for example, indigent hospital patients tried to steer clear from anatomists by emphasising their decency and morality in appeals, hence differentiating themselves from the undeserving poor.[58] In the Habsburg Empire, the anatomical use of the poor accorded with the utilitarian views of Maria Theresa and Joseph II, who saw destitute patients' contribution to the 'greater good' as a righteous repayment for the free medical care they had received.[59]

Historians working on Britain and the United States, the main focus of research until today, have drawn attention to protest against these regulations. They put forward that the dissection of the poor had to happen covertly in order to avoid popular fury. Despite this secrecy, anatomy riots were not unusual. Protests often flared up in times of social unrest, for example during epidemics.[60] Yet opposition was less outspoken in continental Europe. Tatjana Buklijas has convincingly argued that 'the transition from the criminal body to the pauper body was not universally contested'.[61] Protests varied regionally, according to religious beliefs, local customs surrounding death and politics. In Vienna, for instance, there was nearly no resistance against the anatomical use of the poor until the late nineteenth century.[62] In other continental European countries, too, attitudes towards the dissection of the poor were quite permissive. In the German *Länder*, for example, resistance was overcome by offering free burial for dissected corpses.[63] In France and Belgium there is no evidence of violent protest either. The idea that state-funded hospital care had to be compensated by placing one's body at the disposal of medical science appears to have been more or less accepted.

Research suggests, however, that the acquisition of corpses for dissection did become increasingly difficult in continental Europe in the late nineteenth century. In Vienna, protests arose when the anatomical use of paupers' corpses was integrated in the anti-Semitic propaganda of new mass political parties in the tense atmosphere of the *fin-de-siècle*.[64] In Germany, popular resistance against post-mortem interventions grew as cadaver research lost academic prestige from the 1880s onwards, and continued to simmer in the turbulent political context of the Weimar Republic.[65] More generally, historians have shown that the extension of democracy towards the end of the nineteenth century led to criticism against utilitarian and punitive poor relief, which resulted in reforms across

Europe. As the aim of emancipation substituted the wish to discipline and moralise the poor, strict institutional regimes gradually made way for insurance policies.[66]

In the remainder of this chapter, this juncture is the object of analysis. Whereas the beginning of the nineteenth century was characterised by more rigorous attitudes towards the poor and their bodies, the end of the century was marked by an opposite—and opposing—trend. Around 1800, the growing prestige of anatomy combined with harsh attitudes towards the poor led to the forced dissection of hospital patients. Around 1900, however, anatomy had lost much of its scientific standing and poverty became an acute political issue. To tone down social protest, governments started to integrate paupers within society by means of social legislation and (gradual) democratisation. This time, the entanglement between the procurement of corpses and the wider politics of death, the hospital and the destitute turned out less positively for anatomists.

4   CORPSES AND CULTURE WARS

Historians have drawn ample attention to the impact of political tensions on cadaver supply. Both in Britain and in continental Europe, protest against dissection often surged in times of unrest.[67] Given the emotional significance of the dead body, the image of the dissected pauper could mobilise both religious and political pressure groups. On the one hand, dissection conflicted with the sacred importance of human remains and their proper burial, while on the other, the use of the poor for medical purposes could incite class hatred. In Brussels, both kinds of protest came together in the late nineteenth century. Politicians first raised the question of involuntary dissection in conflicts on the secularisation of burial grounds in the 1860s. During the 1880s, the use of the bodies of the poor also became a symbolic issue for social reform movements that advocated class equality and electoral reform. As a result of both the burial question and the social question, the involuntary dissection of the poor gradually became ethically unacceptable.

'Liberal Corpse-Violators'

Discussing the burial policies of asylums in Wales, Pamela Michael and David Hirst have stressed the importance of religious conflicts in the political arena. They argued that political debates on the 1880 Burial Act, which allowed non-conformist burials in Anglican cemeteries, resulted in a growing recognition of the burial wishes of families of psychiatric patients.[68] In a similar vein, political discussions on the control of cemeteries impacted the burial of the institutionalised poor in late nineteenth-century Brussels. As the Liberal city council bore the final responsibility for the burial of the unclaimed bodies from hospitals, the post-mortem treatment of the corpse, from hospital bed to burial ground, was caught up in an ideological struggle with the Catholic Party.

The basis of this rivalry was a disagreement on the structural relationship between the Roman Catholic Church and the State. The Catholic Party's model of society was grounded in religion, whereas the increasingly anticlerical Liberal Party aspired to secularisation.[69] In the 1860s and 1870s, both parties (later followed by the Socialist Party) involved themselves in virtually all areas of society. In this way, they hoped to draw the new middle-class voters into their networks. After all, suffrage had been expanded in 1848 (when the poll tax was lowered to the constitutional minimum) and plural voting was introduced in 1893 (allowing every male citizen over 25 to receive at least one vote).[70] Citizens were expected to participate in the organisations and activities of their respective ideological 'pillar' from cradle to—literally—grave. One of the key issues in the culture war between Catholics and Liberals became the burial, which the Roman Catholic Church traditionally had controlled.[71]

Much like in France, Belgian freethought organised itself around the burial in the second half of the nineteenth century.[72] Civil burials (burials without Catholic ceremony) were no longer imposed one-sidedly by the clergy, as they had been in the first half of the nineteenth century, but became the choice of individuals who wanted to die outside of the Church.[73] Because of their symbolic character, funerals allowed freethinkers to reflect on the organisation of life outside of the traditional patterns of religion and to create shared identity and rituals. Civil burials were often collective events, meant to express and strengthen anticlericalism as a political force. Sometimes these burials went together with mass demonstrations, accompanied by Catholic counter-manifestations.[74] Exhumations were also tools of propaganda. By disinterring the bodies of sympathisers who had received a Catholic burial against their will, Liberals staged the ultimate triumph of freethought over religion.[75]

Mutualist societies brought this ideological struggle to the masses around 1860.[76] To attract members of the working and middle classes into their circles, freethinkers founded anticlerical societies for the promotion of civil burials. These associations asked a moderate fee in exchange for the provision of decent burials without Catholic rituals. For poor members, funerals were provided for free on the condition that they had died and were buried without the interference of clergy.[77] In response to these societies, which were said to lure the poor into damnation by promising 'things for which they care a lot', namely 'a decent funeral and a proper coffin', similar organisations were soon established within Catholic circles.[78]

The burial conflict in essence revolved around a lack of legal clarity. Under nineteenth-century Belgian legislation, cemeteries did not necessarily belong to the municipality in which they were situated. They could also be owned and managed by public organisations, individuals or—most commonly—the parish.[79] In practice this amounted to a Catholic monopoly on burials. Although the Napoleonic Decree of 1804 had transferred the final authority over cemeteries to the municipal council, the Roman Catholic Church continued to control the primary burial grounds of many towns and cities for most of the nineteenth century.[80]

In the second half of the nineteenth century, both Catholics and Liberals interpreted the existing vague legislation to their advantage through an emphasis on freedom and choice. According to a statement from the most important Catholic congress in Belgium, the liberty granted by the constitution implied 'the right to own, consecrate and manage cemeteries as parts of the Church'.[81] For the members of the Liberal city council of Brussels, conversely, freedom signified breaking the Roman Catholic monopoly. They argued that the burial was 'in the first place a civic issue concerning questions of hygiene, health, order and public morals'.[82] They therefore thought that cemeteries should be municipal properties, where the freedom of religion would be guaranteed by allowing priests to consecrate graves of Catholic citizens individually.[83]

This last issue in particular resulted in fierce disputes between the Church and anticlerical politicians. As parochial churchyards were on consecrated ground, citizens deemed 'unworthy' of a Roman Catholic burial—like unbaptised children, suicides or freethinkers—were refused admittance. Mostly, their burial took place outside the walls of the cemetery, in a place disdainfully called the 'dogs' hole' (*trou des chiens*) or the 'corner of the doomed' (*verdoemdenhoek*). Liberal mayors of major cities in Belgium, including in Brussels, publicly renounced this practice. They accused the clergy of separating families after death.[84]

On several occasions, Liberal city councils ordered the exhumation of unbaptised children so as to bury them in consecrated ground, next to their relatives.[85] In the late 1870s and early 1880s, the Belgian Supreme Court repeatedly condemned the compartmentalisation of churchyards as an infraction of the principle of civic equality.[86]

Backed by the Supreme Court rulings and by the Liberal and anticlerical government of Frère-Orban II (1878–1884), Liberal city councils founded new cemeteries independently of the Church in Brussels, Antwerp, Liège and Ghent. The construction of extramural communal cemeteries was above all an answer to hygiene issues, but was also of symbolic and financial significance. Most importantly, the new cemeteries were neither consecrated nor compartmentalised into different sections for each religion or conviction, which implied that Catholics, non-churchgoers and unbaptised people could be buried side by side. On top of that, the revenues from grave concessions in communal cemeteries went to the municipal authorities rather than to the clergy. Facing a loss of cultural authority and financial income, the Roman Catholic Church strongly opposed these changes.[87]

Jolien Gijbels has shown that these heightened ideological tensions had a profound impact on the burial of hospital patients in Brussels. In the St. Jean and St. Pierre hospitals, the Roman Catholic Church held a privileged position for most of the nineteenth century. Although under the constitution all religions were to be treated equally within state-funded institutions, there was only one funeral room: a Roman Catholic chapel. According to complaints, burial services for freethinkers, Jews and Protestants took place in the corridor of the mortuary or even in open air. In the 1880s, the Liberal city council tackled this issue, obliging the hospitals to build new mortuaries with better sanitary conditions and allowing for decent burial services for all hospital patients, regardless of their beliefs. Gijbels put forward that this ultimately led to more freedom for families or burial societies of any denomination to make funeral arrangements for hospital patients. The last wishes of the deceased were increasingly recognised. This led to an enhancement of the burial standards for the poor: as all parties and their respective burial societies claimed moral superiority in the competition for members, the conditions of the pauper burial gradually improved.[88]

Not only the burial of dissected remains, but also the issue of involuntary dissection in itself increasingly came under question against this background. Since city authorities were ultimately responsible for the burial of the unclaimed remains of hospital patients, the association between dissection and the burial policy of Liberal city councils was easily made. The Roman Catholic Church used the image of the dissected pauper to criticise the disrespectful treatment of human remains by, in the words of a newspaper article, 'Liberal corpse-violators'.[89] In Ghent, for example, the Catholic press related the dissection of the corpses of the poor to the introduction of cheaper coffins and to the foundation of a communal burial ground by the Liberal mayors Charles de Kerchove de Denterghem and Hippolyte Lippens. When in 1882 a family discovered that the corpse of their father had been dissected without their approval, a journalist of *Het Handelsblad*, a local Catholic newspaper, seized the opportunity to criticise the policy of the Liberal city council. He represented the foundation of a communal burial ground and the dissection of the poor as two results of the same cause: the mayor's lack of respect for the dead.

> In the first place we notice that the commission of hospitals is planning to support the mayor's plan to bury Catholics in unconsecrated ground. Do they have the slightest authority? In the second place, mutilating corpses, for whatever reason, is very serious. If a citizen dies, no one has the right to dissect the corpse. [...] We do not believe that there is one poor person in the Bijloke hospital who would, if asked, agree to let his or her corpse be cut into pieces.[90]

The establishment of Catholic burial societies for paupers, too, led to discussions on the dissection of the poor. The matter came to the fore during three important Catholic congresses in Belgium, organised in 1863, 1864 and 1867 in Mechelen, the seat of the Belgian episcopate. In 1863, the issue of dissection came up during a discussion on the objectives of burial societies. Should Catholic burial societies claim the bodies of their members and hence avoid their dissection? Even though there is no clear answer to this question in the proceedings—which state that the president of the congress thought 'it was sufficient to draw attention to this point'— it is remarkable that the statement of purpose of burial societies changed after the discussion.[91] In addition to offering the last sacraments and guiding the mortal remains to the cemetery, burial societies had to 'assure the poor, as much as possible, a Mass with the corpse present'.[92] Because dissected remains could not easily be transported, let alone exhibited (open caskets were usual), this amendment suggests that burial societies were to avoid the dissection of members who died in hospital.

This was not always easy. In order to avoid a decrease in the number of available corpses, medical faculties at several moments tried to confine the right to claim bodies to relatives.[93] In this way, members of burial societies who had no family left could still end up on the dissecting table. At the Catholic congress of 1867, representatives of burial societies situated in university cities (in Brussels and Liège) indeed complained that 'special regulations almost completely undermined their charitable work' in hospitals, while others (for instance those from Antwerp or Bruges) posited that 'they buried paupers who died in hospitals in exactly the same way as other paupers'.[94]

In any case, Catholic burial societies posed a problem for members of medical faculties, who repeatedly blamed 'anti-university societies for the redemption of cadavers' for the lack of bodies for dissection.[95] Even with regulations against them in place, in principle depriving them of the privilege to claim bodies, hospital employees often let burial societies claim their members because their 'numerous and continued complaints' gave too much publicity to the dissection of the poor.[96] In Brussels, the Association of Saint-Barbe (*Association de Sainte-Barbe*), founded in 1864, organised Catholic burials for paupers and paid priests to consecrate graves in communal cemeteries. With this initiative, the Church aimed to reduce the number of civic burials in the capital.[97] The Association of Saint-Barbe also claimed the remains of members who died in hospitals, hence preventing their dissection. The Association of Saint-Barbe linked dissections—conducted by students from the Liberal university—to secular pauper burials. The head of the St. Jean hospital, in his own words, had no choice but to let this society claim the bodies of its members, 'because they would cause a scandal in the case of refusal'.[98]

The opposition of the Roman Catholic Church in Belgium partly fits with the findings of Buklijas, who argued that the Austrian Roman Catholic Church tolerated the dissection of corpses as long as the remains received a ritual burial in consecrated ground afterwards.[99] For the Belgian clergy, too, the dissection of the poor only became an issue when burial in consecrated ground was in doubt. These findings suggest that the differing positions of Catholic burial societies in different European countries—some accepting, others opposing dissection—were tied up with the ultimate disposal of dissected remains and with the politics of the pauper burial.[100] For example, the Viennese burial society of Arimathea accepted dissection because corpses could be buried in a wooden coffin, in an individual grave and in consecrated ground afterwards. The Brussels

Association of Saint-Barbe, on the contrary, saved its members from the dissecting table to avoid that they would end up in a communal, unconsecrated cemetery.

In Brussels and other Belgian cities, the acceptance of individual consecrations of graves by Leo XIII eased the burial conflict in the 1890s.[101] Tellingly, Catholic protests against forced dissection were toned down around the same time. In fact, as we will see later in this chapter, Catholics by the early twentieth century no longer represented dissection as an unrighteous and despicable violation of the body, but as an ultimate act of altruism. However, even though the Roman Catholic protest against dissection was circumstantial (since it was connected with larger ideological disputes on burials), the propagandist use of the image of the dissected pauper hints at the importance that was attached to the integrity of remains. The last chapter of this book shows that the poor themselves continued to oppose dissections when the conditions of the pauper burial improved, because the disintegration of the body prohibited burial rituals that revolved around the presence of the corpse.

All Are Equal in Illness and Death

By the time the burial question was resolved, social reform movements that advocated electoral reform and workers' rights had picked up the issue of involuntary dissection. This was not uncommon. In other countries, too, social reformers criticised the dissection of the poor. For example, Nick Hopwood has shown that the anatomical modeller Paul Zeiller made use of the highly politicised environment of the 1848 revolution in Munich to call for the replacement of anatomical instruction on 'proletarian corpses' by anatomical models. The tensions surrounding this controversy were far more explicit than they generally became in German-speaking Europe in this period. Journalists supportive of Zeiller and the revolutionary movement thought that 'to give up the corpses of the destitute without their prior consent' was 'a glaring injustice'.[102] Similar discussions on the social injustices of body procurement arose again in the deeply politically divided Weimar Republic in the 1920s and early 1930s.[103] In Vienna, involuntary dissections were criticised under the banner of equality. In the politically turbulent interbellum, anti-Semitic parties instrumentalised the dissection of poor Christians, allegedly by rich Jewish medical students, to win voters over.[104] In the United States, several Democratic politicians (unsuccessfully) opposed the dissection of the poor in the second half of the nineteenth century, because they saw this, in historian Michael Sappol's words, 'as the ghoulish and illegitimate exercise of class privilege'.[105]

In Brussels, the Socialist movement put the dissection of the poor on the political agenda in the unruly 1880s, when strikes and demonstrations for better working conditions and electoral reform followed on from each other in rapid succession.[106] To encourage class-consciousness, Socialists drew attention to the unequal treatment of the proletariat in public institutions. Two issues in particular acquired symbolic importance in their criticism on class-based society: the treatment of the poor in hospitals and the pauper burial. In both cases, the dissected pauper figured as the epitome of the suppressed working class.

The procurement of corpses for dissection was indeed based on class differences. After all, body procurement in general depended on discrimination against the indigent. Moreover, medical faculties actively pursued associations with hospitals located in working class areas. In response to body shortages, the medical faculty of the Free University of Brussels, for example, wished to strengthen its bounds with the hospital of Molenbeek, an industrial area with a large class of working poor. The importance of this institution was thought 'to grow every day', as 'the large working population from the factories of Molenbeek provides this hospital with very interesting cases'.[107] In 1887, the former professor of anatomy Jean Joseph Crocq emphasised that 'being located in the most industrial region of the capital', the hospital was 'the most interesting for education'.[108] Deroubaix closed a deal in 1891. In exchange for the delivery of unclaimed corpses to the Brussels St. Jean hospital, interns from Molenbeek were allowed to attend university courses for free. Again, the corpses of the poor were traded like commodities.[109]

Yet this conscious use of the bodies of the poor, a reality that had existed for some time, was increasingly problematised by changes in the social composition of the hospital. Medical historians have shown that the introduction of health insurance, medical innovations (most importantly surgeries, which became safer as a result of the use of antiseptics and anaesthetics) and closer contacts with universities started to draw paying patients to hospital wards in the late nineteenth century.[110] As hospitals transformed from charitable institutions for the indigent into respected medical centres for everybody, the subjecting of the patient's body to an authoritarian hospital regime became increasingly difficult. Generally speaking, the admittance of paying patients challenged the idea of a 'natural reciprocity'. Since paying patients could choose their own treatment and did not have to participate in medical education or research, the true 'cost' of a hospital stay became evident for the poor.[111] The emergent Socialist movement quickly integrated the difference between propertied and indigent patients in their political propaganda. In the Socialist newspaper *Le Peuple*, which first appeared in Brussels in 1885, journalists repeatedly argued that the rich oppressed the poor by asking an excessive compensation for care.

In 1887, a scandal gave a strong impetus to their criticism. The police discovered that the body of Cécile Amadou, a woman with Senegalese roots who was involved in prostitution, had not been buried. After exhumation, they found a stovepipe, firewood and stones in the coffin. They quickly discovered that the anatomist and anthropologist Emile Houzé had stolen the body from the morgue of St. Jean. He wanted to keep the skeletal remains of Amadou because they were, in the words of a contemporary journalist, 'entirely of the Negro race'.[112]

Houzé wished to collect and study Amadou's remains because it was his dream to formulate a racial classification theory for Africa. He researched the differences between skulls from northern and southern regions of the continent. In the years 1885–1886, he had published several studies on skulls that had been stolen during 'scientific missions' in the Belgian Congo. Because his results showed large physical variations between different skulls, Houzé believed that scientists needed 'many more data to unveil the mystery surrounding this continent'.[113]

The theft of the corpse, which quickly became known as the 'Coco Affaire' (which referred to Amadou's alias), gave rise to strong reactions. All newspapers closely monitored the case. After all, the scandal made for a good story: the coffin had been dug up after an anonymous tip, written with cut-out letters from newspapers; the key players were a poor prostitute and a successful physician; there was deceit, death, eroticism and exoticism. For several weeks, anatomy and dissection were much-discussed topics in the daily press.

Interestingly enough, the story plucked on the heart strings of the poor. Even though Amadou, as a black woman who worked in prostitution, lived on the fringe of late nineteenth-century society, many people living in poverty identified with her fate. They did not view the theft of her corpse as a result of the popularity of anthropology, but as a typical expression of the exploitation of the poor. Journalists reported that the Coco Affaire 'made a lot of noise in poor neighbourhoods, where the hospital is the sole refuge' and that 'the public opinion demanded a severe punishment for those responsible'.[114] The Socialist newspaper *Le Peuple* grasped the momentum to publish the hospital regulations on the distribution of corpses for dissection in full.[115] The dissection of indigent patients against their wishes, a reality that medical faculties had surrounded with secrecy for decades, became front-page news.

In subsequent years, important pioneers of the Socialist Party continued to represent the dissection of the poor as a revealing example of class inequality, occasionally referring to the Coco Affaire.[116] Louis Bertrand, one of the founding members of the Belgian Labour Party (*Parti Ouvrier Belge*), for instance, recommended to replace the hospital by homecare services.[117] In his opinion, the hospital was a 'temple of suffering', in which the health of the poor was sacrificed to the progress of medicine 'in service of the rich, paying sick'. This injustice was perpetuated in the 'dreaded anatomical theatre', where anatomists and students violated the bodies of the poor once more:

> When dead, their cadavers are brought to the anatomical theatre, where they are subject to more experiments. All this is known, and therefore it is not surprising that the poor are afraid of the hospital and that the expression 'dying in hospital' is used by those who are discouraged and who dread this terrible fate.[118]

The burial of dissected remains and the pauper burial more generally were also critically assessed in the Socialist press. Numerous articles drew attention to the pitiable circumstances in which the anatomised poor were buried: in badly made rectangular coffins, in overcrowded cemeteries and often without their families present.[119] Much like certain Roman Catholics, Socialists argued that every pauper was entitled to a decent funeral. In fact, equality was an important argument for Catholics and Socialists alike. The Catholic thinker Louis Henry, for example, believed that the poor had the same rights to a proper burial—with pomp and without post-mortem—as anyone else, 'for the sentiments and rituals surrounding death are not less vivid for working class families than for rich families'.[120] In nearly the same way as in the Socialist press, Catholic newspapers demonstrated the injustice of involuntary dissection by an emphasis on the emotions of the poor, who cared as much about their dead as the rich. 'Do they think, maybe, that the families of workers are indifferent to the mutilation of their relatives' corpses?'[121] Although Catholics occasionally referred to typical religious rituals—such as laying out the dead or praying in the presence of the corpse—their emphasis on equality was strikingly similar to Socialists' reasoning.

In a parliamentary debate on the working of state-funded teaching hospitals in 1891, progressive politicians linked the dissection of the poor to the quest for universal suffrage. The idea was that the working classes would continue to live (and die) in oppression as long as they were not politically represented. Dissection was seen in a wider perspective, as just one of many expressions of the suppression of the poor.[122] Resentment of the social injustices inherent to body procurement, so it seems, went hand in hand with calls for a more inclusive democracy.

Indeed, all these arguments reflected, and shaped, broader changes in late nineteenth-century Brussels. Attitudes towards the poor had become more emphatic. Research has shown that Socialist agitation, and the resulting universal suffrage introduced in 1893, was the driving force of a general softening of social policy in Belgium. To avoid losing voters to Socialism, other parties started to pay more attention to social justice as well. The general strike of 1886 has especially been discussed as a turning point.[123] As we will see later in this chapter, the extension (and the fear of further extension) of democracy also led local authorities to take measures to avoid discrimination against the indigent.

Historians such as Jo Deferme and Jean-Pierre Nandrin have argued that the fear of social war led to the disintegration of the strictly liberal political culture that had characterised Belgian legislation and social policies since independence. Movements of different ideological make-up emerged, which advocated a different approach towards poverty.[124] Influenced by sociological theories, progressive opponents of all political parties argued that social policies aimed at individual empowerment were insufficient. They no longer believed that all citizens were born equal and free, but argued that governments had to compensate for inequality through structural measures. To give every citizen equal chances in life, freedom had to be imbedded and supported by social legislation.[125] Similar discussions took place in other countries, including in Britain. There was a growing awareness that impoverishment was not the fault of an individual, a thought that according to various historians came to lie at the basis of the early welfare state.[126]

In Belgium, these ideas not only led to the establishment of social legislation on working conditions, but also influenced poor relief more generally. In the last quarter of the nineteenth century, politicians increasingly criticised the state-funded system of assistance for its punitive character. As social determinism replaced the liberal idea of the independent individual, poverty was not considered as one's own fault anymore, and therefore the severity of the workhouse no longer made sense. Progressive politicians posited that punitive institutions had to be reformed because they maintained rather than solved inequality.[127] It is telling that under the new Belgian Poor Law of 1891, poverty ceased to be a crime: begging and vagrancy were no longer punishable offences. Poor assistance gradually became a right (to which one was entitled) rather than a favour (that had to be returned).[128] Against this background, involuntary dissection became harder to defend.

These late nineteenth-century shifts in social policy also accelerated the transformation of hospitals from charitable institutions into more prestigious medical centres. In 1895, a Royal Commission for the Reform of Public Assistance (*Commission royale pour la réforme de l'assistance publique*) was established on the initiative of the Catholic politician Julien Liebaert. The Commission aimed for 'the general and profound reform' of poor relief.[129] Following a political consensus on the need to replace *bienfaisance* (charity) by *prévoyance* (prevention), its main emphasis was on the encouragement of mutualist associations. Much like in Germany, a system of insurance was to substitute charitable institutions. Insurance policies were to prevent severe poverty in cases of unemployment, old age and, most importantly, disease.[130]

In 1894 and 1898, laws on mutualist societies were voted, which established an optional state-supported system of health insurance for workers. The subsidising of recognised insurance funds was a compromise between self-help (reflecting the older ideal of individual precaution) and structural state intervention. As a result of these laws, the number of insured workers increased rapidly, although they remained a minority. In 1890, 54,347 workers were associated with a recognised mutualist society. Ten years later, this number had risen to 185,201. The number of insured workers continued to grow until social insurance became obligatory after the Second World War.[131]

This increased importance of health insurance had a profound impact on the social composition of the hospital, since more and more patients received assistance through their mutualist societies. This separation of care from charitable donations undermined the authoritarian hospital regime. As the number of paying patients grew, the use of the bodies of hospital patients was no longer considered 'natural' or righteous. Instead, hospitals increasingly had to take the wishes of the patient into account.

Most remarkably, the public hospital in Liège was split in two. Whereas the Bavière hospital remained linked to the university, the newly constructed *Hôpital des Anglais* was not. According to a parliamentary discussion, this decision reflected the wish 'to assure the freedom of the poor'.[132]

In fact, the creation of a second hospital, detached from the university, solved a fierce dispute between the Catholic hospital personnel and the university board. The submission of (pregnant) women to clinical courses specifically caused problems, for the obligation to undergo a gynaecological examination or to give birth 'in public, in front of a group of students' allegedly damaged women's modesty (*pudeur*).[133] Yet also dissection was a subject of conflict. The medical faculty of Liège repeatedly complained that the new hospital 'refused to deliver their unclaimed corpses to the university'.[134] The creation of the Anglais hospital apparently gave the indigent of Liège the choice to withdraw themselves not only from clinical courses, but also from dissections. The next pages will reveal that this conflict echoed more general changes in late nineteenth-century Belgium. The clinical encounter gradually began to be based on the principle of consent, even after death.

5   'BY WHAT RIGHT DOES THE SCALPEL ENTER THE PAUPER'S CORPSE?'

As early as the 1870s, criticism on the anatomical use of the bodies of the poor was couched in a discourse of rights and violations. In 1873, for example, a journalist who reported on an involuntary dissection asked his readers: 'By what right does the scalpel enter the pauper's corpse?'[135] Even though official hospital regulations stipulated that corpses of poor patients could be dissected without permission, the question of consent repeatedly came to the fore when social and ideological tensions mounted. The local Catholic newspaper of Ghent, for instance, wrongly pointed out that a corpse could not be sent to the dissecting room 'unless the family had given permission'.[136] In the Socialist press, too, dissections without relatives' approval were called 'fraudulent' and 'unauthorised'.[137] This section shows how the codification of the principle of consent came to influence the distribution of bodies to anatomists around 1900. The codification and reinterpretation of the principle of consent went hand in hand with a legal reassessment of the corpse. Ultimately, poor patients obtained more power to decide over their treatment: they gradually gained ownership of their own bodies, both before and after death.

Consent before Death

Numerous historians have stressed the varied and negotiated nature of nineteenth-century clinical encounters and the importance of patients as autonomous actors. On the competitive private market of medical care, practitioners had to convince patients of their diagnoses and treatments, since they could go elsewhere if they were dissatisfied. Under these circumstances, informing patients of the risks of treatment fostered trust and helped physicians to justify for possible failures, hence protecting their reputation. Clinical relationships in hospitals, however, were different. As indigent patients did not have the freedom to go elsewhere for medical care and were more likely to encounter multiple doctors and nurses (instead of one family physician), they enjoyed a lesser degree of autonomy. Because hospital patients did not have to be convinced to pay for a certain treatment, the provision of information was not routine.[138]

This does not mean that the consent of hospital patients was ignored completely. As early as 1803, Thomas Percival, generally regarded as one of the founding fathers of modern medical ethics, underlined the importance of the patient's agreement and consultation, because medical treatments and experiments could otherwise be seen as coercion or physical assault. As the Manchester Royal Infirmary (a public institution) commissioned his work, his ethical principles also applied for hospital patients. Although Percival considered the preferences of the patient important, he believed that consent would not pose a problem if physicians properly exercised their authority. In the words of Percival, 'the prejudices, caprices and passions of the sick and their relatives' would be easily overcome by the physician's appropriate deportment, which would 'inspire the minds of their patients with gratitude, respect and confidence'. He thought that if the physician carried out his profession with dignity, consent could be assumed.[139]

Percival's ideas attest to what historians have called a 'beneficence-based' or 'paternalist' notion of consent.[140] Within this tradition, physicians and ethicists justified consent practices by considerations of therapeutic benefit rather than by the importance of individual rights. The physician's primary obligation was not to respect patients' autonomy, but to secure their well-being. The need to disclose information or to seek consent was subordinate to medical benefits (or—more accurately—to what physicians believed to be beneficial). Doctors could withhold information if they thought that truthfulness would cause harm, for example, if it would lead to patients giving up hope or refusing a life-saving operation. Consent was above all important because it facilitated cures: physicians sought approval because they assumed that treatments were more likely to be successful if patients agreed to their application. This was especially true for surgeries, since operating without patients' consent was a challenging undertaking in an era without anaesthesia.[141]

This conception of the doctor-patient relationship, in which patients were to put their trust in the physician's judgement for their own good, increasingly came under question through developments in surgery, bacteriology, immunology and pharmacology. Firstly, the introduction of anaesthesia around 1850 made it easier for surgeons to perform operations against their patients' will. Secondly, as mentioned in the previous chapter, experiments on living subjects gradually became more important. This growing demand for clinical trials led to experiments on hospital patients, who were regularly subjected to new methods or treatments. Hospitals were the most advantageous sites for medical experimentation because subjects were numerous and lived under largely the same circumstances. Their confinement to strict regulations of diet and regimen created advantageous conditions for experiments, which were more difficult, if not impossible, to obtain in private practice. Whether it was on living or dead bodies, hospital physicians generally regarded experiments on the poor as a justified compensation for the care provided.[142] In fact, they often justified the subjection of hospital patients' bodies to science by the ideal of beneficence. Because they believed that medical experiments and post-mortem research would lead to progress, they considered the use of hospital patients as something praiseworthy, favourable and 'useful for science, medical education and medical practice alike'.[143]

The admittance of paying and insured patients challenged this kind of reasoning in the late nineteenth century. Whereas historians have noted that the growing population of better-off patients in hospitals mostly signified that there was an overall increase in the proportion of patients who had little or no effective autonomy in their relationships with doctors, the other side of this picture has been overlooked. For Belgium, archival evidence suggests that, whereas hospital admittance might have marked a loss of autonomy for richer patients, their presence did benefit the poor. The admittance of paying patients in Brussels had a moderating effect on the treatment of indigent patients, since it laid bare the aforementioned idea of reciprocity. The different treatment of propertied patients led the poor to realise that they were in fact *paying* for care—though not with financial, but with natural means. This caused a wave of criticism on the strict conditions that governed their hospital stay.

Across Europe and in the United States, physicians were convicted for unauthorised medical interventions, often on hospital patients, in the last decades of the nineteenth century. As early as 1859, the district court of Lyon in France ruled that all kinds of bodily harm, including medical interventions, were punishable under the *Code pénal*.[144] In Germany, two physicians faced a conviction for assault and battery in 1894, because they had conducted surgeries without consent. As judges classed unauthorised medical interventions as physical injury, physicians were forced to adopt practices of seeking consent and informing patients.[145] In Britain, scandals surrounding surgeries on psychiatric patients without their consent gave rise to deontological and legal discussions in the 1860s. The importance of consent came to the fore again in the 1896 case of *Beatty v. Cullingworth*, in which a surgeon had to appear before court because he had removed two ovaries, whereas the plaintiff had only consented to the removal of one. The court underlined the need for patient information and consent, ruling that surgeons had to carefully explain 'what they proposed to do' before operating, and receive 'unequivocal consent'.[146] In the United States, the verdicts of the court cases *Mohr v. Williams* (1905) and *Schloendorff v. Society of New York Hospitals* (1914) both emphasised the importance of patients' consent, as well as their freedom to determine their own treatment.[147]

In Belgium, consent became a legal principle in 1889, in a court case instigated against a surgeon who had performed an osteotomy (a fragmenting of the bone to change its alignment) on a 3-year-old child without seeking the prior consent of the father. The verdict explicitly stated that 'scientific curiosity' could not take priority over the wishes of the patient, not even in the context of the hospital.[148] As an important precedent for Belgian jurisprudence, the court ruling found its way into several legal publications and medical deontological manuals: 'In order to legitimately carry out a surgical procedure, *l'homme de l'art* must be authorised by the patient or by the legal guardian of the patient'.[149]

Historians have argued that these late nineteenth- and early twentieth-century court cases followed from two different legal traditions: negligence and battery.[150] The Belgian case, for example, was not too different from older lawsuits concerning medical negligence, since an important factor in this ruling was that the surgeon had deviated from the medical standard: osteotomies were usually not performed on children younger than 6 years old. The surgeon faced conviction not because he had failed to seek consent, but because the judge did not see his intervention as acceptable from a medical point of view. As such, the case fitted within the tradition of beneficence. The verdict primarily protected patients' well-being, whereas the idea of autonomy was only of secondary importance.

Yet cases instigated on the grounds of battery and assault reflected another way of thinking about consent. First and foremost, the justification for battery liability was not the negligence of the physician, but the basic right of the individual to protect one's bodily integrity from unauthorised intrusions. Additionally, consent in these cases was seen as a full decisional process rather than as a mere permission to perform operations or admit treatments. In other words, the purpose of a consent requirement was not simply to authorise what would otherwise be battery, but to empower the patient to decide on the desirability of a certain treatment. The core idea that underlay these verdicts was not beneficence, but autonomy.[151]

This latter interpretation of consent, however, was still rudimental and rare. Consent mostly remained a passive notion grounded in the ideal of beneficence. In practice, it did not signify that physicians provided hospital patients with information to weigh up the (dis)advantages of possible treatments, but rather that they tried to respect their preferences. They increasingly took into account patients' decisions.[152] This right of refusal did not necessarily imply that patients were free to opt for an alternative treatment. This was true for indigent hospital patients in particular, because their physicians were, in the words of a late nineteenth-century Belgian deontological manual, 'primarily accountable to the institution and not to the sick themselves'.[153]

Even the implications of this largely passive notion of consent were unclear. Consent was only a legal requirement for 'dangerous' procedures. In real terms this meant that the principle only applied for experimental or mutilating (mostly surgical) treatments, such as sterilisations or amputations. Consent was not compulsory in cases of emergency.[154] In reality then, physicians and surgeons were rarely held accountable for professional errors and consent could easily be dodged throughout the nineteenth century.[155] Building on these insights, scholars such as Jay Katz have posited that 'thoroughgoing self-determination', or 'informed consent' as we know it today, did simply not exist.[156]

However, there was one exception to this rule. When it came to non-therapeutic experiments (serving scientific instead of curative purposes), both the legislator and physicians themselves considered the consent of the patient, as well as a preliminary disclosure of information, as indispensable. Since these experiments did not aim to cure patients, it was difficult to fall back on the ethics of beneficence. Judges rather interpreted non-therapeutic experiments on the grounds of assault and battery, offences that were contrary to the individual's right to defend his or her body against harm. As Ruth Faden and Tom Beauchamp have argued, it was this older legal tradition that became the basis for the twentieth-century notion of informed consent. Informed consent did not evolve from the idea of negligence within medical circles, but from the legal principle of self-determination.[157]

In France, the district court of Lyon had forbidden nontherapeutic experiments in 1859. In 1891, purely scientific experiences were condemned again, as the result of a study in which a surgeon had implanted cancer cells in unknowing patients in order to find out if a similar type of tumour would grow.[158] The Prussian parliament established the first official legislation on medical experimentation and consent in 1900. The cause was the Neisser case, in which a physician had injected (and hence infected) uninformed hospital patients with the syphilis virus in an immunological study. In response to this scandal, the Directive of 1900 prohibited nontherapeutic research on incompetents (such as minors or disabled persons) and demanded unequivocal, informed consent in other cases.[159]

In Belgium, the Neisser experiment in particular caused consternation as an 'outrageous' and 'odious' abuse of hospital patients.[160] Deontologists called nontherapeutic medical tests 'grossly unjust', and asserted that 'patients may not figure as experimental subjects for physicians'.[161] Whereas medical ethicists turned a blind eye to experiments that could possibly be advantageous for the patient, they only accepted medical tests for research purposes if the patient cooperated consciously and voluntarily:

> From a moral point of view, it is strictly forbidden to prescribe or use treatments that are probably dangerous, only to know their effects. [...] Indispensable treatments do not deserve as severe a moral judgement.[162]

To sum up, two interpretations of consent coexisted in late nineteenth-century hospitals. On the one hand, surgeons had to seek their patients' consent before they carried out dangerous or mutilating operations. In these instances, consent essentially was a passive notion: even though a bare permission was required, the patient did not participate in the process of medical decision-making. Patients were asked to confirm choices made solely by the surgeon, in order to protect him from the risk of prosecution.[163] On the other hand, nontherapeutic experiments demanded a kind of informed consent *avant la lettre*. Physicians had to inform patients about the possible benefits and risks of an experiment, after which they could decide if they wanted to participate or not. The first interpretation of consent mostly fitted with an older tradition in medical ethics (beneficence) as well as with the older notion of negligence; while the idea of informed consent was new, and came from outside of medicine, specifically from the individual's basic right to protect one's body against unauthorised intrusions (battery and assault).

Interestingly, it appears that these conceptions of consent not only concerned medical interventions on living patients. When it came to their dead bodies, too, consent became increasingly important. In 1889, the same year that consent became a legal principle, the city council of Brussels changed the hospital regulations regarding dissection. From that moment onwards, close relatives could always prohibit a dissection, regardless of their financial capacity. Much like patients could refuse bodily interventions during life, they could now veto dissections after death. The authority of the closest family member to decide on the fate of the deceased reminded of the permission from the legal guardian for surgery on minors or disabled patients. The severity of the mutilations, an important criterion for deciding on the need for consent for surgical procedures, also determined the authorisation of post-mortems. Whereas it became easier for families to claim bodies, and thus prevent their dissection, the less mutilating autopsy was—as will be explained more thoroughly in the next chapter—conducted on both claimed and unclaimed bodies.[164] As De Laet explained in a deontological manual dated 1927:

> With regard to autopsies performed in hospitals, a series of decisions and instructions from the commission of hospitals of Brussels [from the 1880s and 1890s] prescribe that autopsies have to respect the exterior appearance of the corpse, the face in particular, and therefore differ from anatomical dissections, which can only happen on the bodies of the unclaimed dead. [They] authorise that, in hospitals under her direction, every autopsy can happen in spite of the family's advice.[165]

In other words, whereas consent was a requirement for dissection if bodies were claimed (note that claiming a body did *not* have financial implications anymore), autopsies could be performed without families' approval. In the next chapter it will become clear that consent could be neglected precisely because corpses were less severely mutilated during autopsies.

At the same time, a parallel was drawn between involuntary dissections and nontherapeutic experiments. Consent was necessary precisely because dissections are by definition nontherapeutic. As corpses cannot be treated or cured, patients (during life) or their relatives (after death) were to approve of their dissection. Again, the different regulations for autopsies and dissections are telling in this respect. Autopsies could happen without consent for they served scientific interests and because they could possibly be beneficial for families (because they were informed about the cause of death of their relative).[166] Dissections, conversely, were increasingly reduced to mere educational practices, which made it more difficult for anatomists to represent their activities as therapeutically advantageous. Whereas autopsies were essentially seen as *therapeutic* interventions—admittedly not for patients themselves, but for their next of kin or for future generations more generally—dissections were associated with *nontherapeutic* experiments, and therefore required consent.

In general, however, the extension of the principle of consent to the realm of the dead led to disputes between physicians, judges and patients. While consent was broadly recognised as an important criterion for deciding on the justifiability of performing nontherapeutic medical trials on the living, anatomists and students mostly conducted dissections without permission. From a legal point of view, the question of consent not only concerned notions of medical negligence and self-determination, but also raised questions on the ownership of the dead body. As long as the corpse was regarded as a commodity, a dissection was a potential conflict between the relatives of the deceased and medical faculties. The need for consent depended on the respective parties' rights to the dead body. If the pauper's corpse was the rightful property of the hospital, no consent was required; if the family's claim on the cadaver took priority, the physician had to ask for their permission. When the legal interpretation of the corpse changed from a potential property to an extension of the individual at the turn of the century, the need for consent could no longer be denied.

Consent after Death

A dead body is a peculiar thing. As matter that once was a person, it challenges the divide between object and subject. A corpse is both the remains of the living person and a strong reminder of his or her death. In this sense, the deceased is both present and absent in the corpse. A dead body is undeniably human, but only by nature, not by function.[167] The liminal status of the corpse—neither object nor subject—exercised Belgian jurists in the late nineteenth century. Since the status of the corpse was undecided, it was unclear which laws applied to it. In the words of two late nineteenth-century legal scholars, who tried to achieve a synthesis on the legislation concerning corpses:

> When we try to discover the true nature of the laws we can apply to a human corpse, the difficulties are serious. Currently, all the civil rights fall into one of the following four categories: property law, contract law, intellectual property law and personal law. In which of these categories can we classify the right to a cadaver? If we were dealing with an animal, it would obviously be a right of property, a right to a 'res', to a material object outside of the living person, who is the real subject of the rights. But does the same apply for a human corpse?[168]

There existed a complex body of law concerning corpses and their burial in the nineteenth century. Already in the early modern period, laws that required the provision of a reasonable burial, reaching a certain standard of decency and cleanliness, had developed in both religious and civil contexts. Additionally, there was legislation on the disposal of corpses in periods of increased mortality, such as famines or epidemics, when the local government had to ensure burial for those without relatives or means. In the nineteenth century, such provisions led to laws that regulated interments from the point of view of public health, improving hygiene standards.[169] In Brussels, for example, families had to bury their deceased within a certain time frame, outside of the city walls, in a sufficiently big plot, reasonably deep, and preferably without embalmment or leaden coffins in order to ensure that the dead body decomposed rapidly.[170] A central question in all of these laws concerned the ownership of dead bodies. As the quote above suggests, jurists had mostly interpreted claims to a corpse as property rights, similar to the ownership of dead animals. In the same way that an animal was 'the property of the owner of that animal during its life', so the human corpse belonged to the closest family member of the deceased.[171] There was also an emotional component to this principle. Belgian legal repertories from the early nineteenth century argued that the family was entitled to the dead body because they cared the most about the respectful treatment of their deceased relative. Every corpse was therefore 'seen as the property of the family, of which the closest members may dispose of it as they wish, as long as it is in accordance with law'.[172] When the members of the Brussels city council changed the regulation concerning dissection in 1889, they referred to this legal tradition. In the council's opinion, the interests of science 'could not prevail over the rights of the family'.[173]

However, various laws—older laws on epidemics, as well as newer legislation on public hygiene and criminal law—had established exceptions to this rule. Under certain conditions, the greater good took priority over families' claims. Various branches of government could interfere with relatives' funeral arrangements, for example for fear of an epidemic or other hygiene reasons, for needs of investigation and justice, or, more importantly here, in order to enable anatomical education. It is worthwhile to note that in these instances, too, property remained the underlying principle, since the idea was that other parties had a greater claim to take possession of the dead body than the family.

It is not surprising, then, that medical faculties also took on the logic of property rights. At the same time though, they denied the particular emotional and cultural significance of the dead body. As aforementioned, they represented the post-mortem use of the pauper's corpse as a fair compensation for the care or burial provided. Anatomists often referred to a French law of 1809, which was still in force in late nineteenth-century Belgium.[174] Under this law, all the goods indigent patients took with them into hospital became the property of the institution if they died.[175] The hospital was even entitled to the clothes and jewellery on poor patients' corpses, and to the cash that was left in their pockets.[176] Even though this law was applicable for goods (and not patients), it often came to the fore in discussions on the procurement of corpses for dissection. Anatomists in Brussels, for instance, claimed that there was no difference between a dead body and 'inherited jewellery or furniture', objects that could also be confiscated by the hospital despite the sentimental value families might attach to them.[177]

Moreover, physicians argued that their efforts, both before and after death, transformed the corpse into their rightful property, in the same way that natural goods could be turned into property by labour. To put it simply: if a piece of land could become yours by cultivating it, then why should a body not become yours by treating it?[178] Anatomists thought in similar ways about preserved human remains. They argued that they had acquired the right of property by transforming a natural good (a corpse) into a scientific object (an anatomical specimen).[179] Veronique Deblon has shown that these ideas occasionally led to conflicts on the ownership of anatomical specimens between students, teachers and institutions. Students sometimes refused to deposit anatomical specimens in university collections. They claimed that anatomical specimens were *theirs*, for they had personally transformed human remains into preparations through both manual and intellectual labour.[180]

Influenced by the gradual global abolition of slavery in the second half of the nineteenth century, legal experts increasingly claimed that the human body (and corpse) could not be seen as property. In France, an 1843 court order stated that 'the human body, before or after death, cannot be considered as merchandise or classified as an industrial object'.[181] The verdict closed a lawsuit from Marchal de Calvi and Jules Guérin, who had wanted to invalidate Jean-Nicolas Gannal's patent on modern embalming, because his method in their opinion had existed before. In Britain, courts did not interpret the corpse as a thing either, and therefore bodysnatching could only lead to prosecution if one was caught in the act or if a portion of the grave clothes or coffin was taken with the body. Because corpses could not be owned, they could not be stolen. According to this no-property principle, the family did not have the right of ownership to the corpse in a strict sense, but was only entitled to temporarily care for the body until burial.[182] The German Imperial Court was unclear as to whether the corpse was an object or not, but nevertheless asserted that 'no right of property concerning the corpse comes to exist by death alone'.[183] In a similar vein, the well-known Belgian Liberal jurist François Laurent argued that corpses were neither private possessions nor commodities, and that they could therefore not be families' property. In his opinion, corpses could only be claimed by society (and more precisely by medical faculties) in the public interest (to advance the medical sciences).[184]

Yet if property rights only applied to goods, and personal rights were limited to living individuals, the dead body fell outside of the scope of the law. In Belgium this gap in the legislation first came to the fore in a court case in 1868. Hoping to save a soul, a certain Mme. Vandenbussche, a midwife, had cut open the corpse of a recently deceased woman to baptise the unborn child in utero. The plaintiff argued that this was both incautious—how could someone with limited medical knowledge be certain someone was actually dead?—and punishable before the law as a violation of a grave. The Supreme Court, however, decided to acquit Vandenbussche because the law on violation of graves only applied to the desecration of the tomb or the disrespectful treatment of human remains *during* and *after* the burial. Before the funeral, the corpse was not legally protected. Furthermore, other laws were also insufficient to provide an answer. To vindicate its decision, the Supreme Court highlighted the inadequacy of existing laws:

> Manslaughter? No, because nothing allows us to assume that the woman was not dead. Assault? No, one cannot hurt a corpse. [...] Quackery? No, because the confirmed facts and the medical report bear out that there was only one incision, not meant to deliver the foetus but *only* to administer the baptism.[185]

In France, too, the legal vulnerability of the corpse was made clear by baptisms in utero and post-mortem caesarians in the late nineteenth century. Since the recently deceased were not protected either as property or as a person, cutting into their body was not a punishable offence.[186] Similarly, German jurists and physicians argued that an unauthorised post-mortem was not a legal offence: because the dead body was not an object, physicians could not be accused of damaging property, and because the corpse was no longer a living being, a post-mortem could not be classed as medical negligence or assault either.[187] The corpse, in short, was in a legal grey area.

Jurists sought alternatives to avoid abuse of this legal gap. To be able to grant rights, French jurists proposed to protect the corpse as a person rather than a thing.[188] Gabriel Timbal, for example, recommended classifying dead bodies as 'semi-personalities', invested with 'semi-rights' because 'they are persons [...] only without the normal characteristics that belong to the living'.[189] In a similar vein, René Demogue suggested that 'judicially, the dead are semi-persons'.[190] In Belgium, the jurist Théodore Bormans put forward that not only violation of graves, but also 'violation of cadavers' should be included in the law. In this way, judges would be able to convict priests and nuns who conducted post-mortem baptisms and caesarians.[191] This suggestion gave rise to a parliamentary debate on the ownership of the corpse. The Liberal politician (and physician) Jean-François Vleminckx supported Bormans' proposal. In reaction to his statement, Catholic politicians, most importantly Barthélemy Dumortier, defended priests by referring to involuntary dissections:

Does the violation of cadavers also imply the dissection of men and women in anatomical theatres then? To accuse a poor priest, a poor acolyte, who has helped to deliver a child, while you protect the actions of immature medical students [*carabins*] who dissect corpses? This, I confess, is a logic I do not follow. [...] You know, frequently enough, a caesarian is a way to save a child. Even the great Caesar was saved in this way, which is why the operation bears his name. Well, I ask you then, why should we call this operation a violation of cadavers and graves; whilst, in anatomical theatres, we deliver corpses to young physicians who cut them as they seem fit? I think this rather is a violation of cadavers and graves.[192]

Dumortier ended his plea by stating that the corpse, in his opinion, 'was not the property of the physician'.[193] These discussions ultimately led to a new way of thinking about dead bodies. In one of the most important Belgian legal reference works of the late nineteenth century, *Pandectes belges*, a parallel was drawn between the rights of the living and the dead. As the individual was his or her body, he or she also had the right to decide about it, both during life and after death: *notre corps, c'est nous-mêmes—* 'we are our corpse'.[194] The dead body gradually changed from an object to an extension of the living person.

Interestingly then, in an era in which funerals and other death rituals were becoming increasingly personified, the legal status of the corpse was reassessed. As argued above, the burial question led to a growing recognition of the last wishes of the deceased. Especially in circles of freethinkers, the rights of the dead were stressed. By urging members to write down their last wishes in testaments, anticlerical mutualist societies tried to prevent the organisation of Catholic burials. Exhumations, too, were ways to honour the last wishes of the deceased: freethinkers said they rectified unwanted Catholic burials by digging up the corpse.[195] Later in the century, the rights of the individual came to the fore in debates on cremation.[196] Against this backdrop, the funeral became an expression of the identity of the deceased, who could determine his or her post-mortem fate during life. It seems that, as the dead became personified, the idea arose that their bodies belonged to no one but themselves. Gradually, the individual attained the right to decide not only on his or her own burial, but also on the use of his or her remains more generally.

This interpretation, in which the corpse transformed from an object into an extension of the living subject, was formalised in a case concerning the disposal of the mortal remains of one Jules Libert in 1899. The Belgian Supreme Court had to decide whether Libert would receive a civic burial, as he had requested in his will, written ten years before, or whether he would be buried within the Catholic Church following his conversion on his dying day.[197] In the context of heightened Liberal-Catholic conflict over burials, this case was not unique. Civic burials often led to emotional rifts within families, who sometimes took the matter to court.[198] Yet the case of Libert was different because it sparked off a more fundamental debate on the ability to exercise rights after death. Could individuals determine their burial while still living? Referring to a well-known principle from Roman law, namely that 'all rights dissolve with death', and to Laurent's legacy, jurists argued that death marked the end of the legal person. The dead could not own property, and could not have legal faculties or needs either. Consequently, individuals had no power to determine the details of their burial. Other jurists, however, argued that an expression of will continued to have the power of law after death, referring to the right of testate.[199] Here they were following a French judicial decision from 1886, which had suggested that 'the ability to dispose of one's corpse after death does not seem so different from the ability to arrange the use or destination of one's possessions'.[200]

The Belgian Supreme Court ultimately ruled that individuals had the right to decide during their lives on the post-mortem fate of their remains.[201] Significantly, this right was not equal to the right of testate, because wills concerned the transfer of property rights and the corpse was not a thing. Instead the right to the corpse was derived from 'human nature and liberty'. According to the verdict, the right to decide on what happened to one's remains post-mortem was a natural right, which had always been present in the legislation even though it had not been recognised. The rights of the living were extended to the dead: 'Human beings, masters of their own selves during life, have free disposition of their remains after their passing'.[202]

While the case of Libert was above all important for the burial practices of affluent families, it also became a precedent for those wishing to claim the bodies of the poor in hospital. If individuals could decide on the post-mortem fate of their remains, then they should also be able to veto a dissection. In the words of a report on the need for consent for post-mortem examinations, written in 1924, it gradually became 'certain that the right to dispose of a cadaver in the first place belongs to man himself'.[203]

Accordingly, hospitals in Brussels and other Belgian university cities revised their regulations. The right to claim a corpse was gradually stripped of financial implications. As aforementioned, families in Brussels could claim corpses 'even without supporting the cost of burial' from 1889 onwards.[204] In Liège, the city council, in consultation with the commission of hospitals, lowered the price of burial multiple times in the 1880s and 1890s, to the extent that, in the words of a complaining anatomist, 'no family was hindered by such a moderate expense'.[205] Despite the resulting shortage of bodies for education, the low burial price was maintained because the commission of hospitals considered it unrighteous to 'force the poor, by indirect means, to consent to the dissection of a relative's corpse against their intimate wishes'.[206] Significantly, the obligation to pay for funeral expenses was seen as an infraction of individual liberty:

> To avoid that a sick person does not want to enter a hospital because he or she might be subject to anatomical operations in the case of death, we think that his or her liberty and the liberty of his or her family should be respected, and that it is indecent for a charitable administration to stipulate conditions to limit [their freedom], even if this happens in the interest of education.[207]

Even though the situation in Leuven and Ghent is less clear as a result of a lack of primary sources, a few archival traces suggest there were similar developments in these cities. In Leuven, anatomists related a sharp decrease in the number of corpses available for dissection to the increased respect for the wishes of both families and burial societies.[208] In Ghent, a municipal regulation on the procurement of corpses for dissection from 1892 stipulated that families had to claim bodies within a fixed time frame of 36 hours, but did not mention the payment of the burial as a condition anymore.[209]

The evidence presented in this chapter suggests that it was the same idea of self-determination that started to influence the interpretation of consent around the turn of the twentieth century, that impacted the individual's right to decide on his or her burial and on the treatment of his or her corpse more generally. Much like the right of self-determination laid the foundation for patients' right to refuse or accept nontherapeutic experiments, it now became the basis of their right to determine the details of their post-mortem treatment. As a result, individuals obtained the power to decide on their bodies, both during life and after death. Tellingly, hospital patients in Brussels also more often took infractions against the principle of consent to court in the early twentieth century, leading to more severe sentences for physicians.[210] The rights of the hospitalised poor grew. Around 1900, the body and the corpse became inalienable—at least in theory.

6   The Early Days of Anatomical Donation

In 1899, the head of the Brussels St. Jean hospital received a peculiar letter. A certain François Gobbe wanted to donate his body to science:

> November 13, 1899
> 119, Boulevard Anspach
> Brussels
>
> I donate my body to your anatomical theatre in order to facilitate the education of medical students. Nobody or nothing can stop me. My son Ernest, living in 61, Rue des Six Jetons, has received my instructions for this purpose. Afterwards my remains should be thrown in a communal grave without warning anyone. I will die tonight, by my own free will.
>
> F. Gobbe

Startled by this message, the head of the hospital made a few phone calls. The police told him that a man had tried to poison himself with coal gas the night before, but had not succeeded. The head of St. Jean went on to ask the commission of hospitals how he had to deal with 'Gobbe's last wishes, if they would become executable'. After corresponding with multiple municipal institutions and courts, the commission replied that his wishes could be granted if there was a proper testament. Before sending donated bodies to the anatomical theatre, however, he had to warn the Attorney General, presumably to prevent any legal action from families.[211]

The case of Gobbe, the first anatomical gift act preserved in the hospital archives of Brussels, suggests two things: that body donation was still rare at the turn of the twentieth century and that dissection continued to be negatively perceived. The confused reaction of the head of the hospital, as well as the ignorance of both the commission of hospitals and municipal institutions, shows that anatomical donation was very unusual. Despite the existing legal framework, the shift from involuntarily acquired to gifted bodies was slow. Anatomists by and large continued their old ways. Even though families could claim bodies without financial consequences from the turn of the twentieth century onwards, a considerable number of unclaimed corpses continued to find their way to anatomical theatres—in Brussels still an average of 45 bodies per year in the 1920s. As the number of unclaimed bodies dropped persistently because of the new regulations, the medical faculty expanded its reach. All available resources had to be put to use: apart from Brussels institutions, corpses came from the psychiatric institution of Gheel, Antwerp hospitals and the prison of Forest.[212] They were the bodies of individuals who had no close family or written testament, and were not associated with a burial society.

In her research on pauper burials, Gijbels came to similar conclusions. She argued that hospital administrators in Brussels only took into account the funeral wishes of patients 'if there were close family members or burial societies involved'.[213] Patients who died without leaving anyone behind to stand up for them still ended up in the hands of anatomists. As aforementioned, hospitals had confined the right to claim bodies to close relatives. In their opinion, it was impossible to allow 'strangers' to prohibit dissection: if ministers of religion, friends, acquaintances or relatives in the second or third remove could claim corpses without financial obligation, the morgues of anatomical institutes would be empty.[214]

Although a few individuals wished to give their bodies to science in the first two decades of the twentieth century, they remained isolated instances, which gave rise to confusion and discussion, much like in the case of Gobbe. Amongst other things, the university wondered if there was a legal framework for body donation, if the family could object to an individual's donation, and what the requirements for a proper testament were.[215] Albert Brachet, who became professor of anatomy in 1904, was the first to organise a formal body donation programme.[216] His successor, Albert Dalcq, continued his efforts from the 1930s onwards, and by the 1960s all unclaimed bodies had been replaced by donated bodies.[217]

Research on the emergence of anatomical donation in general is scarce, although the findings of a few international studies are similar to the development of body donation in Brussels. For Britain, Ruth Richardson has found a comparable increase in the number of anatomical bequests in the 1940s, although body donation already began on a small scale after the First World War. Bequests counted for 70 to 100 per cent of all bodies used for dissection in the 1960s.[218] In the United States, the shift from unclaimed to bequeathed bodies was gradual, too. Newspapers reported on anatomical donations as early as the 1870s, whereas the first formal willed body programmes were only established in the 1950s.[219] For most European countries, the early history of anatomical donation remains to be written. Preserved legal journals reveal that it became possible to bequeath your body to scientific societies in France in 1886. [220] The initiative for this rule, an amendment to a more general law on the freedom of burials, came from politician and anthropologist Gabriel de Mortillet. It is unclear, however, if (and how) anatomical institutions implemented this principle. A recent article on body procurement in Bayern in Germany reveals that state ministries implemented the absolute transition to body donation against anatomists' wishes in 1960. They did so by stopping any government intervention in the distribution of corpses to medical faculties.[221]

Although a lot is still unknown, from what we do know it seems that the procurement of bodies for dissection developed similarly in different countries. The introduction of body bequeathal was a gradual process, which took over half a century to get off the ground. Until the end of the nineteenth century, anatomists relied exclusively on the bodies of the unclaimed poor, whom they dissected against their wishes; this reliance on the bodies of the socially marginal and secluded came to coexist with body donation from the early twentieth century onwards, until anatomical bequests became the main source of dissectable bodies by the 1960s.

There are several reasons why the introduction of body bequests was slow. First and foremost, it appears that the idea of anatomical donation developed along the same lines as the notion of consent. As aforementioned, consent in the nineteenth century mostly signified the right to refuse medical interventions. Whereas patients were accustomed to the idea that they could deny a proposed treatment, the thought that they could request a medical intervention themselves was, for the poor in particular, new.[222] Regarding surgeries and other medical interventions on the living, Faden and Beauchamp have argued that this more active conception of the patient's role in medical decision-making required a profound, and hence gradual, change of mentality. In their view, patient autonomy only became the bedrock of medical ethics and policy in the 1950s. In this period, the notion of informed consent was lifted from its narrow late nineteenth- and early twentieth-century legal base through lively debates over decisional authority and the doctor-patient relationship. Both case law and medical ethics came to address informed consent in medicine under the influence of various social movements that advocated individual rights. The issues raised by civil rights, women's rights, the consumer movement and so on often included healthcare components, for example regarding human experimentation, reproductive freedom or medical information.[223] Anatomical donation, too, thrived in this context. Much like fifty years before, a new importance attached to self-determination, autonomy and individualism led to a reinterpretation of the principle of consent, accompanied by a shift in the procurement of bodies for post-mortem research.

A more direct reason for the slow increase in the number of bequeathed bodies was the need for a proper testament. Medical faculties were often reluctant to proceed with dissection as the legal validity of bequests was unclear. Although one's will could in principle be drawn up verbally in the presence of witnesses, anatomists for the most part only honoured written wills to avoid lawsuits. Therefore a certain expertise was demanded to donate one's body to science. As the case of Gobbe suggests, a written testament, preferably drawn up with the help of a notary, was the best guarantee for the execution of anatomical bequests. Yet for a large segment of society, the illiterate poor in particular, this kind of testament remained unusual.

In fact, a considerable number of donations were not honoured before the 1950s, despite recurring concerns about insufficient supplies of anatomical material. In her research on the United States, Susan Lederer has shown that not only the fear of prosecution hindered the execution of bequests. Donated corpses could be undesirable for material reasons, too. Anatomists declined bodies because of their weight (too thin/fat) or age (too young/old) or because they had been damaged by disease or death (this happened with suicides especially).[224] Additionally, since anatomical bequests were so unusual, they often led to confusion and protest. The medical faculty of Brussels mostly did not honour bequests if families claimed the body, although the wishes of the deceased took priority from a legal point of view.[225] Another difficulty in this respect was, again, the materiality of the corpse. Donated bodies often decomposed before hospital or university administrators had figured out what to do with them. Similarly, lawsuits of families could delay corpses' departure to anatomical institutes until the point that they had become worthless for medical education and research.

The main obstacle for the introduction of anatomical donation, however, was the negative perception of dissection. The case of Gobbe seems peculiar today, but in fact is a representative illustration of early anatomical donation. In the early twentieth century, individuals mostly bequeathed their bodies for negative reasons. Lederer has found that suicides accounted for as much as one third of all body bequests in the United States between 1870 and 1950. She argued that bequests were a way to make up for a supposedly 'failed' life. First and foremost, they reflected a lack of self-worth. Gobbe's wish to be 'thrown in a communal grave without warning anyone', for example, suggests that he wanted to disappear off the face of the earth without leaving any traces. Rather than being mourned for, he wanted to be forgotten. Gobbe was not alone: Lederer has shown for the American context that the desire to avoid burdening others with the cost of burial was one of the main reasons for anatomical donation until well into the twentieth century. In addition, suicides donating their bodies often desired to procure some utility from their lives. What they, in their view, had lacked in meaning during life, they hoped to gain after death: their dissected remains had to amend for the failures of their living bodies.[226]

Apart from a lack of self-esteem, also a fear to be buried alive drove individuals to donate their bodies to science. The precise criteria of death were a topic of intense debate in the second half of the nineteenth century in Belgium, both in medical circles and in society more broadly. For some, death and its illegibility gave rise to anxiety. Various forensic physicians, for instance, represented dying as 'a process, which does not happen at once, but which has a beginning and an end'. They believed in 'an intermediary stage between life and death that always occurs'.[227] Discussions on the volatility of death were, next to hygienic and religious reasons, an important motive for the construction of morgues. Since 'decomposition was the only certain sign of death', morgues had to figure as waiting rooms in between the worlds of the death and the living.[228] The invention of new embalming techniques further problematised the clarity of death. It might have seemed as if the uncertain border between life and death could be overcome from both sides. The fear of apparent death (someone looks dead, but is alive) as well as modern embalming (someone looks alive, but is dead) changed the face of death, and had repercussions on popular culture.[229] The liminality and deceptiveness of corpses inspired writers, playwrights, poets and painters, who imagined how it was to be buried alive.[230]

In Brussels, this perceived haziness of death led a few individuals to donate their bodies in order to be on the safe side. In 1909, for example, a certain Alfred Prume wrote a letter to the head of the St. Jean hospital. For the sake of his 'tranquillity of mind', he asked to be 'dissected in service of education, and not buried, or at least autopsied, in order to detract heart and brains from my body'.[231]

The head of St. Jean forwarded his letter to the commission of hospitals. He noted that 'his offer might interest the university' and that 'this gentleman probably is afraid of being buried alive'.[232]

Yet as long as negative emotions, such as self-hatred, self-effacement or fear, inspired anatomical donations, they remained unusual. Bequests had to be reframed before they could become the main source of cadaveric material. In order to encourage individuals to donate their bodies, anatomy had to be redefined as something worth giving to. From this perspective, it is not surprising that anatomists increasingly occupied themselves with their public image in the first half of the twentieth century, from the 1930s onwards in particular. Through their efforts, body donation gradually transformed from a deplorable way to punish oneself to an admirable act of altruism.

Stories of physicians donating their own bodies to science increasingly shaped the public face of dissection in the first half of the twentieth century. In 1912, for example, 200 New York physicians publicly pledged themselves as body donors in an effort to overcome public prejudice against dissection.[233] Their examples provided an offset for typical nineteenth-century 'saved from the dissecting table stories', in which families were urged to claim their deceased in order to avoid the fragmentation of their bodies. Over time, these stories, in which dissection was something progressive, dignified and altruistic rather than something shameful, disrespectful and unrighteous, became the dominant narrative on body donation in the United States.[234] In her study on Britain, Richardson came to similar conclusions. She posited that the decision of the Inspector of Anatomy, the person responsible for the distribution of bodies to anatomical institutes, to bequeath his own body to science in 1938 marked a more general change in attitudes towards dissection. His positive representation of anatomical donation prompted others to give up their bodies as well.[235] In a similar vein, Dalcq, who was responsible for the generalisation of anatomical donation at the Free University of Brussels, donated his own remains to science. In doing so, he enhanced the public image of his institution. The fact that he was willing to bequeath his body himself was supposed to convince others of the value of dissection.[236]

Physicians' willingness to be scrutinised after death was not entirely new. In France, anatomists urged each other 'to set a good example by asking for an autopsy in their testament' as early as the 1880s.[237] Even earlier, medical obituaries occasionally included autopsy reports, particularly during the heyday of pathological anatomy in the 1830s and 1840s.[238] Belgian medical journals, for example reported on the well-known French anatomist and surgeon Guillaume Dupuytren, who left his body for dissection to the Parisian anatomists François Broussais and Jean Cruveilhier. Their examination not only revealed the cause of death (an effusion of pus in his right lung) but also showed that Dupuytren's 'brain was remarkable because of its volume'.[239] Similarly, the Ghent anatomist Adolphe Burggraeve published the autopsy results of the 95-year-old physician Pierre-Englebert Wauters, who had predicted that his colleagues would find a stone in his bladder. In the report, Burggraeve linked the excellent condition of most of Wauters' organs, despite his old age, to his 'austere and regular life' in service of science.[240] These autopsy reports were more about heaping praise on the deceased than about medical progress. They above all showcased physicians' devotion to their profession by emphasising that their first priority was the advancement of the medical sciences, even after death.[241]

Whereas these earlier examples were primarily aimed at the creation of a shared identity and morality within the medical corps, historians have shown that late nineteenth- and early twentieth-century tales on heroic physicians also provided an answer to contemporary anxieties. In the face of criticism on animal and human experimentation, there was a professional investment in improving medicine's image, which led to the publication of almost mythological reports and necrologies in medical journals and newspapers. The ideal of self-sacrifice became a political and emotional strategy. The ability to point at researchers' own willingness to be subjected to experiments enabled them to take the moral high ground and to earn public support.[242] Physicians were represented not only as heroes, but also as martyrs: not only successful medical trials, but also failed self-experiments could bolster their reputation. In the fields of bacteriology and radiology especially, medical pioneers were portrayed as martyrs, who took life-threatening risks, or even lost their lives, for the benefit of medical progress. In these narratives, medicine was a calling and a duty, which demanded practitioners to put their self-interest aside.[243]

A similar development took place in the field of anatomy. In the same way that stories of heroic medical researchers could encourage patients to cooperate in medical trials, reports of physicians bequeathing their bodies to medical faculties emphasised the respectful and useful nature of dissections. In Belgium, newspapers published stories of anatomical martyrs, who were injured because of their strenuous efforts in the dissecting room. Students who died from the effects of a *piqûre anatomique* (infections from a cut with the dissecting knife) were called 'heroes of science', 'martyrs of science' or 'victims of duty'.[244] Since the handling of corpses increased the likelihood of dangerous infections, dissections, much like other medical (self-) experiments, were deemed heroic and altruistic activities:

The list of names of medical martyrs has grown by one. Every year, the medical community loses a number of their members in this way, whether it is through a *piqûre anatomique*, a poisoning following the preparation or administration of certain drugs, or an epidemic disease caught by the physician at the patient's sickbed. [...] Both science and humanity feel very honoured by the dedication and the courage with which professional duties are executed [by physicians].[245]

These kinds of reports coincided with renewed efforts to commemorate the work of anatomical pioneers. Joris Vandendriessche has posited that the early modern anatomist Andreas Vesalius, who was born in Brussels in 1514, 'became the icon with which Belgian physicians identified and which could represent the medical profession in society'.[246] As his life and work could excite both patriotism and admiration for the medical profession, Vesalius became the subject of numerous biographies and eulogies that appeared from the mid of the nineteenth century onwards. In his 'Studies on Andreas Vesalius' (*Etudes sur André Vésale*), for example, Adolphe Burggraeve, professor of anatomy at the University of Ghent, enhanced Vesalius' reputation as the founder of anatomy by charting the extent of his knowledge and by underlining his contribution to the progress of science. Burggraeve used the figure of Vesalius to convey a positive image of modern science and a feeling of Belgian pride.[247] Vesalius' memory was also picked up in the literary and artistic world. His dissections were visually attractive topics for painters, and his remarkable course of life inspired writers.[248]

Tellingly, Vesalius' life became a story of medical martyrdom. The legal obstacles and religious 'prejudices' he had to overcome to obtain cadavers were represented as signs of his dedication and perseverance. His flee from Europe after he was sentenced to death also reflected the ideal of self-sacrifice, showcasing renowned physicians' willingness to suffer for science. In a work entitled 'The Martyrs of Science' (*Les martyrs de la science*), for example, Vesalius was portrayed as an altruistic and heroic physician, who risked his life multiple times to acquire anatomical knowledge: not only by facing the Inquisition, but also by 'fighting dogs for an already decomposed cadaver' at the cemetery (Illustration 3.1).[249] As Vesalius' hard life illustrated anatomists' dedication, and his contribution to medical progress showed the usefulness of dissection, it is no wonder that the newly constructed anatomical institute of Leuven was called after him in 1877.[250]

Vésale disputant à des chiens une proie déjà putréfiée... (Page 302.)

Illustration 3.1 'Vesalius competes with dogs for an already decomposed prey.' From: Gaston Tissandier, *Les martyrs de la science*, 2nd ed. (Paris: Maurice Dreyfous, 1882), 287. © Collection Erfgoedbibliotheek Hendrik Conscience, Antwerp, G 46379

Apart from stressing their own sacrifice and the contribution of anatomy to medical progress, anatomists also emphasised the respectful treatment of the corpse. In a case study on the London teaching hospitals, Keir Waddington has shown that medical schools tried to counter popular stereotypes in the late nineteenth century. In order to enhance the status of medicine and assert its credentials, they attempted to reinvent their students, who were popularly regarded as foolish or vicious pariahs, as industrious gentlemen. Around the same time, Waddington asserted, medical schools established a disciplinary apparatus aimed at giving this new image of the hardworking and respectable student some substance.[251]

Students' behaviour in the dissecting room became a thorn in the flesh of medical faculties. As mentioned in the previous chapter, historical research has revealed that dissections went together with shabby behaviour and black humour. Students suppressed the discomfort and fear caused by the cadaver through jokes and pleasantries.[252] Brussels newspapers in the second half of the nineteenth century occasionally reported on inappropriate student jokes with human remains, or on students stealing body parts from the anatomical theatre in order to continue their studies at home. During exam periods specifically, students seem to have stolen anatomical remains, after which they dumped them on the street.[253]

Medical faculties were increasingly annoyed with students' misconduct. Because 'feeble jokes', in the words of the medical faculty of Brussels, 'evoked upheaval amongst the population', they took disciplinary measures to regulate students' behaviour.[254] Medical faculties made sure to communicate to newspapers that they did not accept tomfooling in the anatomical theatre, and emphasised that disrespectful students 'would be severely punished'.[255] The honourable behaviour of students became even more important when donation became the primary source of dissectable bodies. When medical students had stolen an ear as a 'bad joke' in 1950, for instance, the fervent supporter of donation (and future donor) Albert Dalcq asked the rector of the Free University of Brussels 'to fully assign him with the disciplinary authority to prevent an ignominious use of anatomical material'. The reason for his unforgiving stance was that he believed that the respectable treatment of bodies in the anatomical theatre was quintessential for the success of bequeathal programmes.[256] Bodies were also to be treated with dignity after dissection. As I will show in more detail in the fifth chapter of this book, which discusses the disposal of dissected remains, anatomists stressed that body donation did not result in the denial of a proper burial.

Concerns over the public image of dissection also influenced the architecture of anatomical institutes. In late nineteenth-century Belgium, anatomical theatres were modelled after Churches, buildings which were associated with a respectful treatment of the dead body. In 1899, for example, architect Louis Cloquet designed the Rommelaere Institute for Hygiene, Bacteriology and Forensic Medicine in Ghent in a neo-Gothic style (Illustration 3.2). Families who wanted to pay their last respects entered the building via an entrance made especially for this purpose, which led them to a room that resembled traditional funeral chapel. If they looked up, they saw a black ceiling.[257] In 1875, the University of Leuven began constructing a new anatomical institute with financial support of Monseigneur De Montpellier, the ultramontane archbishop of Liège. Two years later, the Vesalius institute, designed in a neo-Gothic style by the Catholic architect Joris Helleputte, was inaugurated with great pomp and circumstance.[258]

In Brussels hospitals, precious materials were used to impress families who came to visit deceased patients in the morgue.[259] Liège anatomists,

Illustration 3.2 Postcard depicting the Rommelaere Institute, date unknown. ©
City Archives of Ghent, SCMS_PBK_3302

too, wished to give their post-mortem rooms 'a certain luxury and religious ostentation'. In their opinion, replacing 'bare, wretched walls' with
'paintings, stained glass, curtains, etcetera' would emit a sense of respect,
and therefore would tone down protests from patients and their families.
They thought that the architecture and interior of anatomical institutions
were important, as they 'had a profound influence on families' decisions
about claiming cadavers':

> It is clear that when relatives notice that the corpse has been treated with
> decency and when the mortal remains are surrounded with a certain splen-
> dour, which does not have to be expensive to look luxurious in the eyes of
> the poor, they have less to fear when they entrust the care for their relatives'
> interment to us.[260]

In the 1920s and 1930s, Catholic authors represented dissection as an
ultimate act of altruism in biographies and eulogies, often published with
the support of Catholic universities. A willingness to undergo dissection
symbolised the final stage of a life in service of others, for example in

biographies and tales about the Catholic businessman and social activist Philibert Vrau or the early modern bishop and saint François de Sales.[261] Similarly, in a study dedicated to Charles Nelis, professor of anatomy at the University of Leuven from 1919 until 1935, the physician and historian Jean-Joseph Tricot-Royer used examples from Christian history, from antiquity until recent times, to prove that the Roman Catholic Church did not oppose to, and had never been opposed to, anatomical research:

> We conclude that the Church has not hindered anatomical studies, but has encouraged them; she has made sure that dissections were practised with the necessary respect for the cadaver of one's fellow man; and in the end, she has taken crucial precautions to spare the feelings of the friends and family of the deceased.[262]

Not only could dissection be reconciled with traditional religious beliefs; it could also figure in more alternative conceptions of the afterlife. In the last two decades of the nineteenth century, the Society for Mutual Autopsy (*Société d'autopsie mutuelle*) created a furore in France. Members of this society, mostly physicians and left-wing politicians, declared themselves prepared to undergo an autopsy after death in order to contribute to scientific progress. In their view, post-mortem research could give free-thinkers an alternative, secular life after death. An eternal existence in scientific results gave meaning to death without interfering with materialist convictions.[263] Belgian newspapers reported repeatedly on the French initiative. Autopsy reports of members were published and explained, the goals of the society were both praised and ridiculed.[264] Either way, the idea that post-mortem research could give meaning to, and procure utility from, death gained visibility.

Yet, as Richardson has rightly argued, this reconciliation between dissection and last rites could only be fully realised when the social meanings of the corpse and its spiritual associations shifted. The rise of anatomical donation only became possible when both rituals surrounding the dead body and the perception of the dead body itself changed. As long as burial rituals, such as the laying out of the corpse, wakes or last visits, were essential to the culture of death, and as long as dead bodies were seen as possibly sensitive, anatomical donation could not become the main source of cadaveric material for medical faculties. The increase in bequests in the 1930s and 1940s was incited not only—and perhaps not even primarily— by a more positive view of dissection and medical science, but also by a growing disbelief in the spiritual power and animated nature of the corpse. Tellingly, the number of body donations rose parallel with the popularity of cremation in the United Kingdom.[265] In Brussels, too, the development of anatomical bequeathal went hand in hand with the growing acceptance of cremation, a practice that was legalised in 1931.[266]

Despite different conceptions of consent, science and the corpse, numerous individuals donating their bodies to science today are driven by similar motives as François Gobbe in 1899. All has changed, but nothing has changed. On the one hand, the positive wish to mean something to others lives on. The aspiration to benefit the living when dead has inspired donors from the nineteenth century until today. The significance of the maxim *in morte vita quaeritur* ('in death, we search for life'), which was engraved on the walls of the dissecting room in the Brussels St. Pierre hospital in the early nineteenth century, has only been enforced by the development of organ donation from the 1940s onwards. The core idea remains the same: what is useless to you when dead, might save someone's life.[267] On the other hand, the negative wish to make up for a failed life and to avoid burdening others has continued to push people to anatomical donation. Anthropologists and ethicists in the United Kingdom have argued that escaping burial costs is still one of the main reasons in favour of body bequeathal. As evidence suggests that the number of body donations rises parallel with average funeral expenses, one might wonder if poverty still is at the heart of present-day donors' motivation.[268] Those giving their bodies to science today might not be so different from their nineteenth-century predecessors as we would like them to be.

7   Conclusion

In 1923, Mme. Lankester, who had been a hospital nurse in Brussels, died from a rare neurological condition. The psychiatrist and neurologist Auguste Ley removed her brain for research. When the nurses found out, they communicated the misconduct (it was not allowed to conduct post-mortems on employees of the institution) to the press.[269] In his defence, Ley stressed the importance of the last wishes of Lankester. As a nurse who had assisted at many autopsies, she on multiple occasions had expressed her belief in the need for post-mortems for the advancement of neurology. She had also consented to her own autopsy, because she believed it would lead to interesting scientific results. In Ley's opinion, the prohibition of performing autopsies on hospital staff was subordinate to Lankester's own wishes and beliefs: 'I consider the personal decision to be more important than regulations'.[270]

The sharp contrast between Ley's defence in 1923 and Deroubaix' proposal in 1888, with which I began this chapter, shows an important historical evolution. Even though we cannot be sure if Ley was sincere, his argumentation does attest to the cultural importance attached to the consent of the individual. As a result of political and legal debates on the treatment of the poor and the ownership of the dead around the turn of the century, individual patients had obtained the right to decide over their bodies. This marked the beginning of the end of the nineteenth-century transaction that involved the provision of assistance in exchange for bodies. As an inalienable right of the individual, the body, whether dead or alive, could no longer be used as a means of payment. As the case of Lankester suggests, anatomists increasingly complied with the principle of consent.

Although more research is needed to fully grasp the links between ideological conflicts, the extension of democracy, changing attitudes towards the poor, the codification of the principle of consent and the introduction of anatomical donation, the few studies that exist seem to confirm the conclusions of my analysis. Throughout this chapter, it has become clear that the Brussels debate on involuntary dissection fitted with broader trends. It therefore seems likely that the beginning of the twentieth century was a juncture with regard to the acquisition of anatomical material. Much like the bodies of paupers replaced the bodies of criminals in the early nineteenth century, the bodies of paupers appear to have been very gradually replaced by anatomical donation from the early twentieth century onwards.

At least many of the circumstances that gave rise to anatomical donation in Brussels existed in other European countries as well. Firstly, historians such as Christopher Clark and Wolfram Kaiser have identified nineteenth-century culture wars as a European phenomenon, with outbursts in Central Europe (for example in Belgium, France, the Netherlands, Germany and Austria), Southern Europe (for example in Spain and Italy) and even across the North Sea, in the United Kingdom. In many countries, these conflicts led to an increased emphasis on the last wishes of the deceased.[271] Secondly, attitudes towards the poor became more emphatic in other countries, too. This evolution was tied up with the gradual extension of democracy and debates on the 'dignity' of poor relief. The 'modernisation' of Belgian poor relief was modelled after Germany, where Otto von Bismarck had introduced health insurance in 1883 in order to reduce support for Socialism.[272] In Britain, the Poor Law system began to decline with the emergence of other forms of assistance, such as friendly societies and trade unions. In 1905, a Royal Commission was established to reform poor relief.[273] Elizabeth Hurren has shown that democratisation, too, had a profound influence on the transformation of poor law politics. By means of case studies on pauper burials and dissection, she has argued that few politicians could afford to ignore 'ordinary people's negative reaction to the worst excesses of poor law retrenchment' in an era of the coming of democracy.[274] Thirdly, it appears that the principle of consent, which was codified across Europe in the last decades of the nineteenth century, prepared the ground for formal willed body programmes. From the early twentieth century onwards, body donation slowly became the main source of anatomical material for European and American medical schools.[275]

Not only is the shift from involuntarily acquired to donated bodies an important evolution in the social history of anatomy. It also offers new insights into the significance of the corpse and the funeral around the start of the twentieth century.

By discussing the debate on involuntary dissection from multiple perspectives, this chapter has related the idea that everyone can decide on the post-mortem fate of his or her remains during life to changes in the Western culture of death, which increasingly revolved around the individual. As the funeral became a personal ritual that expressed the identity of the deceased, the meaning of the dead body also shifted. In a sense, the individual could live on a little longer through his or her corpse: either by choosing his or her own burial ritual or by donating his or her body to science. By embodying the will of the deceased, the dead body could become a subject rather than an object—even in the anatomical theatre.

NOTES

1. Based on: "Détournement de cadavre," *Journal de Bruxelles*, March 4, 1887; "In St. Jansgasthuis," *Het Handelsblad*, March 3, 1887 "Détournement d'un cadavre à l'hôpital St Jean," *Indépendance belge*, March 3, 1887; "L'affaire de l'hôpital St Jean, à Bruxelles," *La Meuse*, March 3, 1887; "L'affaire Coco," *La Meuse*, March 24, 1887; "A propos de l'histoire du cadavre de la négresse Coco," *Le Peuple*, March 5, 1887; "Cadavre disparu," *Le Peuple*, March 4, 1887; "Le cadavre de Coco," *Le Peuple*, April 23, 1887.

2. Procès-verbaux des séances du Conseil d'Administration, March 17, 1888, book 4, UAB.
3. Procès-verbaux des séances du Conseil d'Administration, November 10, 1888, book 4, UAB.
4. This chapter draws largely on the following article: Tinne Claes, "By What Right does the Scalpel Enter the Pauper's Corpse? Dissections and Consent in Late Nineteenth-Century Belgium," *Social History of Medicine* 31, no. 2 (2018): 258–77.
5. The standard work on the history of consent is: Ruth R. Faden and Tom L. Beauchamp, *A History and Theory of Informed Consent* (Oxford: Oxford University Press, 1986).
6. Katherine Park, "The Life of the Corpse: Division and Dissection in Late Medieval Europe," *The Journal of the History of Medicine and Allied Sciences* 50, no. 1 (1995): 111–32, on p. 129.
7. Helen MacDonald, "A Body Buried is a Body Wasted: The Spoils of Human Dissection," in *The Body Divided: Human Beings and Human 'Material' in Modern Medical History*, eds. Sarah Ferber and Sally Wilde (Farnham: Ashgate, 2012), 9–27; Elizabeth T. Hurren, *Dying for Victorian Medicine: English Anatomy and its Trade in the Dead Poor c. 1834–1929* (Houndmills: Palgrave Macmillan, 2012), 41–73.
8. Helen MacDonald, *Possessing the Dead. The Artful Science of Anatomy* (Melbourne: Melbourne University Press, 2010), 100–2.
9. Maurice De Laet, "Le droit à l'autopsie et à prélèvements post-mortem," *Bulletin de l'Académie royale de médecine de Belgique* 107 (1948): 84–92, on p. 84.
10. Edmond Picard and Napoléon d'Hoffschmidt, "Cadavre," in *Répertoire général de législation, de doctrine et de jurisprudence belges donnant pour toutes les matières du droit belge*, vol. 15 (Brussels: Ferdinand Larcier, 1885), 244–65, on pp. 259–60.
11. Pierre Nuard, *Sciences, médecine, pharmacie, de la révolution à l'empire* (Paris: Dacosta, 1970), 140.
12. Toby Gelfand, "The 'Paris Manner' of Dissection: Student Anatomical Dissection in Early Eighteenth-Century Paris," *Bulletin of the History of Medicine* 46, no. 2 (1972): 99–130.
13. Note that the supply of cadavers in France was in fact regulated two months later than in the Southern Netherlands, by the law of 29 Brumaire, an VII (November 19, 1798). See: Dora B. Weiner, *The Citizen-Patient in Revolutionary and Imperial Paris* (Baltimore and London: Johns Hopkins University Press, 1993), 183.
14. In 1796, the French revolutionary regime allowed local authorities to appoint a board to administer the institutions for the intake of the infirm poor (the sick, the elderly, orphans or neglected children, etc.). This reg-

ulation was based on municipal responsibility and was originally aimed at the control of religious charity. As the members were elected by the city council and all of the commission's decisions had to be approved, local politics came to influence the relationship with the religious who were active in public medical institutions. See: Stijn Van de Perre, "Public Charity and Private Assistance in Nineteenth-Century Belgium," in *Armenfürsorge und Wohltätigkeit. Ländliche Gesallschaften in Europa, 1850–1930/Poor Relief and Charity. Rural Societies in Europe, 1850–1930*, eds. Inga Brandes and Katrin Marx-Jaskulski (Frankfurt am Main: Peter Lang, 2008), 93–123, on p. 95–9.

15. Picard and d'Hoffschmidt, "Cadavre," 259–60.
16. Cay-Rüdiger Prüll, "No Law, No Rights? Autopsy in Germany since 1800," in: *Coping with Sickness: Medicine, Law and Human Rights— Historical Perspectives*, eds. John Woodward and Robert Jütte (Sheffield: EAHMH Publications, 2000), 30–53, on p. 35.
17. For example: Peter King, *Crime and Law in England, 1750–1840: Remaking Justice from the Margins* (Cambridge: Cambridge University Press, 2006), 1–2.
18. Circular letter sent to Belgian hospitals concerning the distribution of corpses for dissection, 1881, folder 1, Gasthuis allerlei, ASSL.
19. "Hôpitaux: mode d'information d'un décès aux membres de la famille," *Bulletin communal de Bruxelles* 55 (1904), I, 897.
20. Form letter for notification of death, 1887, folder 38, Inhumations, Affaires générales, ASSB; P. Cuypers to C. Buls, January 19, 1887, folder 38, Inhumations, Affaires générales, ASSB.
21. C. Buls to Conseil des hospices, January 19, 1887, folder 38, Inhumations, Affaires générales, ASSB.
22. Dewael to Tirlemond, February 29, 1887, folder 38, Inhumations, Affaires générales, ASSB.
23. Dossier concernant la plainte de la veuve Pouillon relative à l'enterrement de son mari décédé à l'hôpital de St. Jean, sans avoir été prévenue à temps, 1885, folder 113, Hôpital St. Jean: généralités, Affaires générales, ASSB.
24. Quotes from: Charles Firket, *Du but et de l'organisation des services d'autopsies* (Liège: Vaillant-Carmanne, 1883), 41.
25. Vandervelde to Conseil des hospices, May 18, 1900, folder 107, Autopsies, Affaires générales, ASSB.
26. Firket, *Du but et de l'organisation*, 37.
27. Charles Firket and Félix Putzeys, *Rapport sur le projet de loi proposé par la commission des Hospices civils de Liège, relativement à l'interprétation de l'art 8 de la loi du 15 juillet 1849, organique de l'enseignement supérieur*, 1891, folder 254, Fonds du secrétariat central, UAL.

28. Procès-verbaux des séances du conseil d'administration, November 10, 1888, book 4, UAB.
29. A similar argument is made in: Elizabeth T. Hurren, "Abnormalities and Deformities: The Dissection and Interment of the Insane Poor, 1832–1929," *Journal of the History of Psychiatry* 23, no. 1 (2012): 65–77, on p. 66.
30. Eugène Bouchut and Armand Després, *Dictionnaire de médecine et de thérapeutique médicale et chirurgicale* (Paris: Librairie Germer Ballière, 1883), 725.
31. Rapport sur l'organisation du service des autopsies dans les hôpitaux de Bruxelles, [1878], folder 107, Autopsies, Affaires générales, ASSB.
32. Michel Foucault, *La naissance de la clinique* (Paris: Presses universitaires de France, 1963), 85.
33. Grégoire Chamayou, *Les corps vils: expérimenter sur les êtres humains au XVIIIe et XIXe siècle* (Paris: La découverte, 2011), 178–80.
34. Séverine Icard, *La constatation des décès dans les hôpitaux en France et à l'étranger et la nécessité de la pratique hâtive des autopsies* (Paris: Maloine, 1910), 17.
35. Nicolas-Joseph Ansiaux, "Discours d'ouverture du cours de clinique chirurgicale à l'Université de Liège," *Le Scalpel*, February 5, 1849, cited in: Carl Havelange, "L'hôpital à la croisée des chemins: la question des malades payants," *Annales belges d'Histoire des Hôpitaux et de la Santé Publique* 25 (1987): 83–94, on p. 84–5.
36. Valérie Leclerq, "Une histoire des interactions hospitalières avant l'ère du 'patient autonome', Bruxelles 1870–1930" (PhD diss., Université Libre de Bruxelles, 2017).
37. For example: Proceedings of the plenary sessions of the Belgian Chamber of Representatives of May 13, 1891. https://sites.google.com/site/bplenum/proceedings/1891/k00161208/k00161208_00. Accessed February 27, 2017.
38. Dossier concernant la plainte du Sieur J. Maeck, s.d., folder 123, Généralités, Affaires générales, ASSB. For more examples: "Une histoire des interactions."
39. Dossier concernant l'autorisation de recevoir à l'hôpital les cadavres des personnes qui meurent pendant leur transport en civière, 1893, folder 76, Généralités, Affaires générales, ASSB.
40. For example: Dossier concernant un article des 'Nouvelles du Jour' à charge de l'hôpital St Pierre, 1880, folder 130, Hôpital St. Pierre: Divers, Affaires générales, ASSB.
41. Dirk Van Damme, "Divergerende wegen van sociale beheersing," in *Op vrije voeten? Sociale politiek in West-Europa, 1450–1914*, eds. Catharina

Lis, Hugo Soly and Dirk Van Damme (Leuven: Kritak Uitgeverij, 1985), 171–8; Van de Perre, "Public Charity," 95–101.

42. Register of bodies brought to the Institut d'anatomie, 1920–1968, Private Archives of Stéphane Louryan, Brussels; Dossier 'Ter beschikking stellen van lijken veroordeelden voor onderwijs', 1887, folder 4A 43 (71), UAG.

43. Jo Deferme, *Uit de ketens van de vrijheid. Het debat over de sociale politiek in België, 1886–1914* (Leuven: Leuven University Press, 2007), 59–88.

44. Two works provide an overview of this network of private charitable organisations in Belgium around 1900: Molly de Spoelberch de Lovenjoul, *Belgique charitable. Bruxelles: charité, bienfaisance, philanthropie, etc., etc* (Brussels: Larcier, 1893) and Ludovic Saint-Vincent and Charles Vloeberghs, *Belgique charitable* (Brussels: Dewit, 1904). On Catholic charity in Belgium more generally: Leen Van Molle, "Social Questions and Catholic Answers: Social Reform in Belgium, c. 1780–1920," in *Charity and Social Welfare. The Dynamics of Religious Reform in Northern Europe, 1780–1920*, ed. Leen Van Molle (Leuven: Leuven University Press, 2017), 101–21. On Catholic healthcare in Belgium: René Stockman, *Pro Deo. De geschiedenis van de christelijke gezondheidszorg* (Leuven: Davidsfonds, 2008).

45. André Tihon, "Les religieuses en Belgique du XVIIIe siècle. Approche statistique," *Belgisch Tijdschrift voor Nieuwste Geschiedenis 7*, no. 1–2 (1976): 1–54, on p. 40–1. See also: Jan De Maeyer and Jo Deferme, "Vrouwelijke religieuzen in de openbare en private gezondheidszorg in het België van de negentiende en twintigste eeuw: tussen traditie en moderniteit," in *Bezielde zorg. Verpleging door katholieke religieuzen in Nederland en Vlaanderen (negentiende—twintigste eeuw)*, eds. Liesbeth Labbeke, Vefie Poels and Rob Wolf (Hilversum: Uitgeverij Verloren, 2008), 10–28.

46. Firket, *Du but et de l'organisation*, 38.

47. Faculty of medicine to Conseil des hospices, December 13, 1873, folder 107, Autopsies, Affaires générales, ASSB.

48. "Discussion sur la désinfection des sépultures: Société royale des sciences médicales et naturelles de Bruxelles: séance du 2 mars 1874," *Journal de médecine, de chirurgie et de pharmacologie* 58 (1874): 278.

49. On post-mortem punishment in Britain: Elizabeth T. Hurren, *Dissecting the Criminal Corpse: Staging Post-Execution Punishment in Early Modern England* (Basingstoke: Palgrave Macmillan, 2016).

50. Dossier 'Ter beschikking stellen van lijken veroordeelden voor onderwijs', 1887, folder 4A 43 (71), UAG.

51. See for example: Tinne Claes and Veronique Deblon, "When Nothing Remains: Anatomical Collections, the Ethics of Stewardship and the

Meanings of Absence," *Journal of the History of Collections* 30, no. 2 (2017): 351–62.

52. On popular anatomical museums in the Low Countries: Tinne Claes and Veronique Deblon, "Van panoramisch naar preventief. Populariserende anatomische musea in de Lage Landen," *De Negentiende Eeuw* 39, no. 3–4 (2015): 289–306.

53. Proceedings of the plenary sessions of the Belgian Chamber of Representatives of May 13, 1891. https://sites.google.com/site/bplenum/proceedings/1891/k00161208/k00161208_00. Accessed February 27, 2017.

54. On France: Gelfand, "The 'Paris Manner' of Dissection," 99–130; On the German *Länder* and Austria: Karin Stukenbrock, *'Der zerstückte Cörper': Zur Sozialgeschichte der anatomischen Sektionen in der frühen Neuzeit 1650–1800* (Stuttgart: Franz Steiner Verlag, 2001) and Tatjana Buklijas, "Cultures of Death and Politics of Corpse Supply: Anatomy in Vienna 1848–1914," *Bulletin of the History of Medicine* 82, no. 3 (2008): 570–607. On Britain: Ruth Richardson, *Death, Dissection and the Destitute*, 2nd ed. (London and Chicago: University of Chicago Press, 2000), 30–74; MacDonald, *Possessing the Dead*; Hurren, *Dying for Victorian Medicine*; Fiona Hutton, *The Study of Anatomy in Britain, 1700–1900* (London: Pickering & Chatto, 2013).

55. Richardson, *Death, Dissection and the Destitute*, 266.

56. Ibid., 261–81.

57. Werner Piechocki, "Zur Leichenversorgung der halleschen Anatomie im 18. und 19. Jahrhundert," *Acta Historica Leopoldina* 2, no. 1 (1965): 67–105, on p. 75.

58. Irmtraut Sahmland, "Verordnete Körperspende: Das Hospital Haina als Bezugsquelle für Anatomieleichen (1786–1855)," in *An der Wende zur Moderne. Die hessischen Hohen Hospitäler im 18. und 19. Jahrhundert*, eds. Friedrich Christina Vanja Von Arnd and Irmtraut Sahmland (Petersberg: Michael Imhof Verlag, 2008), 65–105, on p. 77–80.

59. Buklijas, "Cultures of Death," 579–80. A similar argument was made in Denmark: Maria Olejaz, "Anatomical Dissection Revisited: Exchanging the Dead Body in 18th Century Denmark," in *(aus)tauschen: Erkundungen Einer Praxisform*, eds. Sebastian Mohr, Lydia Maria Quart, and Vetter Andrea (Berlin: Berliner Blätter, 2013), 96–106.

60. Richardson, *Death, Dissection and the Destitute*, 226–30 and 263–4; Sean Burrell and Geoffrey Gill, "The Liverpool Cholera Epidemic of 1832 and Anatomical Dissection: Medical Mistrust and Civil Unrest," *Journal of the History of Medicine and Allied Sciences* 60, no. 4 (2005): 478–98; Michael Sappol, *A Traffic of Dead Bodies: Anatomy and Embodied Social*

Identity in Nineteenth-Century America (Princeton and Oxford: Princeton University Press, 2002), 136–7.

61. Buklijas, "Cultures of Death," 575–6.
62. Ibid., 570–85.
63. Stukenbrock, *Der zerstückte Cörper*, 26–78.
64. Buklijas, "Cultures of Death," 598–605. A similar development took place in Poland: Natalia Aleksiun, "Jewish Students and Christian Corpses in Interwar Poland: Playing with the Language of Blood Libel," *Jewish History* 26, no. 3–4 (2012): 327–42.
65. Cay-Rüdiger Prüll and Marianne Sinn, "Problems of Consent to Surgical Procedures and Autopsies in Twentieth-Century Germany," in *Historical and Philosophical Perspectives on Biomedical Ethics: From Paternalism to Autonomy?*, eds. Andreas-Holger Maehle and Johanna Geyer-Kordesch (Farnham: Ashgate, 2002), 73–93; Sabine Hildebrandt, "Insights into the Freiburg Anatomical Institute during National Socialism, 1933–1945," *Annals of Anatomy* 205 (2016): 90–102.
66. On France: Paul V. Dutton, *Differential Diagnoses: A Comparative History of Health Care Problems and Solutions in the United States and France* (Ithaca and London: Cornell University Press, 2007), 31–64. On England and Germany: Peter E. Hennock, *The Origin of the Welfare State in England and Germany, 1850–1914: Social Policies Compared* (Cambridge: Cambridge University Press, 2007). On England: Elizabeth T. Hurren, *Protesting about Pauperism: Poverty, Politics and Poor Relief in Late-Victorian England, 1870–1900* (Woodbridge: Boydell Press, 2007).
67. Burrell and Gill, "The Liverpool Cholera Epidemic," 478–98; Nick Hopwood, "Artist versus Anatomist, Models against Dissection: Paul Zeiller of Munich and the Revolution of 1848," *Medical History* 51, no. 3 (2007): 279–308; Buklijas, "Cultures of Death," 598–605.
68. Pamela Michael and David Hirst, "Recording the Many Faces of Death at the Denbigh Asylum, 1848–1938," *History of Psychiatry* 23, no. 1 (2011): 40–51.
69. On this phenomenon in Europe more generally: Christopher Clark and Wolfram Kaiser, "The European Culture Wars," in *Culture Wars: Secular-Catholic Conflict in Nineteenth-Century Europe*, eds. Christopher Clark and Wolfram Kaiser (Cambridge: Cambridge University Press, 1993), 1–10. On Belgium: Els Witte, "The Battle for Monasteries, Cemeteries and Schools: Belgium," in *Culture Wars*, 102–28.
70. In the system of plural voting, introduced in 1893, every male citizen over 25 received one vote for legislative elections. A few electors received up to 2 supplementary votes according to education and financial capability.

71. Witte, "The Battle for Monasteries," 113–18.
72. On France: Thomas Kselman, "Funeral Conflicts in Nineteenth-Century France," *Comparative Studies in Society and History* 30, no. 2 (1988): 312–32.
73. Christoph De Spiegeleer and Jeffrey Tyssens, "Secularizing funerary culture in nineteenth-century Belgium: A product of political and religious controversy," *Death Studies* 41, no. 1 (2017): 14–21.
74. Witte, "The Battle for Monasteries," 113–18.
75. Christoph De Spiegeleer, "Tussen banketten en begrafenissen: de radicaal-liberale burgerlijke cultuur rond Charles Potvin in Brussel tijdens de tweede helft van de 19e eeuw," *De Negentiende Eeuw* 34, no. 4 (2010): 289–308; Gita Deneckere, *Geuzengeweld. Antiklerikaal straatrumoer in de politieke geschiedenis van België 1831–1914* (Brussels: VUB Press, 1998), 63–86.
76. Similar societies also existed in the United States, see: Sappol, *A Traffic of Dead Bodies*, 134–5.
77. For example: *L'Affranchisement* (1854), *Les Solidaires* (1857) and *La Libre Pensée* (1863). On these organisations, see De Spiegeleer and Tyssens, "Secularizing funerary culture," 14–21.
78. *Assemblée générale des catholiques en Belgique, première session à Malines, 18–22 Août 1863*, vol. 2 (Brussels: Goemaere, 1864), 10.
79. R.J. Bonjean, J.B. Bivort, J.J. Cloes, and E.A. Dubois, *Revue de l'administration et du droit administratif de la Belgique* (Liège: Imprimerie de Veuve Verhoven-Debeur, 1837), 28.
80. Christoph De Spiegeleer, "Secularisering van stedelijke begraafplaatsen in de tweede helft van de negentiende eeuw in België," in *R.I.P. Aspecten van 200 jaar begrafeniscultuur in Vlaanderen*, ed. Tamara Ingels (Ghent: Academia Press, 2015), 5–23, on p. 6.
81. *Assemblée générale 18–22 Août 1863*, vol. 1, 106.
82. Quote from: "Développement de la proposition de M. Funck relative à la propriété des cimetières," *Bulletin communal de Bruxelles* 11 (1861): 433.
83. De Spiegeleer, "Secularisering van stedelijke begraafplaatsen," 5–23.
84. Ville de Gand, *Règlement sur les inhumations. Rapport de la commission du contentieux* (Ghent: C. Annoot, 1865), 3–7; De Spiegeleer, "Secularisering van stedelijke begraafplaatsen," 9.
85. For example: Proceedings of the plenary sessions of the Belgian Chamber of Representatives of July 23, 1862. https://sites.google.com/site/bplenum/proceedings/1862/k00063913/k00063913_06. Accessed February 23, 2017.
86. De Spiegeleer, "Secularisering van stedelijke begraafplaatsen," 5.
87. Ibid., 5–21.

88. Jolien Gijbels, "Reassessing the Pauper Burial: The Disposal of Corpses in Nineteenth-Century Brussels," *Mortality: Promoting the Interdisciplinary Study of Death and Dying* 23, no. 2 (2018): 184–98.
89. "Nieuws over de kerkhofkwestie: liberale lijkenschenners," *Het Handelsblad*, January 30, 1873.
90. "Wat is er deze morgend in het gasthuis gebeurd?," *Het Handelsblad*, April 22, 1882.
91. *Assemblée générale 18–22 Août 1863*, vol. 2, 17.
92. *Assemblée générale 18–22 Août 1863*, vol. 1, 465.
93. Dossier concernant les observations faites à l'université relativement à la reprise des cadavres, folder 107, Autopsies, Affaires générales, ASSB. See also: Gijbels, "Reassessing the Pauper Burial," 184–98.
94. *Assemblée générale des catholiques en Belgique, troisième session à Malines, 2–7 septembre 1867*, vol. 2 (Brussel: Victor Devaux, 1868), 373.
95. Behaeghel to Vanderlinden, January 22, 1906, folder 107, Autopsies, Affaires générales, ASSB.
96. Head of St. Pierre to Conseil des hospices, July 21, 1885, folder 66, Affaires générales, ASSB.
97. Els Witte et al., *Nieuwe geschiedenis van België 1830–1905*, vol. 1 (Tielt: Lannoo, 2005–2009), 303–8.
98. Head of St. Jean to Conseil des hospices, July 22, 1885, folder 66, Université, Maternité, Amphithéâtre etc., Affaires générales, ASSB.
99. Buklijas, "Cultures of Death," 582–5.
100. On burial societies in Austria: Buklijas, "Cultures of Death," 582. On the German federal states: Piechocki, "Zur Leichenversorgung," 76. On Britain: Richardson, *Death, Dissection and the Destitute*, 276.
101. De Spiegeleer, "Secularisering van stedelijke begraafplaatsen," 14.
102. Hopwood, "Artist versus Anatomist," 290–8.
103. Hildebrandt, "Insights into the Freiburg Anatomical Institute," 94.
104. Buklijas, "Cultures of Death," 598–605.
105. Sappol, *A Traffic of Dead Bodies*, 131.
106. The protest culminated in the general strike of 1886, leading to the first social legislation in Belgium. Deferme, *Uit de ketens van de vrijheid*, 89–101.
107. Procès-verbaux des séances du Conseil d'Administration, July 9, 1887, book 4, UAB.
108. Ibidem.
109. Procès-verbaux des séances du Conseil d'Administration, March 12, 1891, book 4, UAB.
110. For example: Alfons Labisch, "From Traditional Individualism to Collective Professionalism: State, Patient, Compulsory Health Insurance and the Panel Doctor Question in Germany 1883–1931," in *Medicine and Modernity: Public Health and Medical Care in Nineteenth- and Twentieth-Century Germany*, eds. Manfred Berg and Geoffrey Cocks

(Cambridge: Cambridge University Press, 1997), 18–34; Keir Waddington, "Unsuitable Cases: The Debate over Outpatient Admissions, the Medical Profession and Late-Victorian London Hospitals," *Medical History* 42, no. 1 (1998): 26–46. On Belgium, see: Havelange, "L'hôpital à la croisée des chemins," 83–94.

111. The hospital regulations for paying patients can be found in: Règlement pour les hôpitaux de St Pierre et de St Jean à Bruxelles, [1889], folder 131, Hôpital St. Pierre: Divers, Affaires générales, ASSB.

112. "L'affaire de l'hôpital Saint-Jean," *Indépendance belge*, 4 March 1887.

113. Maarten Couttenier, *Congo tentoongesteld: een geschiedenis van de Belgische antropologie en het museum van Tervuren (1882–1925)* (Leuven: Leuven University Press, 2005), 70–80.

114. "A propos de l'histoire du cadavre de la négresse Coco," *Le Peuple*, March 5, 1887.

115. Ibidem.

116. For example: César De Paepe, "A l'Amphithéâtre," *Le Peuple*, November 18, 1889.

117. Louis Bertrand, "Bienfaisance publique et assurance sociale," *Revue de Belgique* 25 (1894): 5–33.

118. Louis Bertrand, "L'hôpital," *Le Peuple*, December 7, 1891.

119. For example: "Farce lugubre," *Le Peuple*, March 27, 1895. "Service des inhumations," *Le Peuple*, 11 June 1904.

120. Louis Henry, "La crémation," *Revue Catholique* 14 (1875): 545–72, on p. 566.

121. "Nieuws over de kerkhofkwestie: liberale lijkenschenners," *Het Handelsblad*, January 30, 1873.

122. For example: Proceedings of the plenary sessions of the Belgian Chamber of Representatives of May 13, 1891. https://sites.google.com/site/bplenum/proceedings/1891/k00161208/k00161208_00. Accessed February 27, 2017.

123. Deferme, *Uit de ketens van de vrijheid*, 89–101. See also: Jean-Pierre Nandrin, "Les libéraux et la genèse du droit social en Belgique. Peut-on parler d'un modèle paternaliste libéral?," in *Liberalism and Paternalism in the 19th Century*, eds. Erik Aerts, Claude Beaud and Jean Stengers (Leuven: Leuven University Press, 1990), 94–103.

124. Namely the Christian Democrats, the progressive Liberals and the (Proudhonian) Socialists. Note that the Socialist movement was already established in the years around 1848.

125. Deferme, *Uit de ketens van de vrijheid*, 145–206.

126. See, for example: Elizabeth T. Hurren, "World without Welfare. Pauper Perspectives on Medical Care under the Late Victorian Poor Law 1870–1900," *Obligation, Entitlement and Dispute under the English Poor Laws*, eds. Peter Jones and Steven King (Newcastle upon Tyne: Cambridge

Scholars, 2015), 292–320; Labisch, "From Traditional Individualism to Collective Professionalism," 18–34; Paul V. Dutton, *Origins of the French Welfare State: The Struggle for Social Reform in France, 1914–1947* (Cambridge: Cambridge University Press, 2002); Hennock, *The Origin of the Welfare State.*

127. For example: Bertrand, "Bienfaisance publique," 18; Cyrille Van Overbergh, *Réforme de la bienfaisance en Belgique* (Brussels: Lesigne, 1900).

128. Van de Perre, "Public Charity," 99.

129. Proceedings of the plenary sessions of the Belgian Chamber of Representatives of January 10, 1895. https://sites.google.com/site/bplenum/proceedings/1895/k00180422/k00180422_00. Accessed March 1, 2017.

130. Deferme, *Uit de ketens van de vrijheid,* 323–82.

131. Griet Van Meulder, "Mutualiteiten en ziekteverzekering in België (1886–1914)," *Belgisch Tijdschrift voor Nieuwste Geschiedenis* 27, no. 1–2 (1997): 83–134.

132. Proceedings of the plenary sessions of the Belgian Chamber of Representatives of May 13, 1891. https://sites.google.com/site/bplenum/proceedings/1891/k00161208/k00161208_00. Accessed February 27, 2017.

133. Ibidem.

134. Letter Faculty of medicine to Louis Trasenster, June 15, 1882, folder 254, Fonds du secrétariat central, UAL.

135. "Nieuws over de kerkhofkwestie: liberale lijkenschenners,"*Het Handelsblad*, January 30, 1873.

136. "Lijken in het gasthuis," *Het Handelsblad*, April 30, 1882.

137. "Cadavre disparu," *Le Peuple*, March 4, 1887.

138. Sally Wilde, "Truth, Trust, and Confidence in Surgery, 1890–1910: Patient Autonomy, Communication and Consent," *Bulletin of the History of Medicine* 83, no. 2 (2009): 302–30; Kathleen E. Powderly, "Patient Consent and Negotiation in the Brooklyn Gynaecological Practice of Alexander J.C. Skene: 1863–1900," *Journal of Medicine and Philosophy* 25, no. 1 (2000): 12–27; Nancy Theriot, "Negotiating Illness: Doctors, Patients and Families in the Nineteenth Century," *Journal of the History of the Behavorial Sciences* 37, no. 4 (2001): 349–68; Regina Morantz-Sanchez, "Negotiating Power at the Bedside: Historical Perspectives on Nineteenth-Century Patients and their Gynaecologists," *Feminist Studies* 26, no. 2 (2000): 287–309.

139. Lisbeth Haakonssen, *Medicine and Morals in the Enlightenment: John Gregory, Thomas Percival and Benjamin Rush* (Amsterdam and Atlanta: Rodopi, 1997), 122–72; Robert Baker, "Deciphering Percival's Code," in *The Codification of Medical Morality: Historical and Philosophical*

Studies of the Formalisation of Western Medical Morality in the Eighteenth and Nineteenth Centuries, eds. Robert Baker, Dorothy Parker and Roy Porter, vol. 1 (Dordrecht: Springer, 1993), 179–212.

140. Faden and Beauchamp, *A History and Theory*, 59–60; Andreas-Holger Maehle, *Doctors, Honour and the Law. Medical Ethics in Imperial Germany* (Houndmills: Palgrave Macmillan, 2009), 69–70.

141. Faden and Beauchamp, *A History and Theory*, 74–7.

142. On nineteenth-century medical experimentation, see: Marion Maria Ruisinger, "Geschichte des Humanexperiments. Zur Entwicklung der Forschung am Menschen," in *Standards der Forschung. Historische Entwicklung und ethische Grundlagen klinischer Studien*, eds. Andreas Frewer and Ulf Schmidt (Frankfurt am Main: Peter Lang Verlag, 2007), 19–35; William Bynum, "Reflections on the History of Human Experimentation," in *The Use of Human Beings in Research. With Special Reference to Clinical Trials*, eds. Stuart Spickert et al (Dordrecht: Springer, 1988), 29–46.

143. Rapport sur l'organisation du Service des autopsies dans les hôpitaux de Bruxelles, [1878], folder 107, Autopsies, Affaires générales, ASSB.

144. Robert Nye, "Médecins, éthique médicale et État en France 1789–1947," *Le Mouvement Social* 214, no. 1 (2006): 19–36; Christian Bonah, *Histoire de l'expérimentation humaine en France* (Paris: Les Belles Lettres, 2007), 114–25; Alex Dracobly, "Ethics and Experimentation on Human Subjects in Mid-Nineteenth-Century France: The Story of the 1859 Syphilis Experiments," *Bulletin of the History of Medicine* 77, no. 2 (2003): 332–66.

145. Barbara Elkeles, "The German Debate on Human Experimentation between 1880–1914," in *Twentieth Century Ethics of Human Subject Research*, eds. Volker Roelcke and Giovanni Maio (Stuttgart: Franz Steiner Verlag, 2004), 18–33; Maehle, *Doctors, Honour and the Law*, 69–94.

146. Phil Fennell, *Treatment without Consent. Law, Psychiatry and the Treatment of Mentally Disordered People since 1845* (London: Routledge, 1996), 66–77, on p. 72.

147. Faden and Beauchamp, *A History and Theory*, 120–3.

148. "Demarche contre Deschambre," Tribunal civil de Liège, November 27, 1889, in: *La Belgique judiciaire* 46 (1890): 474.

149. Fernand Héger-Gilbert, *Manuel de Déontologie médicale: résumé du cours professé à l'Université Libre de Bruxelles* (Brussels: Imprimerie médicale et scientifique, 1928), 41–3.

150. Faden and Beauchamp, *A History and Theory*, 114–50; Maehle, *Doctors, Honour and the Law*, 69–94.

151. Faden and Beauchamp, *A History and Theory*, 117–25.
152. Ibid., 77–83; Martin S. Pernick, "The Patient's Role in Medical Decision-Making: A Social History of Informed Consent in Medical Therapy," in *Making Healthcare Decisions*, President's Commission for the Study of Ethical Problems in Medicine and Biomedical and Behavioural Research, vol. 3 (Washington: The Commission, 1982), 1–35.
153. Alfred Moreau, *La responsabilité médicale* (Brussels: Bruylant-Christophe, 1891), 50.
154. Maehle, *Doctors, Honour and the Law*, 69–94.
155. Faden and Beauchamp, *A History and Theory*, 140–3. On Belgium specifically, see: Julie de Ganck, "Cultiver la différence: Histoire du développement de la gynécologie à Bruxelles, 1870–1935" (PhD diss., Université Libre de Bruxelles, 2016), 337–45.
156. Jay Katz, *The Silent World of Doctor and Patient* (New York: Macmillan, 1984), 79.
157. Faden and Beauchamp, *A History and Theory*, 105–11.
158. Bonah, *Histoire de l'expérimentation humaine*, 114–30.
159. Michael A. Grodin, "Historical Origins of the Nuremberg Code," in *The Nazi Doctors and the Nuremberg Code*, eds. George J. Annas and Michael A. Grodin (Oxford: Oxford University Press, 1992), 121–44; Thomas G. Benedek, "Case Neisser. Experimental Design, the Beginnings of Immunology and Informed Consent," *Perspectives in Biology and Medicine*, 57, no. 2 (2014): 249–67.
160. "Un scandale universitaire," *Journal de Bruxelles*, February 19, 1900; "Odieuses expériences sur des malades," *Journal de Charleroi*, June 20, 1900.
161. *Responsabilité médicale: notes de jurisprudence* (Brussels: Ch. Miguet, s.d.), 31.
162. Jozef Salsmans, *Geneeskundige plichtenleer* (Leuven: Vlaamsche Boekenhalle, 1919), 44.
163. Emmanuel Betta, "Between law and profession: the origins of informed consent (1840–1900)," in *Doctors and Patients. History, Representation, Communication from Antiquity to the Present*, ed. Maria Malatesta (San Francisco: University of California Medical Humanities Press, 2015), 108–33.
164. See Chap. 4.
165. Maurice De Laet, *Les responsabilités du médecin* (Brussels: Imprimerie médicale et scientifique, 1927), 45.
166. De Laet, "Le droit à l'autopsie," 84–92. I will return to this matter in the next chapter.

167. Margaret Schwartz, "An Iconography of the Flesh: How Corpses Mean as Matter," *communication +1*, 2, no. 1 (2013): 1–16, on p. 4.

168. Edmond Picard and Napoléon d'Hoffschmidt, "Cadavre," *Pandectes belges: Répertoire général de législation, de doctrine et de jurisprudence belges donnant pour toutes les matières du droit belge. Les lois en vigueur les décisions judiciaires, les circulaires administratives et la doctrine des auteurs*, vol. 15 (Brussels: Larcier, 1885), 247.

169. On this complex body of law in the Anglo-American context: Susan Lawrence, "Beyond the Grave. The Use and Meaning of Human Body Parts: A Historical Introduction," in *Stored Tissue Samples: Ethical, Legal and Public Policy Implications*, ed. Robert F. Weir (Iowa: University of Iowa Press, 1998), 111–42, on pp. 117–23.

170. Règlement sur les inhumations et les transports funèbres, May 3, 1880, folder 1866, ASB Cultes, Archives administratives, CAB.

171. Picard and d'Hoffschmidt, "Cadavre," 245.

172. Charles De Brouckère and Franciscus Thielemans, "Cadavre," *Répertoire de l'administration et du droit l'administratif de la Belgique*, vol. 4 (Brussels: Weissenbruch, 1838), 61.

173. Conseil des hospices to city council, May 19, 1890, folder 107, Autopsies, Affaires générales, ASSB.

174. Charles De Gronckel, *Hospices civils et bureaux de bienfaisance. Précis du régime légal de l'assistance public* (Brussels: L. Bourlard et V. Havaux, 1884), 325.

175. Règlement pour les hôpitaux de St Pierre et de St Jean à Bruxelles, [1889], folder 131, Hôpital St. Pierre: Divers, Affaires générales, ASSB.

176. "Hospices de Bruxelles contre la veuve Leemans," Cour d'appel de Bruxelles, March 14, 1849, in *La Belgique judiciaire* 7 (1849): 425–7.

177. Rapport sur les autopsies et les dissections, 1924, folder 107, Autopsies, Affaires générales, ASSB.

178. Rohan Hardcastle, *Law and the Human Body: Property Rights, Ownership and Control* (Oxford and Portland: Hart Publishing, 2007), 28–40.

179. Lawrence, "Beyond the Grave," 129.

180. Veronique Deblon, "Commercialized Bodies: the Trade in Anatomical Preparations." Paper presented at the EAHMH Conference, Cologne, September 2015.

181. Vidal de Cassis, "Procès intenté par MM. Guérin et Gannal," *Annales de la chirurgie française et étrangère* 10 (1844): 126.

182. Hardcastle, *Law and the Human Body*, 27.

183. Prüll, "No Law, No Rights?," 37–8.

184. François Laurent, *Droit civil international*, vol. 4 (Brussels: Bruylant, 1880–1881), 154.

185. "Vandenbussche et Legein," Cour de Cassation, November 2, 1868, in *Pasicrisie belge. Recueil général de jurisprudence de la royaume de Belgique* (Brussels: Bruylant, 1869): 8.

186. Hélène Popu, *La dépouille mortelle, chose sacrée. A la découverte d'une catégorie juridique oubliée* (Paris: L'Harmattan, 2009), 161–2.

187. Prüll and Sinn, "Problems of Consent," 83; Prüll, "No Law, No Rights?," 35–8.

188. Popu, *La dépouille mortelle*, 66. In the United States, jurists found a different solution to this problem. In her research on the United States, Susan Lawrence has shown that the corpse came to be seen as 'quasi-property': not true property, a point of view inherited from English common law, but with some of the characteristics of a property relationship with the living: Lawrence, "Beyond the Grave," 120.

189. Gabriel Timbal, *La condition juridique des morts* (Toulouse: Edouard Privat, 1903), 182.

190. René Demogue, "La notion de sujet de droit: caractère et conséquences," *Revue trimestrielle de droit civil* 3 (1909): 611–55, on p. 631.

191. Théodore Bormans, "De la violation des cadavres," *La Belgique judiciaire*, April 17, 1870. Note that Bormans' work was published two years after his actual research, which explains the time difference between the parliamentary discussion and this publication.

192. Proceedings of the plenary sessions of the Belgian Chamber of Representatives of December 8, 1868. https://sites.google.com/site/bplenum/proceedings/1868/k00082654/k00082654_00. Accessed March 3, 2017.

193. Ibidem.

194. Picard and d'Hoffschmidt, "Cadavre," 248. Note that the word 'corps' in French can refer to a living as well as to a dead body.

195. See also: Gijbels, "Reassessing the Pauper Burial," 184–98. On similar events in France: Kselman, "Funeral Conflicts," 327.

196. Karel Velle, *Begraven of cremeren: de crematiekwestie in België* (Ghent: Stichting mens en kultuur, 1992), 74–8.

197. "Procureur général en cause de Libert," Cour de Cassation, June 3, 1899, in *Pasicrisie belge. Recueil général de jurisprudence de la royaume de Belgique* (Brussels: Bruylant, 1899), 318–22.

198. Witte, "The Battle for Monasteries," 113–18.

199. "Procureur général en cause de Libert," 320.

200. "De la Tour Landry c. époux d'Origny," Cour de Cassation, March 30, 1886, in *Jurisprudence générale. Recueil périodique et critique de jurisprudence, de législation et de doctrine en matière civile, commerciale, criminelle, administrative et de droit public* (Paris: Dalloz, 1886), 451.

201. The court ruling of 1899 continued to influence Belgian legislation. Until today, the law considers the last wishes of the deceased with regard

to his/her burial more important than the wishes of the family. See: Arianne Vervliet, "Het juridische statuut van een lijk" (Unpublished MA diss., University of Antwerp, 2013), 15–16.

202. "Procureur général en cause de Libert," 321–2.
203. Rapport au conseil sur les autopsies et les dissections, January 28, 1924, folder 107, Autopsies, Affaires générales, ASSB.
204. Conseil des hospices to the Head of St. Pierre, March 11, 1898, folder 292, Décès—autopsies 1852–1922, Fonds du directeur de l'hôpital St. Pierre, ASSB.
205. Firket, *Du but et de l'organisation*, 38–9.
206. Conseil des hospices to city council, July 19, 1883, folder 255, Faculté de médecine 1891–1914, Fonds du secrétariat central, UAL.
207. Ibidem.
208. Circular letter from Commission des Hospices de Louvain sent to several universities, June 25, 1901, folder Gasthuis Allerlei 1, Autopsie 1839–1901, ASSL.
209. Arrête sur la déclaration de décès et l'inhumation, September 6, 1892, folder 4A150 (700), UAG.
210. De Ganck, "Cultiver la différence," 344.
211. Dossier concernant la lettre de M. Gobbe portant qu'il lègue son corps à l'amphithéâtre, folder 116, Hôpital St. Jean: Généralités, Affaires générales, ASSB.
212. Register of bodies brought to the Institut d'anatomie, 1920–1968, Private Archives of Stéphane Louryan, Brussels.
213. Gijbels, "Reassessing the Pauper Burial," 194.
214. Dossier concernant la remise des cadavres aux seules personnes de la famille non aux ministres des cultes, folder 113, Hôpital St. Jean: Généralités, Affaires générales, ASSB.
215. Dossier concernant la question posée par l'université quant aux mesures à prendre pour prendre possession d'un cadavre qui lui a été légué, 1899, folder 66, Université, Maternité, Amphithéâtre, etc., Affaires générales, ASSB.
216. Stéphane Louryan and Nathalie Vanmuylder, "L'Institut d'Anatomie Raoul Warocqué de l'ULB (1893–1928)," in *Le Pôle Santé de l'ULB: Histoire de lieux, de personnages, de découvertes*, eds. Stéphane Louryan and Paul Kinnaert (Brussels: Memogrames, 2009), 107–24, on p. 117.
217. Jacques Mulnard, "Albert Dalcq," in *Le Pôle Santé de l'ULB*, 149–58; Register of bodies brought to the Brussels Institut d'anatomie, 1920–1968, Private Archives of Stéphane Louryan, Brussels; Various documents in folder 1 FD 703, Corps légués ou mis à la disposition du service d'anatomie 1945–1968, UAB.

218. Richardson, *Death, Dissection and the Destitute*, 258–60. Note that forth-coming research by Elizabeth Hurren claims that body donation in Britain only really began in 1954, a few years after the National Healthcare System was established. In her view, there was considerable longevity in the use of former Poor Law institutions (renamed as care homes and cottage hospitals) from which corpses for dissection continued to be generated long after the New Poor Law formally ended in 1929. See: Elizabeth T. Hurren *Dead Ends? Disputed Bodies in Modern Medical Research 1945–2000* (Cambridge: Cambridge University Press, forthcoming).

219. Ann Garment, Susan Lederer, Naomi Rogers and Lisa Boult, "Let the Dead Teach the Living: The Rise of Body Bequeathal in 20th-Century America," *Academic Medicine* 82, no. 10 (2007): 1000–5; Susan Lederer, "Bodies for Science." Paper presented at the Bodies Beyond Borders Conference, Leuven, January 2015.

220. "Déliberations sur la proposition de loi relative à la liberté des funérailles," in *Journal officiel de la République française. Débats parlementaires. Chambre des députés*, March 30, 1886, 607–25, on p. 609; "De la Tour Landry c. époux d'Origny," Cour de Cassation, March 30, 1886, in *Jurisprudence générale. Recueil périodique et critique de jurisprudence, de législation et de doctrine en matière civile, commerciale, criminelle, administrative et de droit public* (Paris: Dalloz, 1886), 451.

221. Mathias Schütz, "Erzwungener Wandel. Die Transformation der anatomischen Leichenbeschaffung in Bayern nach 1945," *Medizinhistorisches Journal* 54, no. 1 (2019), 70–92.

222. Wilde, "Trust, Truth and Confidence," 327.

223. Faden and Beauchamp, *A History and Theory*, 86–8.

224. Lederer, "Bodies for Science."

225. This was different in the United States, where the final decision about the disposition of the body continued to rest with the family until 1968. Because the law did not consider corpses as property, individuals could not transfer their bodies by will. In Belgium, an alternative for this legal situation was already established in 1899 (see above). As opposed to the Belgian situation, the Uniform Anatomical Gift Act of 1968 made sure that the donor's wishes took priority over those of next of kin in court precisely by establishing the human body as property. See: Garment et al, "Let the Dead Teach the Living," 1000–5.

226. Lederer, "Bodies for Science."

227. Icard, *La constatation des décès*, 9.

228. For example: André Fritz, *Du danger des inhumations précipitées* (Brussels: Jamin et Coosemans, 1879); Dossier dépôts mortuaires des hôpitaux, folder S 107.2, Bienfaisance, CAB.

229. On the impact of modern embalming on visual representations of death: John Troyer, "Embalmed Vision," *Mortality: Promoting the Interdisciplinary Study of Death and Dying* 12, no. 1 (2007): 22–47.

230. For example: "Un homme enterré vivant," *Le Bien Public*, March 28, 1862; "Un affreux réveil," *Journal de Bruxelles*, July 28, 1874.

231. Alfred Prume to Head of St. Jean, August 14, 1909, folder 107, Autopsies, Affaires générales, ASSB.

232. Head of St. Jean to Conseil des hospices, August 17, 1909, folder 107, Autopsies, Affaires générales, ASSB.

233. Garment et al., "Let the Dead Teach the Living," 1002.

234. Lederer, "Bodies for Science."

235. Richardson, *Death, Dissection and the Destitute*, 258–9.

236. Mulnard, "Albert Dalcq," 149–58.

237. Désiré-Magloire Bourneville and Paul Bricon, *Manuel de technique des autopsies* (Paris: Librairie du progrès médical, 1887), 25.

238. Joris Vandendriessche, *Medical societies and scientific culture in nineteenth-century* Belgium (Manchester: Manchester University Press, 2019), 230–8.

239. "Biographie médicale," *Bulletin médical belge* 2 (1835): 149–51.

240. Adolphe Burggraeve, *Discours sur le médecin P. E. Wauters; prononcé le jour de son enterrrement* (Ghent: Gyselynck, 1840), 12–13.

241. Vandendriessche, *Medical societies*, 230–8.

242. Rebecca M. Herzig, *Suffering for Science: Reason and Sacrifice in Modern America* (New Brunswick: Rutgers University Press, 2005); Susan E. Lederer, *Subjected to Science: Human Experimentation in America Before the Second World War* (Baltimore and London: Johns Hopkins University Press, 1995), 126–38; Ian A. Burney, *Bodies of Evidence: Medicine and the Politics of the English Inquest, 1830–1926* (Baltimore and London: Johns Hopkins University Press, 2000), 137–64.

243. Lederer, *Subjected to Science*, 126–38; Lawrence K. Altman, *Who Goes First? The Story of Self-Experimentation in Medicine* (Berkeley, Los Angeles and London: University of California Press, 1986); Daniel S. Goldberg, "Suffering and Death among Early American Roentgenologists: The Power of Remotely Anatomizing the Living Body," *Bulletin of the History of Medicine* 85, no. 1 (2011): 1–28.

244. "Une victime de son devoir," *La Meuse*, October 26, 1898; "Les héros de la science," *Vingtième Siècle*, February 11, 1899; "Martyr de la science," *Le Courrier de l'Escaut*, February 13, 1886; "Ils sont nombreux," *Journal de Bruxelles*, December 26, 1902.

245. "Martyr du devoir," *Avenir du Luxembourg*, May 1, 1912.

246. Vandendriessche, *Medical societies*, 215.

247. Adolphe Burggraeve, *Etudes sur André Vésale, précédées d'une notice histo-rique sur sa vie et ses écrits* (Ghent: Annoot-Braeckman, 1841). See also: Vandendriessche, *Arbiters of Science*, 215.

248. Vandendriessche, *Medical societies*, 215–17. On nineteenth-century paintings depicting Vesalius: Maurits Biesbrouck, Luc Missotten and Omer Steeno, "De Vesalius-schilderijen van E.J.C. Hamman (1819–1888)," in *Heel-meesters. Befaamde artsen en figuren uit de geschiedenis van de geneeskunde*, eds. Bob Van Hee and Cornelis Van Tilburg (Antwerp: Garant, 2014), 19–39.

249. Gaston Tissandier, *Les martyrs de la science*, 2nd ed. (Paris: Maurice Dreyfous, 1882), 292–3.

250. Kaat Wils, "Institut Vésale," in *Album van een wetenschappelijke wereld/ Album of a Scientific World*, eds. Geert Vanpaemel, Marc Derez and Jo Tollebeek (Leuven: Leuven University Press, 2012), 162–9.

251. Keir Waddington, "Mayhem and Medical Students: Image, Conduct and Control in the Victorian and Edwardian London Teaching Hospital," *Social History of Medicine* 15, no. 1 (2002): 45–64.

252. Emmanuelle Godeau, *L'esprit de corps: Sexe et mort dans la formation des internes en médecine* (Paris: Maison des sciences de l'homme, 2007), 27–39; Sappol, *A Traffic of Dead Bodies*, 74–97.

253. For example: "Jeudi soir," *Journal de Bruxelles*, November 15, 1868; "Une découverte singulière," *Indépendance belge*, February 20, 1869; "Crime ou farce?," *Le Peuple*, March 23, 1890; "A qui le bras," *Le Peuple*, August 25, 1892.

254. Dossier concernant la trouvaille faite d'une main humaine rue Lacaille, 1876, folder 66, Université, Maternité, Amphithéâtre etc., Affaires générales, ASSB.

255. "L'oreille coupée," *Le Peuple*, February 20, 1887.

256. Albert Dalcq to Marcel Barzin, December 1, 1950, folder 1 FD 703, Corps légués ou mis à la disposition du service d'anatomie 1945–1968, UAB.

257. Elienne Langendries and Anne-Marie Van der Meersch, *Het Rommelaere complex: onderdeel van het gebouwenmasterplan voor de Gentse universiteit op het einde van de 19de eeuw* (Gent: RUG Archief, 1999).

258. Wils, "Institut Vésale," 162–9.

259. Specifications concerning the construction of the morgue of St. Pierre, April 9, 1886, folder 34, Cultes 1811–1895, Affaires générales, ASSB.

260. Firket, *Du but et l'organisation*, 46–51.

261. See, for example, Louis Baunard, *Les deux frères: Cinquante années de l'Action Catholique dans le nord : Philibert Vrau, Camille Feron-Vrau 1829–1908* (Paris: Maison de la bonne presse, 1926), 365–6.

262. Jean-Joseph Tricot-Royer, *L'Eglise et la mutilation du cadavre humain. Décarnisation, dissection pour enseignement, embaumement, autopsie* (Leuven: Nova et Vetera, 1936), 16.

263. Jennifer Hecht, "French Scientific Materialism and the Liturgy of Death: The Invention of a Secular Version of Catholic Last Rites (1876–1914)," *French Historical Studies* 20, no. 4 (1997): 703–35.

264. "À propos de la mort de M. Asseline," *Journal de Bruxelles*, April 14, 1878; "La société d'autopsie mutuelle," *L'Indépendance belge*, March 13, 1894.

265. Richardson, *Death, Dissection and the Destitute*, 260.

266. Velle, *Begraven of cremeren*, 97–8.

267. Édouard Suau de Varennes, *Die Mysterien von Brüssel* (Stuttgart: Franck Verlag, 1846), 373.

268. Richardson, *Death, Dissection and the Destitute*, 260; Sophie Jamieson, "British people donating bodies to science to avoid funeral costs," *The Telegraph*, January 8, 2016.

269. "Un scandale aux hôpitaux de Bruxelles," *La Libre Belgique*, February 10, 1924; "Aux hôpitaux de Bruxelles: un scandal!," *Midi*, February 10, 1924; "Grave incident dans un hôpital bruxellois," *Le journal de Paris*, February 10, 1924.

270. Dossier concernant l'autopsie pratiquée clandestinement du cadavre de l'infirmière Lankester, 1923–1924, folder 107, Autopsies, Affaires générales, ASSB.

271. Clark and Kaiser, "European Culture Wars," in *Culture Wars*, 1–10.

272. Labisch, "From Traditional Individualism to Collective Professionalism," 18–34.

273. Donnacha Lucey, *The End of the Irish Poor Law? Welfare and Healthcare Reform in Revolutionary and Independent Ireland* (Manchester: Manchester University Press, 2015), 148–88; Hurren, *Protesting about Pauperism*, 159–247.

274. Hurren, "World Without Welfare," 317.

275. Garment et al., "Let the Dead Teach the Living," 1000–5.

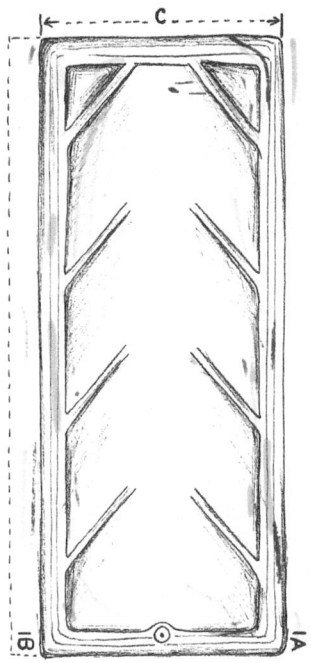

CHAPTER 4

Under the Scalpel: Dividing the Body

1 PREAMBLE: 'IT ALWAYS HAPPENED THAT WAY', 1902

Mrs. De Ro buttons up her coat as she rushes to the hospital of St. Pierre. It is that time of the year in which cold temperatures bring people together in heated living rooms. The heat from stoves permanently steams up the windows of Brussels houses. Her husband, Henri De Ro, has been in the hospital for months. Things do not look good for him. It might be their last Christmas together.

As usual, Mrs. De Ro arrives around 2 in the afternoon. The hospital personnel greet her by name. They have given her special permission to visit her husband every day because of the severity of his condition. Mrs. De Ro remembers that she felt strange when she thanked them for this privilege, as if she accepted her husband's imminent death too easily. It is 4.30 when she takes her coat and walks home.

* * *

Around 8 in the evening, one of the hospital nuns notices that Henri De Ro has developed a raging fever. Before she ends her round, she asks the hospital attendant to call Mrs. De Ro to warn her that the time has come to say her last goodbyes. Yet the hospital attendant does not call. He thinks it is exaggerated to have Mrs. De Ro return to the hospital only a few hours after she left. He decides to contact her at the break of dawn.

T. Claes, *Corpses in Belgian Anatomy, 1860–1914*, Medicine and Biomedical Sciences in Modern History, https://doi.org/10.1007/978-3-030-20115-9_4

Around 3 in the morning, the ward is suddenly awakened. One of the patients later describes that his sleep was disturbed by 'an exclamation of the dying'. Delirious with fever, Henri De Ro cried out for his wife and child. Two hours later, he yelled 'Adieu, my wife, my child, my friends'. Peace returned to the ward. Henri De Ro died at 5 in the morning on Sunday, the 21st of December 1902.

*　*　*

Mrs. De Ro wakes up when her phone rings. It is the call she has been both expecting and dreading for days. A few hours later, she arrives at the hospital with her husband's most beautiful clothes. The hospital attendant tells her that her husband's last thoughts were for her. She walks up to his deathbed and dresses him for the funeral. Then she returns home to her child.

*　*　*

Right after she has left, the interns of the autopsy service arrive. They put the body of Henri De Ro on a cart and cover it up carefully with a piece of black cloth. The corpse has to be hidden from view: they do not want to disturb patients who are still fighting for their life. They roll the cart through the corridors of the hospital until they reach their working place, close to the morgue. They undress the body and put him ready for the autopsy course.

Students arrive, dressed in white coats. Their teacher tells them to open the corpse carefully, using but one straight incision. They cut out various organs, dissect them a little. Bodily fluids leak from Henri De Ro's body straight into the gutter system of the autopsy table.

When the course is done, the teacher urges his students to make sure the body looks decent again. All organs are put back into place; here and there the students refill the body with wads of cotton wool. Then they wash the body with plenty of water. All dressed again, the body is rolled through the heavy, sound-proof doors of the autopsy room, back to the morgue.

*　*　*

The next day Mrs. De Ro returns to the hospital carrying a wreath to place on her husband's coffin. She enters the morgue through the separate entrance from the Rue aux Laines. As her husband is not lying in the chapel yet, one of the hospital personnel guides her to the room where the bodies are stored. Curtains separate the dead from each other. She asks behind which curtain her husband rests and shoves the firm fabric aside.

When she walks closer, Mrs. De Ro notices that the collar of the shirt that she had carefully selected for her husband looks wet. While she opens the buttons, she discovers faded blood stains and recent stitches on his body. She calls the hospital personnel and asks for an explanation. She demands to view the state of the cadaver in detail. The hospital employees firmly discourage her from undressing her husband's corpse. A heartbreaking scene unfolds in the morgue.

*   *   *

After the burial, Mrs. De Ro sends a letter to the mayor of Brussels. She complains about the hospital's lack of respect for the dead. The mayor orders the hospital to carry out an investigation. After a few days, however, he dismisses the widow's complaint. The commission of hospitals explained to him that nothing exceptional was done to the corpse of Henri De Ro. In their words, 'it always happened that way'. Ultimately, only the hospital attendant who had refused to call Mrs. De Ro when her husband was dying is punished for his 'reprehensible' behaviour.[1]

2   Introduction

In 1872, 47 students requested the faculty of medicine of the Free University of Brussels to swiftly find a solution for 'the outrageous shortage of bodies' that hindered their education. 'Students often cannot dissect a single bit for over two weeks.'[2] Their professors also worried about the lack of corpses. They regularly postponed or cancelled classes on anatomy, or illustrated them with 'old, deteriorated specimens preserved in alcohol' out of sheer necessity, although they believed that 'fresh material' offered students 'a more profound and useful insight' into the human body.[3] At all Belgian universities, regular body shortages led to initiatives for the reform of the distribution of cadaveric material to anatomical theatres in the late nineteenth century. Medical faculties, for example, limited the right to claim bodies to relatives of the deceased, or arranged to transport unclaimed bodies from other institutions (such as prisons, homes for the elderly or psychiatric hospitals) to them.[4]

The changes discussed in the previous chapter further intensified the dissatisfaction of anatomists with the number of available corpses for dissection. As indigent patients or their families obtained the right to decide on their post-mortem fate, fewer and fewer bodies ended up on the dissecting table. The years in between the lifting of financial duties to claim bodies (around 1890) and the organisation of donation programmes (in the 1930s and 1940s) in particular were times of scarcity for medical faculties. As the rising number of claims kept corpses out of the hands of anatomists and the body donation movement was yet to develop, the morgues of anatomical institutes were worryingly empty. In Brussels, for example, the city council's decision to give all families the right to claim the bodies of their deceased without financial consequences led the faculty of medicine to write multiple complaints on the 'quickly decreasing number of corpses harming education'.[5] In 1900, a university administrator deplored that Also 'sometimes an entire week passes by without a transfer of a cadaver'. At the moment he wrote the complaint, the mortuary of St. Jean contained 'not a single body for the university'.[6]

Also before changes in regulations made it easier for the poor to claim their deceased, insufficient supplies of bodies had bothered anatomists. The abolition of financial duties only worsened the problem. Shortages had been frequent throughout the nineteenth century as a result of a rising demand. As mentioned in the introductory chapter, hands-on dissection became an obligatory part of medical education in Belgium in 1876 and had already been usual in the decades before. Because hands-on dissections were routine, an increase in the number of students went hand in hand with a greater need for corpses. And student numbers grew indeed: Pieter Dhondt has shown that the number of enrolments at Belgian medical faculties more than doubled from the 1860s until the 1890s.[7] It is no wonder that in 1898 the medical faculty of Brussels reported that the shortage of cadavers was becoming more and more acute as a result of the increased normality of claims by relatives and the 'ever growing number of students'.[8] On top of that, dead bodies were increasingly needed outside of dissecting rooms, for cadaveric material became an essential didactic object and research tool for other disciplines: most importantly for pathological anatomy, but also for newer branches of medicine, such as gynaecology and ophthalmology.[9]

This chapter shows that the growing shortage of bodies deeply influenced anatomical practices and the formation of disciplines in the late nineteenth and early twentieth centuries. Through an analysis of the establishment and working of the autopsy service in Brussels, the argument is made that medical faculties in Belgium solved the lack of bodies for education through a redefinition of the autopsy.[10] The Free University of Brussels was not only the first Belgian university to establish an autopsy service; it was also the university where student numbers increased most remarkably. Whereas the Brussels faculty of medicine was the smallest of all Belgian universities in 1850 (with some fifty student enrolments), it had become the biggest by 1880 (with almost four hundred students).[11] The autopsy service responded to the result of this growing number of students: a glaring scarcity of bodies.

This chapter argues that the autopsy was a compromise, accommodating educational, social and scientific changes. By distinguishing autopsy from dissection, the medical faculty of Brussels regained access to the bodies of the poor. They were able to convince the city council that autopsies could happen without patients' or families' consent, as the intervention was invisible for the untrained eye. At the same time, the new emphasis on autopsies echoed the shifting ambitions of the university and the hospital. Autopsy services became the first stop for cadavers because pathologists had persuaded the university of the scientific nature of their activities. Dissections, conversely, were increasingly reduced to mere educational practices.

By studying dissections and autopsies in relation to each other, I respond to the calls of various scholars, who have recently drawn attention to the historical meanings of different post-mortem procedures, which have all too often been confused with each other.[12] Elizabeth Hurren, for instance, has shown that anatomisation and dissection were two separate procedures in the early modern period. The first was associated with the cross-shaped incision made down the torso and across the chest for examining the major organs in order to check whether someone was medically dead; the latter meant the fragmentation of the body to its extremities, implying extensive cutting and dismemberment.[13] Whereas autopsy originally meant to look at or observe the surface of a corpse externally, it came to resemble anatomisation in the eighteenth and nineteenth centuries:

dead bodies were opened up, often through a cross-shaped incision, in order to search for the cause of death. Unlike in the case of whole-body dissections, however, the corpse was sewn up after the procedure.[14] In addition, Sandra Menenteau has shed light on the differences between forensic and clinical autopsies.[15]

Even though these works, especially when compared with each other, implicitly show that the meanings, scope and functions of post-mortem examinations are unstable and interdependent, their authors mostly studied different procedures separately. Hurren, for example, discussed anatomisation and dissection as two distinct steps in the punishment of criminals in order to show that their frequent elision as 'dissection' in crime historiography does not do justice to the complexities of post-mortem harm.[16] In a similar vein, Simon Chaplin argued that 'ottamisation', which signified the preservation and display of ones' remains as a moral lesson to others, came *after* anatomisation.[17] In these cases, it was valuable and appropriate to distinguish different post-mortem procedures from each other, as bodies undergoing anatomisation, dissection or ottamisation were moved to other places (from public to private and the other way around) and were handled by different actors (e.g. anatomists, students or museum visitors).

Yet, the histories of dissection and autopsy are more difficult to disentangle. Medical students, for example, handled bodies both in anatomical theatres and in autopsy rooms, often in order to learn roughly similar skills. Still, historians have mostly tried to grasp the meanings of these examinations by categorising them. Lynsey Cullen, for instance, recently argued that autopsies in late nineteenth-century British asylums might better be defined as dissections because they were not aimed at determining the cause of death.[18] This chapter goes one step further. The aim is not only to distinguish the actions of medical students and practitioners into existing categories—were they performing autopsies or dissections?—but also to look at how distinctions between post-mortem examinations came about, evolved over time and worked in practice.

In so doing, this chapter follows the approach of Andrew Cunningham. In his study on the eighteenth-century disciplinary identities of anatomy and physiology, Cunningham argued that attempts to read back modern ideas about disciplines or research practices had all too often hindered historians to get a grasp on the actual nature of the investigations that were carried out.[19] Much like historians of physiology could not, according to Cunningham, contemplate the thought that there were experiments being conducted on the live animal without jumping to the conclusion that such practices were early instances of physiology, historians of anatomy have largely been unable to recognise that autopsies might have consisted of different practices, and might have served different functions, in the nineteenth century than today. By approaching autopsy and dissection as actor's categories that shifted meanings over time, this chapter aims to shed light on overlapping and cross-cutting areas between them, and on their interrelated evolution.

In the following three sections of this chapter, it will become clear that the members of the medical faculty of Brussels reassured themselves of a steady supply of cadavers by consciously distinguishing autopsy from dissection. First, I put forward that autopsies took on several didactic functions from dissections when corpses were in short supply (Sect. 3). This was possible because the medical faculty had established more lenient regulations by representing autopsies, unlike whole-body dissections, as examinations that were directly useful for families, as well as respectful towards the dead and their bodies (Sect. 4). In addition, the last part of this chapter argues that the emergent distinction between autopsy and dissection also reflected and shaped the relationships between medical disciplines. Studying the, sometimes successive, partition of cadaveric material allows historians to shed light on the relative prestige of post-mortem examinations and the medical disciplines associated with them, as well as on the priorities of, and conflicts between, the university and the hospital (Sect. 5). This chapter hence provides an insight not only into the significance of clinical autopsies in the late nineteenth century, but also into changing sensitivities with regard to dead bodies, social relationships within hospitals and the interplay between medical disciplines.

3   How Autopsy Replaced Dissection

In 1866, Guillaume Rommelaere proposed that the Brussels medical faculty establish a specialised autopsy service.[20] Rommelaere, who would become professor of anatomy and dean at the Free University of Brussels, had visited hospitals in Britain, Prussia, Austria and France after his medical studies. The British autopsy services in particular had left an impression, because they organised applied courses in pathological anatomy for students.[21] The travels of Rommelaere were not unique. Charles Firket, professor of pathological anatomy in Liège, also praised British medical education after having visited schools in England, Ireland and Scotland.[22] Generally speaking, international comparison formed an important inspiration for Belgian advocates of the 'scientification' of medicine, who often travelled to leading medical centres for internships.[23] Since medical centres abroad were an incentive to reform education and research, it is no wonder that the Belgian case was part of broader evolutions, as we will see throughout this chapter.

In 1872, the faculty of medicine brought Rommelaere's request to establish an autopsy service to the university board. The project finally received support in 1878, during Rommelaere's own deanship. As a shared initiative of the university and the hospital, the newly founded autopsy service had a twofold mission: pathological-anatomical education and diagnostic surveys.[24] After the British example, students could attend practical courses in pathological anatomy and clinicians could check their diagnosis by inquiring a specialised pathologist to perform an autopsy on their deceased patient.[25] Other Belgian universities followed suit. In the early 1880s, the teaching hospitals of Ghent and Liège established specialised autopsy services. Leuven followed ten years later, when hands-on exercises in pathological anatomy became an obligatory part of the medical doctorate.[26]

The Brussels autopsy service from the very beginning occupied itself with the shortage of bodies for dissections and other post-mortem examinations. In fact, the wish to 'place many more bodies at the disposal of students' was one of the reasons for its existence.[27] Right after its inauguration in 1878, the autopsy service produced a bulky report on the most important impediments for its working. This report discussed the shortage of bodies as a result of an 'exaggerated importance attached to patients' wishes' and 'deplorable prejudices'. Particularly the fact that burial societies could claim the bodies of their members, and thus avoid their dissection, was deemed 'a crying abuse'. The report posited that only close relatives should be able to claim the bodies of their deceased after they had paid for their burial and within a strictly limited timeframe of 24 hours, 'as the changes that happen in the cadaver very quickly are in a great number of diseases the most important to know'.[28]

It is not hard to see that these statements clashed with the changing sensitivities towards the poor and their bodies discussed in the previous chapter. Whereas the autopsy service wished to limit patients' autonomy in order to supply anatomical theatres with a sufficient number of bodies, the Brussels city council decided that the wishes of patients and their families prevailed over scientific and educational needs. The right to claim bodies was stripped of financial implications in 1889. Georges Feron, head of the Brussels autopsy service, explained the result of these changes in 1903:

The city council has instructed us that we have to inform all interested parties that claiming bodies is absolutely free of charge for indigent families. Year after year, the statistics show us the influence of this measure on the number of bodies claimed by the poor.[29]

According to Feron, the regulation had cost anatomists about 136 cadavers in the year before. Yet in the same letter Feron also noted that 'the number of autopsies had doubled'. While less and less hospital patients ended up in the anatomical theatre, more and more patients were placed on the autopsy slab.

Preserved figures, which are unfortunately rare and incomplete, suggest that the decrease in the number of dissectable bodies coincided with an increase in the number of autopsies. In Brussels, the number of autopsies rose spectacularly in the years between 1878 and 1908. In 1878, approximately 40 per cent of all deceased patients were autopsied in the hospital of St. Jean. Twenty years later, this number had risen to over 60 per cent, and thirty years later even to 80 per cent.[30] More detailed figures show a steady increase in the number of autopsies in the Brussels St. Jean and St. Pierre hospitals from 1888 until 1897, while the number of deaths remained quite stable (Illustration 4.1).

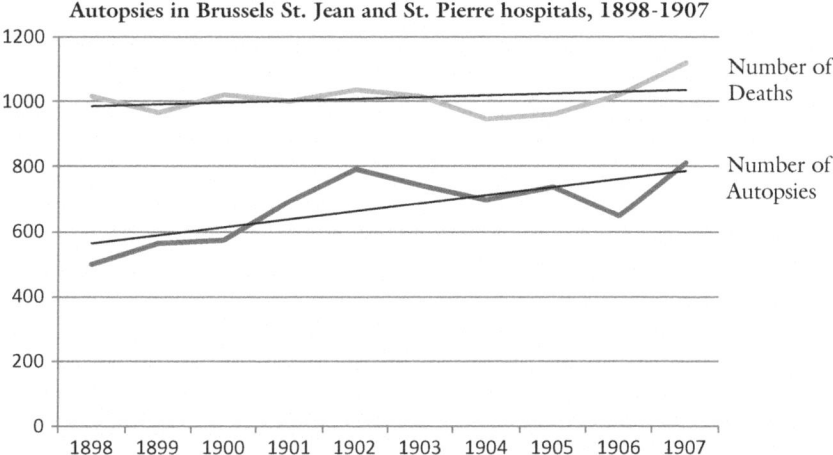

Illustration 4.1 Based on: Dossier concernant l'augmentation du nombre de chefs de service d'autopsies, folder 107, Autopsies, Affaires générales, ASSB

As the Brussels autopsy service published these figures in evaluations of its own working or in requests for more personnel, one should perhaps take them with a grain of salt. The general upward trend of these statistics is credible nonetheless, because the Brussels case would not have been exceptional. Historians have found that the number of autopsies conducted in hospitals augmented in other European countries, too.[31] In the history of psychiatry, several studies have shown that the number of autopsies rose to sensational heights in the last decades of the nineteenth century.[32] In the early 1880s, for example, 49 to 65 per cent of deceased mental patients ended up in the autopsy room in England and Wales. This number had risen to more than 70 per cent ten years later.[33]

Belgian pathologists even dreamt of autopsying all deceased patients. Six years before the foundation of the autopsy service, members of the Brussels faculty of medicine looked up to teaching hospitals in Vienna and Germany, where autopsies were the standard:

> In our country, and in France in particular, the autopsy is the exception; in Vienna, and almost everywhere in Germany, the autopsy is the rule. [...] Without a doubt, we immediately have to establish an anatomical-pathological institution modelled after Vienna or Berlin in Brussels.

In 1879, just one year after the foundation of the autopsy service, the president of the Society for Pathological Anatomy of Brussels (*Société anatomo-pathologique de Bruxelles*) expressed his hope that 'the autopsy, like in neighbouring countries, would become obligatory for every person dying in hospital'.[34] Similarly, Firket complained that he and his colleagues in Liège could 'only' perform approximately 100 autopsies per year. In his opinion, this was both deplorable and astonishing because even the smallest university towns in Germany outnumbered them.[35] It appears that medical faculties tried to turn the autopsy into a frequent, almost standard, procedure in the last two decades of the nineteenth century, and in fact succeeded at it.

In the historiography of psychiatry, which historians of anatomy have unfortunately largely overlooked, the rising number of autopsies has led to research on the function of these examinations. In their study on mental hospitals in late nineteenth-century Canada, for example, David Wright, Laurie Jacklin and Tom Themeles posited that 'the line between necessary assessment to determine cause of death and exploratory pathological inquiries for medico-scientific research was a murky one, further complicated by the question of how many autopsy subjects may have then been used for the very different purpose of pedagogic dissection'.[36] In case studies on asylums in colonial Victoria (Australia) and Oxford (Britain), respectively, Dolley MacKinnon and Lynsey Cullen reached similar conclusions. In psychiatric institutions, so they argued, autopsies were used for extensive pathological research as well as for teaching students internal anatomy.[37]

Autopsies might have fulfilled these purposes in teaching hospitals, too. Preserved archival documents suggest that the number of autopsies was inversely proportional to the number of dissections precisely because autopsies took over several didactic functions of dissections. There are clear indications that the rising number of autopsies in Brussels compensated for the lack of regular dissecting opportunities. First and foremost, the rise in the number of autopsies went hand in hand with a reorganisation of autopsy courses. Before the late 1870s, physicians in training learned how to conduct an autopsy by observing their professors' hands perform the procedure and by listening to his explanations. Autopsies were learned by eyes and ears, not hands. Only *élèves internes*, selected students who stayed in hospital permanently during an internship, on rare occasions had the opportunity to assist the professor. With the establishment of specialised hospital services, however, these demonstrative lessons became hands-on exercises, in which all students handled the scalpel.[38]

Autopsy exercises enabled students to train skills that historians usually associate with the anatomical theatre. Much like during dissections, students used knives, saws and scissors to open the body, after which they removed organs and cut them into sections. In Liège, students dissected internal organs during autopsy courses. The heart, for example, was handled 'with long scissors' and cut 'beginning from the top of the heart, through the left septum of the ventricle and through the left side of the aorta'.[39] In Ghent, the professor of pathological anatomy Daniel Van Duyse also taught his students how to dissect internal organs in some detail. The brain, for instance, was opened up entirely. To get a better look at the lobes, the corpus striatum (a part of the cerebrum) or the thalamus, Van Duyse urged his students to 'multiply the number of incisions' (Illustration 4.2).[40]

In addition, autopsies in part replaced dissections as grab bags for anatomical museums. Organs collected during autopsies were later prepared and used for anatomical demonstrations, or kept for the museum. In Brussels, one of the main reasons for the establishment of the autopsy

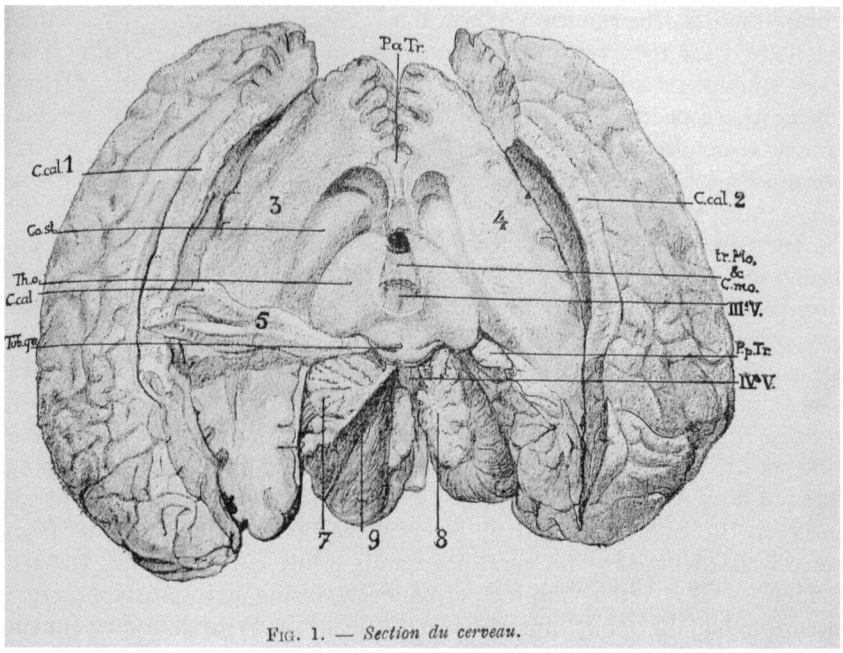

FIG. 1. — *Section du cerveau.*

Illustration 4.2 The 'dissection of the brain' was a part of the autopsy course by Daniel Van Duyse at Ghent University. As the brain was cut in order to uncover its main anatomical structures, one could wonder what the difference was between internal anatomy taught during autopsies and dissections. From: Daniel Van Duyse, *Technique des autopsies* (Ghent: Annoot-Braeckman, 1896), 7

service was that it would allow for an expansion of the anatomical collection. In a request dated 1872, members of the faculty of medicine stated that the possibility to remove 'scientific treasures' during autopsies would 'in a small amount of time and without great expenses enrich the existing anatomical museum'.[41] When the autopsy service was inaugurated six years later, its regulations explicitly stated that 'it was obligatory to remove all scientifically or educationally interesting anatomical or pathological parts, and hand them over to the university'.[42] All of this was not without result: within the first year of its foundation, the employees of the autopsy service had prepared over eighty specimens.[43] In Ghent, too, 'interesting pieces collected during autopsies' were frequently used for anatomical demonstrations or became part of the university museum.[44]

Teachers also urged their students to prepare interesting pieces themselves. In Brussels, the Society for Pathological Anatomy organised an annual competition from 1865 onwards. Students could win a considerable cash prize by preparing and presenting either the most or the best pathological specimens. Through the competition, pathologists wanted to encourage students to 'make the most of the hospital's resources' and to 'contribute to science'.[45] The criteria were severe. Students had to prepare anatomical specimens perfectly. Meticulous clinical information was a requirement. Towards the end of the nineteenth century, additional microscopic and chemical research became a criterion.[46] As a result of these high standards, the prize was often not rewarded.[47] Nevertheless, physicians in training, trainees from the autopsy service in particular, retrieved anatomical specimens from autopsied bodies every year. In some years, students accounted for as much as two thirds of the specimens presented at the meetings of the Society for Pathological Anatomy.[48] By participating in the society's competition, they polished up their preservation methods and enriched the university's collections.

Significantly, the autopsy could only take over didactic functions of the dissection, such as accustoming students to the scalpel and providing them with opportunities to prepare specimens, because the scope of the examination had been expanded. In the words of a frequently used French manual by Désiré-Magloire Bourneville and Paul Bricon, 'the autopsy always had to be complete': a phrase that became usual in the late nineteenth century and to which we will return in a moment.[49] Rommelaere and Léon Stiénon, both teachers at the Brussels autopsy service, stated that only 'complete autopsies' could make a contribution to the field of pathological anatomy.[50] In Liège, Firket and Constant Vanlair, then professor of pathological anatomy, argued that autopsies should not be limited to 'supposedly diseased' organs, but had to take into account the general condition of the intestines and the main organic systems.[51]

These 'complete' autopsies were not just aimed at determining the cause of death, but also at collecting as much information as possible on the course of a certain disease. Belgian pathologists argued that they could only uncover the development of diseases if they related different lesions to each other. Whereas anatomical lesions *an sich* were snapshots, the body as a whole could provide insights into the progress of an illness. Through complete autopsies, pathologists aspired to round out medical knowledge on the course of disease by revealing so-called 'secondary lesions', disorders that were caused by the initial disease but had spread elsewhere in the body. This kind of research was inspired by studies on inflammation, which had shown that inflammatory cells could migrate from afar, and affect different parts of the body.[52] By scrutinising all organs, late nineteenth-century pathologists wanted to distinguish causes from effects, and interrelate lesions.[53] In the words of Firket:

> In order to produce everything that can reasonably be expected [from the field of pathological anatomy], the autopsy should not confine its role to a simple observation of lesions that exist at the moment of death, as these deadly lesions are in most cases nothing but the ultimate outcome of a long series of alterations. [...] It is the connection between lesions we need to know; it is, in the words of Laennec, *non pas la lésion toute faite, mais la lésion se faisant.* For this, more than ever, we feel the need for a complete autopsy dedicated to the relationships between organs and systems.[54]

In real terms, this meant that pathologists had to study the entire internal anatomy of the body. Autopsy manuals show that during university courses, students did not examine bodies on the basis of clinical histories or the assumed cause of death, but scrutinised them *systematically*. In the 1920s, for example, Eugène Picard explained to his students at the Catholic University of Leuven that 'an absolutely complete autopsy' consisted of five phases: the autopsy of the skull, the autopsy of the abdomen, the autopsy of the thorax and neck, the autopsy of the genitourinary organs and the autopsy of the spinal column.[55] Manuals used at other universities also divided the autopsy in distinct steps, including at least the dissection of head, abdomen (and containing organs) and neck.[56]

The Belgian case was not unique. French autopsy manuals, too, stated that the examination 'required a methodical approach' and a fixed order.[57] Anne Carol has suggested that body shortages impacted the scope of autopsies in France. When anatomical theatres were empty, students spent more time in autopsy rooms. Carol asserted that autopsies in those moments encompassed quite spectacular mutilations even if there was no dismemberment. All body cavities and organs—abdomen, neck, skull, spines, etcetera—could be examined and dissected.[58] In her study on anatomical education in Oxford and Manchester, Fiona Hutton has posited that a gradual growth of autopsies might have provided teaching opportunities when bodies were in short supply through the provisions of the Anatomy Act.[59] Much like in Belgium, the boundaries of the autopsy were pushed in order to provide students with opportunities to train their hands.

Interestingly, the changing scope of autopsies coincided with shifts in the practice of dissection. While autopsies were elaborated, dissections were cut back. As argued in the introductory chapter, topographic dissections increasingly replaced detailed, systematic dissections at Belgian universities in the last two decades of the nineteenth century. The systematic fragmentation of bodies into muscles, joints, vessels, nerves and internal organs was in part superseded by an approach based on regions in the body. Whereas systematic anatomy studied the body's individual systems, for instance, the skeletal or cardiovascular system, topographic anatomy looked at parts of the body, such as head or neck, with a focus on the spatial relationships between organs and structures. Charles Ledresseur, professor of topographic anatomy in Leuven, for example, split his course in five regions: head, neck, abdomen, arms and legs. Significantly, three out of these five steps (head, neck and abdomen) were distinct parts of complete autopsies as well.[60] Moreover, a topographic approach was said to replace 'an emphasis on details' by 'a complete overview' of the human body. Much like complete autopsies, topographic dissections had to shed light on the interrelations between organs and other, surrounding body parts.[61]

As a result of these developments, the differences between autopsies and dissections at the end of the nineteenth century were vague to say the least. Ironically though, medical faculties increasingly distinguished autopsy from dissection in the same period. The next section argues that the autopsy could only take over the functions of the dissection because a clear distinction between the two practices had been introduced. As a result of the more stringent rules that came to govern dissections, the employees of the autopsy service of Brussels emphasised that autopsies had nothing in common with 'the large and pointless mutilations of the anatomical theatre' in order to regain access to patients' bodies. They divided the post-mortem examination in two, contrasting the 'cruel' dissection to the 'respectful' autopsy.[62] While dissections and autopsies grew towards one another in real terms, they were represented as growing apart.

4   'IT IS IMPORTANT TO ESTABLISH A DISTINCTION
      BETWEEN DISSECTION AND AUTOPSY'

In 1890, the Brussels commission of hospitals wrote a letter to the city council on behalf of the autopsy service. The reason was the recent change in regulations, which had enabled poor families to prevent their relatives who died in hospital from being subjected to post-mortem examinations. In the view of the personnel of the autopsy service, this more stern regulation was the result of 'a deplorable confusion between two very different things: dissecting and autopsying bodies'. The letter explained that whereas dissection often incited feelings of repugnance and horror because it implied a drastic *morcellement du corps* (a complete disintegration of the body), autopsies did not. On the contrary, autopsies were decent and humane examinations that were beneficial for families and society alike. If the procedure happened with respect, the letter concluded, families would not reject autopsies but embrace them.[63] In Liège, too, Firket argued that 'it was important to establish a distinction between dissection and autopsy', because 'these interventions are very different in terms of their goals, their methods and their effects on the bodies under scrutiny'.[64]

Autopsies indeed had a better reputation than dissections. Katharine Park has shown that in early modern Europe, the autopsy was a privilege for aristocratic and highly esteemed citizens. Back then, the examination often happened at the family's request: sometimes relatives wished to reassure themselves that they and their doctors had done everything in their power; more commonly, parents wanted to protect their children from hereditary diseases. Additionally, family physicians often performed autopsies, unlike dissections, in private, domestic contexts, hence leaving the modesty of the deceased unharmed.[65] In a similar vein, Russell Maulitz has put forward that whereas dissection was abhorrent to a sizeable part of nineteenth-century British society because it was perceived as a disgraceful and disrespectful treatment of the body, autopsy was tolerated at least in principle. In Maulitz' view, this was in part due to the character of the examination. While dissection signified the complete fragmentation of the body, autopsies were limited to interesting anomalies and did not imply a methodic dissection.[66] In Belgium, the autopsy was still a common procedure for aristocratic and royal families in the nineteenth century.[67] Whereas the dissection was a stigma of poverty, the autopsy was a status symbol.

Even though the actual differences between dissections and autopsies were becoming smaller in late nineteenth-century Belgium, medical faculties and hospitals tried to bolster the more positive reputation of the latter.

In order to regain their access to the bodies of the poor, they built on an old distinction that had almost disappeared in real terms. For instance, as early as 1881, Brussels pathologists established a voluntary autopsy service where everyone could request an examination of a relative's body, even if he or she had died outside of the hospital.[68] The initiative came from members of the Liberal city council, from the new mayor Charles Buls specifically. Buls wanted to enhance hospitals' reputation and turn them into centres of knowledge production. He thought that specialised postmortem services in particular could facilitate the political use of science, for example, by drawing up detailed mortality statistics.[69] Pathologists greeted the city council's proposal to establish a voluntary autopsy service with enthusiasm, because they believed the initiative could show post-mortems in the best possible light.[70] The service was established in the same year.

The foundation of the voluntary service was communicated with enthusiasm and in all candour. Newspapers and posters explained the procedure (Illustration 4.3). Specialised pathologists could perform autopsies at home or at the hospital, and always informed relatives about the results. Most significantly, indigent families could request autopsies for free, whereas others had to pay a considerable sum of money (20 francs and additional fees of up to 100 francs for individual pathologists). Both Belgian and foreign pathologists lauded this rule, which consciously transformed the autopsy from burden to privilege in the eyes of the poor. Firket wanted to organise a similar service in Liège. He thought the Brussels initiative was an 'excellent measure' because it 'sought to popularise the practice of cadaveric examinations by publicly representing it as an advantage'.[71] The French pathologists Bourneville and Bricon also heaped praise on the voluntary service for it facilitated the 'general acceptance' of autopsies.[72]

In order to promote voluntary autopsies, Brussels pathologists underlined the direct benefits of the procedure. In advertisements and public statements, they represented the autopsy as a useful examination because it could trace hereditary conditions. The idea was that the voluntary service allowed for an exchange of specific knowledge for a more general contribution to science: families could learn about the disease of their deceased relative, whereas pathologists received access to the knowledge contained in his or her body. In the words of the head of the autopsy service, Jean Wehenkel:

> These measures [the foundation of a voluntary autopsy service] first and foremost allow families to possibly require information of paramount importance, providing them with objective knowledge on which they can base

their decision to take precautionary action at a chosen moment in time, and thus maybe prevent the development of serious diseases. These measures also result in interesting facts for science and education; facts that would otherwise be irrevocably lost.[73]

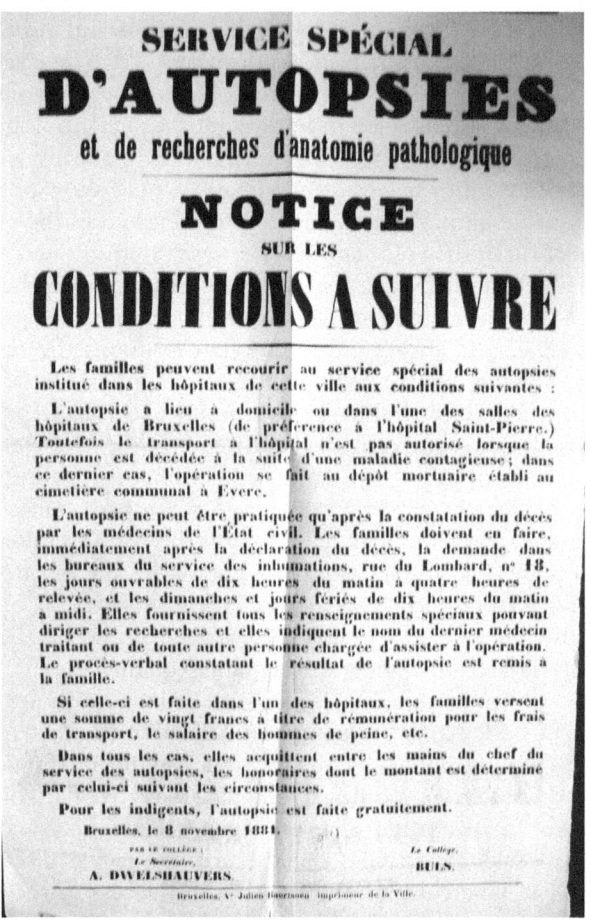

Illustration 4.3   Poster on the establishment of the voluntary autopsy service, 1881. © Archives of the Social Services of Brussels. Dossier concernant l'organisation d'un service d'autopsie des cadavres des personnes décédées en dehors des établissements de l'administration 1881–1896, folder 107, Autopsies, Affaires générales

The voluntary service responded to broader worries about heredity and degeneration that tormented late nineteenth-century Belgian society. The idea was that the autopsy could single out 'degenerated' individuals before they negatively influenced the evolution of society as a whole. As was typical for hygienist beliefs in Belgian medical circles at the time, this kind of knowledge was phrased positively. Thanks to the autopsy, hereditary defect individuals would be given 'objective knowledge' through which they could make an informed decision about possible preventive treatments (which were rare) and, more importantly, about their reproductive behaviour. In this hygienist discourse, the autopsy figured as a form of empowerment.[74] In practice, however, the voluntary service was hugely unpopular. In several years of its rather short existence, from 1881 until 1896, not a single autopsy was requested.[75]

The voluntary service was not the only way in which pathologists tried to bolster the reputation of the autopsy. From the 1880s onwards, medical faculties and hospitals wanted to make clear that the arguments which would eventually lead to the new regulations governing dissection were invalid when it came to autopsies. In their view, the *raisons d'être* of social protest and the resulting stricter regulations were not applicable for this examination. They posited not only that autopsied were directly beneficial for families, but also that they did not intervene with patients' burial wishes. After all, the procedure was limited in time and the body remained largely intact.

Belgian autopsy services, for instance, allowed families to pay a last visit to the deceased and to claim and bury the body after the procedure.[76] As will be argued in more detail in the next chapter, the autopsy, unlike the dissection, did not imply a denial of funeral rites. Pathologists made sure that autopsies did not result in a disrespectful interment of disintegrated remains; on the contrary, the body would be 'neatly presented' in a separate coffin, so that burial rituals could take place in its presence.[77] In this way, pathologists reconciled the autopsy with the burial wishes of the deceased. As they returned the body afterwards, autopsies did not imply that relatives had to assign the rights on the cadaver to medical faculties. They could still fulfil the lawful wishes of the deceased, only with a few hours delay.

The return of the corpse was possible because the autopsy was not mutilating. Whereas dissections turned bodies into a ragbag of human material, autopsies, in the words of the Ghent pathologist Van Duyse, always came to an end with 'the closure of the open body cavities and *la toilette du cadavre*'.[78] Late nineteenth-century autopsy manuals—in

Belgium, but also in France—indeed paid a lot of attention to the exterior of the body.[79] The outward appearance of the corpse mattered from the first cut: instead of the well-known Y-incision, manuals recommended to make only one middle incision from chin to pubis, because it was easier to stitch two skin scraps than pulling together three, four or five of them.[80] Once the body was opened, the examination was continued 'with the greatest reserve'. All incisions had to be 'straight and clean, without rips'.[81]

The most important distinctive feature of the autopsy was its conclusion. The term *obduktie*, a Dutch synonym for autopsy, stemmed from the Latin word *obducere*, which means 'to close' or 'to put back into place'.[82] Whereas the dissection ended in decapitation and fragmentation, pathologists carried out the autopsy with respect for its etymological roots. First, they restored the body's original shape by properly rearranging organs that had been temporarily removed or by filling the bodily cavities with 'wads of cotton wool and towels'; then, they sewed the corpse back together; and finally, they washed the body thoroughly in order to remove blood stains or other liquids.[83]

This emphasis on the outward appearance of the corpse seems opposed to the importance attached to completeness. Yet outer integrity could go hand in hand with inner destruction. A photograph that depicts a complete autopsy in the Brussels St. Pierre hospital in 1892 shows that a considerable number of organs could be taken out of the body without leaving visible traces (Illustration 4.4). Even the brain could be removed without mutilating the corpse. By making an incision underneath the hairline, pathologists could cover up the wound by rearranging the hair of the deceased (Illustration 4.5). Firket claimed that he in this way could research the brain without hurting the feelings of the family, for 'the hair masks absolutely every trace of the intervention, and the face, of which the preservation is essential, is not deformed at all'.[84]

Pathologists from different universities even promised city councils that families would not notice that an autopsy had taken place at all. Relatives would 'not find a trace of the autopsy' because all physical traits through which they knew and recognised their deceased relative were 'absolutely preserved'.[85] Face and hands were of utmost importance, because they were both the bastions of identity and the most visible parts of the body.[86] Again in the words of Firket, the autopsy could be 'covered up completely with a shirt' by holding the scalpel at a distance from face and hands.[87]

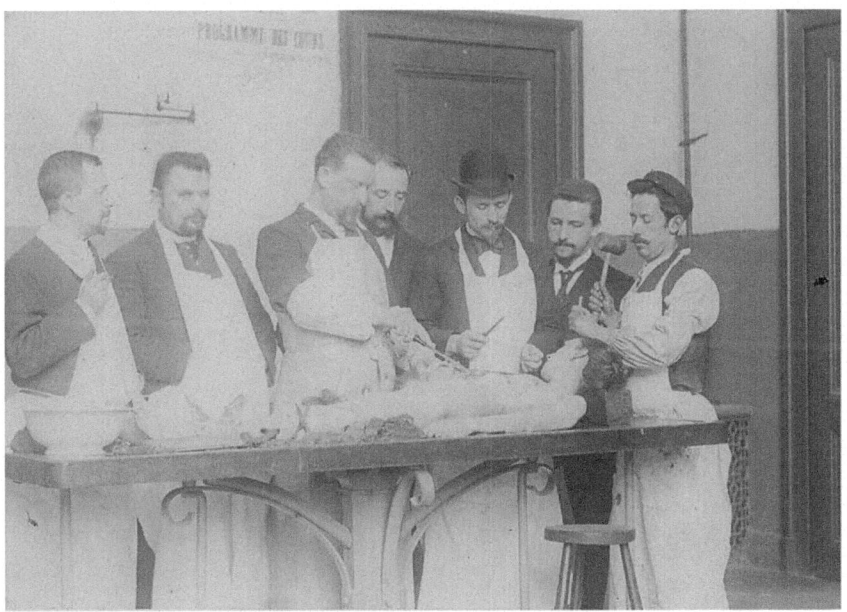

Illustration 4.4   An autopsy in the Brussels hospital of St. Pierre, 1892. © Archives of the Free University of Brussels, 2Y2.11. This photograph, which depicts a young woman surrounded by seven bearded men, is staged and highly gendered. More important for this chapter's argument, however, is the number of organs exhibited on the autopsy table. It seems that the group of students and their teacher were performing a complete, methodic autopsy. The outward appearance of the corpse is striking: although the body was ransacked on the inside, it was not cut extensively on the outside

In her research on France, Sandra Menenteau has put forward that this increased respect for the outward appearance of the corpse was part of a new 'moral protocol' for clinical autopsies. Menenteau argued that pathologists changed their ways in order to spare the family's feelings and to bolster the reputation of cadaveric examinations.[88] Yet the archives reveal that they had a more direct reason to use their scalpel differently. Although the protocol for clinical autopsies changed in real terms, this was not due to the morality of pathologists. For Belgium, archival evidence shows that the respect for the integrity of the corpse rather was a creative way to procure bodies for

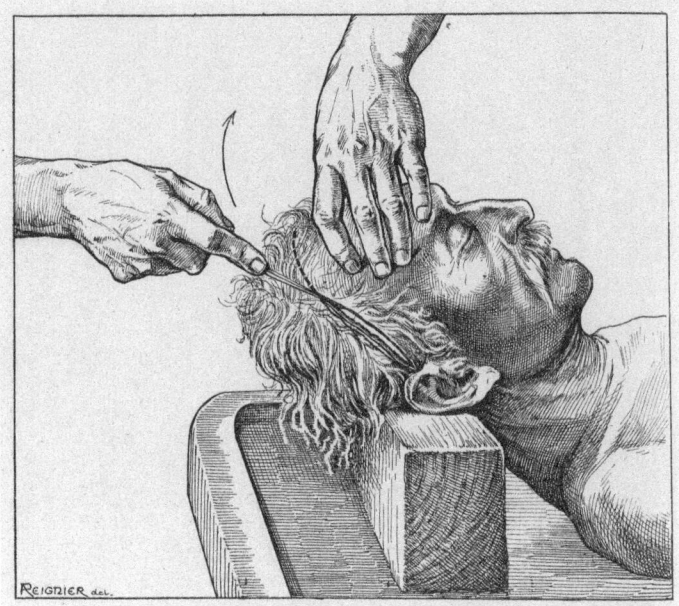

Fig. 93. — Autopsie de l'encéphale. Incision du cuir chevelu.

Illustration 4.5   By making the incision underneath the hairline, the removal of the brain was hidden from families' view. In this way, pathologist reconciled their wish for completeness with relatives' sensitivities. From: Maurice Letulle, *La pratique des autopsies* (Paris: Naud, 1903), 380. © University of Amsterdam, Bijzondere Collecties, UBM: 1320 B 24

researchers and students. On the one hand, medical faculties tried to convince families of the usefulness and decency of the autopsy; on the other hand, they hid the examination from their sight as much as possible.

In fact, the emphasis on corporal integrity had been quintessential for the establishment of a separate regulation for autopsies. In response to the aforementioned complaint by the autopsy service, the city council of Brussels decided that autopsies could be authorised by tacit consent. From 1890 onwards, not even a year after the more stringent rules on post-mortems had been put in place, *all* patients entering teaching hospitals could be autopsied, unless they spontaneously lodged an objection. Both claimed and unclaimed bodies could end up on the autopsy table, even if families had reimbursed the costs of treatment and burial. Dissections, conversely, could only be performed on unclaimed bodies. This distinction was a useful compromise for the members of the city council, who wished to reconcile public grievances with scientific needs.[89] In Ghent, too, physicians could always request an autopsy, for claimed as well as unclaimed bodies.[90] In Liège, all hospital patients could be autopsied from 1885 onwards.[91]

Belgium was not an exception. Helen MacDonald has shown that dissections and autopsies were not only regarded, but also regulated differently in the United Kingdom. While the procurement of corpses for dissection was mostly subject to law, rules concerning autopsies were vague, embedded in unclear and conflicting legislative traditions. As a result, rules concerning autopsy were more lenient in real terms. In Guy's Hospital in London, for instance, bodies were autopsied automatically even though there was no law authorising this.[92] Other historians have reached similar findings in case studies on hospitals in important medical centres, in more peripheral cities and in colonial settings.[93] In psychiatric institutions, both in the United Kingdom and in continental Europe, autopsies also appear to have been executed routinely without clear-cut legal authorisation or consent from patients or their guardians.[94]

Although historians have drawn attention to these lenient regulations and the resulting rising frequency of autopsies, they have overlooked an important prerequisite for this evolution. In Brussels, the city council attached one condition to the new regulation:

> We do not anticipate on any inconveniences by permitting, as a general rule, that all corpses of persons who died in hospital can be autopsied, even if their families have paid for the burial expenses. This examination, however, has to respect the face of the cadavers and may not hinder the return of the body, in a decent way, to those who claim it.[95]

Similar rules were in place in other Belgian and European hospitals. In Liège, autopsies had to happen 'discretely' and 'with respect for the corpse', meaning that the body had to be sewn back together, dressed and laid in state for the family's visit after the procedure.[96] In Paris, pathologists had to restrict themselves to the organs that were related to disease and death.[97] In the words of the Parisian pathologist Emile Goubert, the integrity of the corpse had to be respected 'both because of regulations and because the body is often visited by the family afterwards'.[98] In Vienna, where the number of autopsies almost equalled the number of hospital beds, pathologists 'tried to avoid useless disfigurements of cadavers as much as possible'.[99]

In fact, it appears that the autopsy could become a standard procedure precisely because the appearances of the body were taken into account. The requirement of bodily integrity fitted with ideas about consent to medical interventions more broadly. When it concerned living patients, too, surgeons only had to actively seek consent for mutilating interventions. All other examinations and treatments could be—and were indeed, particularly in hospitals—authorised by tacit agreement.[100] Tellingly, also the other arguments discussed above questioned the need for consent. As consent was only necessary for scientific interventions that did not contribute to patients' well-being, the appeal to indirect therapeutic benefit for families was a powerful argument to establish a more lenient regulation for autopsies. Additionally, by returning the body after the procedure, medical faculties detached the autopsy from the rights of the individual and his or her family on the body. As the corpse was given back afterwards, the autopsy did not require a transfer of rights.

Admittedly, one could argue that dissections were also authorised by tacit consent. The members of the Brussels city council might have given patients and their families the right to refuse dissections without financial consequences in 1889, but they had not required anatomists to actively seek consent from dying patients. Dissections were also only prevented through individuals' or families' spontaneous objection, in the sense that a body needed to be claimed. The main difference was that families *knew* they had to claim a body in order to avoid dissection, as social protests and public scandals had given this procedure a certain publicity. Against the background of culture wars and social conflicts, the Catholic and Socialist press regularly printed stories on families who had saved their deceased from the dissecting table. In 1887, a Socialist newspaper in Brussels had even published the hospital regulations in its entirety, following the 'Coco Affaire', a scandal in which a professor of anthropology had stolen the body of a black woman.

Autopsies, conversely, were successfully hidden from patients' and families' view. It is telling, in this respect, that the Brussels city council did not formally ratify the new regulations, and hence did never inform the public about the new rules. Only the employees of the autopsy service were notified. When one of them reflected on this lack of public communication a few decades later, he stated that 'the city council without a doubt did not want to provoke useless controversies'.[101]

The idea was that families did not need to be informed because the autopsy was invisible for their untrained eyes. In 1902, for example, the widow of Henri De Ro opened the collar of her deceased husband's shirt and discovered recent stitches. When she thereupon demanded to view the state of the cadaver in detail, a heartbreaking scene unfolded in the morgue. A hospital employee and eyewitness explained that she had been convinced that 'bodies could not be autopsied without families' approval'. In reality, the commission of hospitals and the mayor of Brussels quickly dismissed the widow's complaint because, in their opinion, nothing illegal or improper had happened.[102]

To avoid 'useless controversies', pathologists not only changed their incisions, but also held their tongues. They only talked about autopsies in medical jargon and euphemisms. In Liège, Vanlair used a chalkboard in the central entrance hall of the hospital to communicate practical information regarding autopsies to students. In his own words, he employed a 'strictly medical language' for this, which was unintelligible for patients and their families.[103] Similarly, autopsy manuals recommended students to replace the word 'amphitheatre' by 'the usual expression *Chez Morgagni*', a reference to one of the founding fathers of pathological anatomy. 'Autopsies' were better called 'necropsies'.[104] This deceptive, secretive language contrasted sharply with the open communication on the inauguration of the voluntary service in Brussels. The reason for this was pragmatic. Through the voluntary service, pathologists could receive access to corpses on which hospital regulations did not apply. The more visible the voluntary service, the more bodies they could win. Within the hospital, however, silence was preferable as bodies could only be lost.

Autopsies were also hidden by spatial means. Post-mortem rooms were mostly situated in a separate wing inaccessible to patients at the rear of the hospital, close to the morgue and far away from the wards. Ideally, families could visit their deceased without entering the main building by using a separate entrance, often through a chapel and with direct access from a public road.[105] This kind of architecture was modelled after pathological institutes in Vienna and Berlin. In Belgium, the autopsy service of Ghent was exemplary.[106] In Brussels, the situation improved after 1886, when new mortuaries were built on the edge of the hospital complexes of St. Jean and St. Pierre (see Illustrations 4.6 and 4.7).[107]

The underlying principle of this architecture was the separation of the dead from the living. According to a report from the Brussels autopsy service, 'wards in more or less direct contact with repositories for organic tissue in decomposition' posed a serious threat for patients.[108] Without

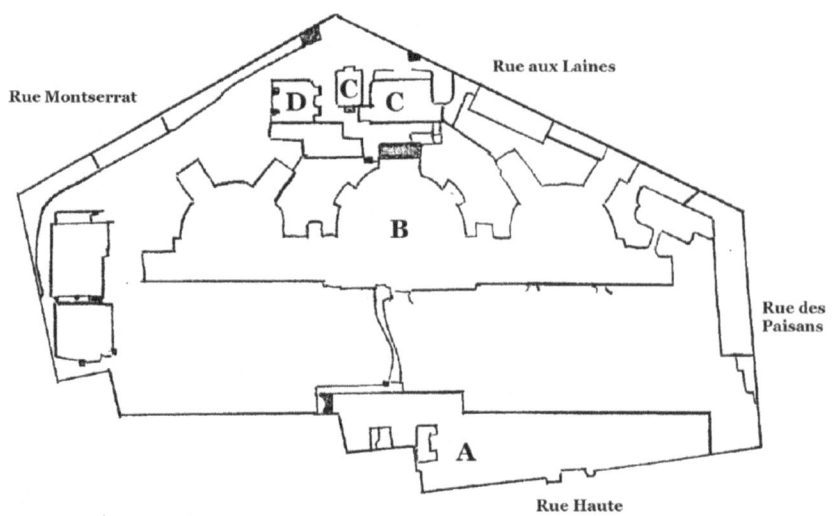

Illustration 4.6 The autopsy service (C) of the Brussels St. Pierre hospital was situated at the rear of the hospital, close to the morgue (D) and far away from the wards (B) and the central entrance (A). Own sketch based on the original plan (which is unfortunately in disrepair). © Archives of the Social Services of Brussels. Plan général des bâtiments, 1917, folder 12, Projet de transformation et d'agrandissements, cartes et plans

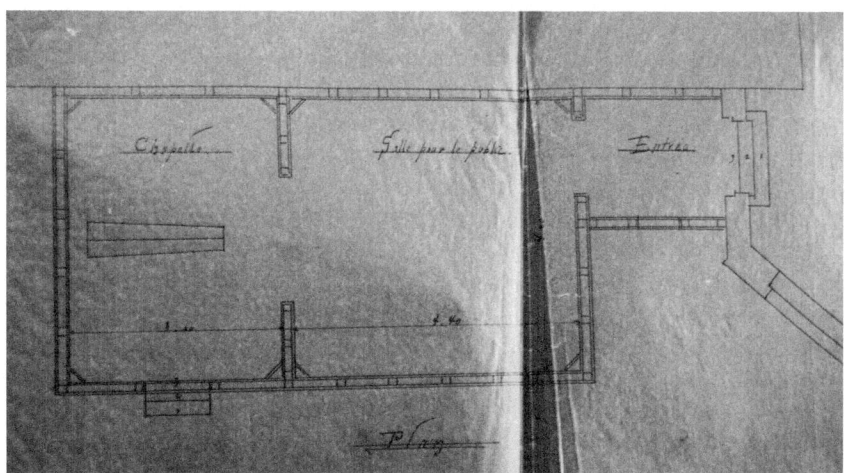

Illustration 4.7 In the Brussels St. Jean hospital, the room where mourners could pay their last respects to the deceased could be reached through a separate entrance. Adolphe Samyn, plan of the mortuary of St. Jean, July 19, 1886. From: Dépôts mortuaires des hôpitaux: construction, 1886, folder S107/2, ASB Bienfaisance, CAB

proper constructional divisions, diseases could spread from dead to living bodies through 'pestilent odours' or by the infectious presence of the employees of the autopsy service who had to traverse the hospital on their way home.[109] The confrontation with death was also believed to have a negative impact on patients' moral. The idea was that those fighting for their lives should not be confronted with their potential, dreaded fate. In the words of a newspaper article, hospital patients should not 'see death' as this caused 'a moral suffering' that was often worse than their physical condition.[110]

Most significantly, the remote location of the autopsy service allowed for the concealment of the examination from patients' senses. In the words of a French autopsy manual, it was 'simultaneously humane and careful' to make sure that patients did not see pathologists and their students when they left the autopsy room dressed in bloody aprons.[111] In Liège, Firket complained that the autopsy room was ill-aired because the curtains were always closed, because it was otherwise possible to observe the opening of cadavers from the syphilis ward.[112] The new morgue of the St. Pierre hospital did not have any windows on the street side despite the need for ventilation.[113] In addition, autopsies were performed at a certain distance from the chapel and from the room in which the body was laid out in order to avoid mourners from hearing 'painful noises', such as hammers knocking on bone or saws slicing skin, and from smelling 'nauseating scents'. For the same reason, the doors of the autopsy room were made from heavy wood and provided with secure locks.[114]

Despite the benefits of an isolated location, the autopsy service needed to be attached to the hospital because the transport of cadavers had important disadvantages. In the words of Firket, the transport of bodies 'on bumpy roads' could produce 'changes in the relations between organs, perforations, etcetera, which may falsify the results of the anatomical examination'.[115] The coming and going of hearses between the hospital and the autopsy service could also move people's emotions. In 1874, for example, neighbours from the Leuven hospital complained about their 'sad view': hearses accompanied by grieving women and crying children.[116] When the Brussels anatomical theatre moved to Parc Léopold (a scientific site outside of the city centre) in 1896, the commission of hospitals decided that the autopsy service had to stay where it was. The reason for this was twofold. First and foremost, moving the autopsy service would give publicity to the procedure. As early as 1883, following a negative report from the Brussels hygiene service, the commission of hospitals worried that 'the repulsion for the hospital would grow' if the post-mortem service was relocated, because 'the public would learn there is a special establishment for the study of cadavers'.[117] Additionally, the move would result in delays, as transports of bodies via the public road could only happen before the first break of day or after sunset in order to avoid public protest.[118] For the same reason, Leuven pathologists moved bodies from sickbed to autopsy slab 'through the enclosed garden of the hospital, mostly at 1 AM, when all patients have left'.[119]

Although mortuaries were marginal, in the sense that they were built at the edges of hospital complexes out of patients' sight, they also represented the public face of the institution, since they were mostly established at a public road so that families could enter the building via a separate entrance. In her study on asylum mortuaries in Central Europe, Leslie Topp has linked the ambiguous status of the house for the dead, at the same time isolated and central, to its different functions. On the one hand, the building was a workspace for pathologists; on the other hand, it was a stage for the various processes and rituals undertaken on the dead body. In this last capacity, mortuaries could emphasise the respectful treatment of patients within the institution.[120]

Within Belgian hospitals, too, mortuaries were multifaceted spaces, which consisted of morgues and cool rooms for the storage of bodies, well-lit theatres for autopsies and chapels for funeral rites.[121] The design of these rooms reflected their function. The private spaces of the mortuary, such as laboratories and post-mortem rooms (the pathological institute), had a clinical look, with concrete and steel as popular materials. Thick walls without windows and sound-proof doors hid them from sight. Conversely, the public spaces, such as the chapel and the façade, were designed with care and they often, much like anatomical theatres, resembled churches. In order to enhance the prestigious reputation of the autopsy, buildings were made of precious materials, such as blue limestone, and were decorated with several ornaments.[122]

The design of the mortuary hence was a game of hide and seek, of secrecy and openness. As Thomas Gieryn has asserted for scientific buildings in general, and Allan Brandt and David Sloane have shown for hospitals in particular, architectural features influenced not only how people acted and moved inside buildings, but also how they were perceived by those outside. Much like was the case for laboratories and hospitals, control and order were at the heart of the design of pathological institutes. Whereas the façade enhanced the public image of medicine by integrating cultural values and expectations regarding both scientific progress and the treatment of the dead, the spatial arrangements segregated insiders from outsiders and shaped normative encounters between doctors, patients and their families.[123] By managing the spaces between the autopsy room, the morgue, the hospital and the city, pathologists sought to control the symbolic and social boundaries of the autopsy.[124] Whereas the public spaces had to uphold an image of respect, pathologists consciously covered up the core of their activities, both spatially and sensory.

To sum up, even though medical faculties placed the respect for the individual and his or her family at the heart of their argumentation in their quest for more lenient regulations, they did not really inform them about their activities. They concealed the autopsy with clothing, discussed it in unintelligible jargon and performed it in an inaccessible wing of the hospital. Both in practice and in regulations, they hid the autopsy from view in order to avoid spontaneous objections. By rendering the autopsy invisible—physically, legally, rhetorically and spatially—medical faculties regained access to the bodies of the poor.

Whatever their motivation, pathologists did change their ways. The requirement of bodily integrity changed the institution of the autopsy profoundly. In the late nineteenth century, autopsied patients were only objectified as long as they were on the table. Afterwards their bodies were restored as individuals, both physically, through stitches and clothes, and socially, as they were visited, buried and remembered by their friends and family. In a way, the distinct spaces of the mortuary reflected the changing status of the cadaver. As the body was stitched back together and moved from the autopsy slab to the funeral chapel, it changed from an object of study to a mourned individual.

Yet this was not true for unclaimed bodies. Whereas claimed bodies were returned to their families after their passage through the autopsy service, unclaimed corpses were further divided. While claimed bodies transformed into individuals with ongoing social relationships again, the unclaimed dead remained socially isolated. As valued scientific and didactic objects (rather than social subjects), they became the stake of a disciplinary struggle between various medical practitioners.

5   A Scientific Examination

In 1900, Auguste Brunin, who was responsible for the teaching of hands-on dissections, alerted the board of the Free University of Brussels that the anatomy department received no more than 140 to 150 bodies per annum anymore. To make matters worse, he could not even use this 'less than brilliant' number of corpses in his lessons, since he had to share them with others. Brunin complained that not only relatives of the deceased, but also his colleagues took bodies out of his hands. In the year before, the available cadaveric material had, for instance, been used by students who had worked in the dissecting room during four and a half months, in courses on arthrology, myology, splanchnology, angiology, neurology, topographic anatomy and surgery, in the course on anatomy at the Academy of Fine Arts and in surgical exercises.[125] On top of that, gynaecologists, forensic physicians, ophthalmologists and hygienists regularly requested (parts of) corpses.[126] This competition for cadavers, Brunin concluded, severely threatened the workings of his 'nonetheless important' department.[127]

Historians have drawn attention to the impact of medical specialisation on cadaver supply. As the disciplinary landscape expanded, more and more interested parties requested their fair share of cadaveric material for research and education.[128] In this section, such internal competition over bodies is studied in order to gain an insight into the relationships between medical disciplines, universities and hospitals. The scarcity of corpses made tacit assumptions about the relative importance of different post-mortem examinations explicit, which led to conflicts between researchers, teachers and clinicians. The increased number of autopsies, in particular on *unclaimed* bodies, gave rise to what Thomas Gieryn has called 'disciplinary boundary work', as different medical practitioners sought to underline the value of their work in contrast to one another.[129] By looking at the content and persuasiveness of their arguments, the next pages shed light on the standing of different medical practices within the hospital and university in the late nineteenth century.

In Brussels, the autopsy service was the starting point of a well-considered distributive system for cadavers. When the service was established in 1878, it immediately became the first stop in the afterlife of unclaimed bodies. From 1890 onwards, most claimed bodies also passed through the autopsy service before burial. From there, corpses were divided according to a hierarchic logic. First, *chefs de clinique*, heads of department who were responsible for the clinical education of university students, could request both claimed and unclaimed bodies for scientific research. The head of the autopsy service, who customarily was professor of pathological anatomy at the Free University of Brussels (and hence a *chef de clinique*), had the first access to cadavers. Then, bodies could be used for other university courses, such as operative medicine, or could serve to teach autopsy techniques to students. Only in a third stage, those corpses that were not claimed by their families could be transported to the anatomical theatre for dissection. *Chefs de service*, heads of department who exclusively worked for the hospital and not for the university, came last in line: they were only able to request an autopsy when none of the parties above had taken possession of the body.[130]

It is important to keep in mind that individual bodies were not just used for autopsies or for surgeries, dissections etcetera, but were used *successively* for different aims. Corpses were consumed not only by different actors but also inch by inch. In 1924, for example, over half of the bodies that reached the anatomical theatre had been autopsied before.[131] Many unclaimed corpses were first autopsied for research purposes or to check a clinical diagnosis, then handled by students in courses on autopsy, surgery or gynaecology, and ultimately dissected to their extremities in the anatomical theatre. Although their successive use was both possible and usual, corpses' didactic value did decrease along the way. Because different postmortem examinations took time, dead bodies either decomposed, leading to health risks for researchers and students, or had to be preserved, altering the colour, touch and biochemical composition of tissues. Despite the improvement of preservation techniques in the late nineteenth century, clinicians and researchers commonly agreed that nothing could replace or surpass the *pièce fraîche*.

The priority of autopsies was therefore not a mere practical matter, in the sense that autopsies could happen before dissections but not the other way around (since there was no body left to autopsy anymore). The priority of the autopsy rather followed from its relative prestige. Autopsies could happen before dissections because they were deemed more important. The Brussels autopsy service had been able to take the lead in the competition for cadavers because the autopsy fitted with the ambitions of both the hospital and the university board. On the one hand, pathologists underlined the clinical importance of the service as a diagnostic arbiter; on the other hand, they emphasised the scientific value of autopsies.

The idea was that diagnoses could be checked more objectively by taking the autopsy out of clinicians' hands. Through the establishment of a specialised service, the treatment of living patients was separated from their post-mortem examination. In Brussels, for example, 'specialised physicians' performed autopsies without prior knowledge of the case. Because of its independence from the clinic, the autopsy service was able to function as a kind of quality control within the hospital. Autopsies mostly confirmed diagnoses, thus proving that patients had received the correct treatment and that death had been inevitable; in other cases, the autopsy service informed the doctor who had attended the patient during life, so that he could learn from his mistakes.[132]

In order for the autopsy to function as an objective assessment, however, pathologists needed the time to perform thorough examinations. Advocates of the autopsy service argued that because clinicians had to spend the bulk of their working day in the sick ward, they mostly limited themselves to checking if post-mortem lesions fitted with their diagnosis. Specialised pathologists, conversely, could spend hours studying the dead body and could assess lesions with 'complete freedom of mind' and 'rigid impartiality' because they had not seen the symptoms during life.[133]

The working of the autopsy service reflected a notion of objectivity, according to which truly 'scientific' results required the elimination of the individual researcher. Much like mechanically produced observations had to, in the words of Lorraine Daston and Peter Galison, 'quiet the observer so that nature could be heard', an independent, will-less pathologist was supposed to conduct post-mortem research without allowing the clinician to impose his assumptions onto the corpse.[134] In fact, it was argued that judging lesions with an open mind was only possible if the corpse was a blank page. If the pathologist knew the case history and the assumed cause of death beforehand, he would not think of alternative scenarios. Significantly, this lack of knowledge and expectations went hand in hand with a need for extensiveness: since uninformed pathologists had no clue where to look first, they had to examine the entire body, and hence perform a complete, methodic autopsy.[135]

Yet the 'primary goal' of specialised autopsy services was not checking clinical diagnoses, but 'developing the medical sciences'.[136] According to Brussels pathologists, the strict division of tasks between clinicians and pathologists would lead not only to an advancement of therapeutics and clinical knowledge, but also to scientific progress in general. They argued that pathologists in Germany and Austria, most importantly Carl von Rokitansky in Vienna, had been able to turn their cities into leading medical centres 'by raising the autopsy to a strictly scientific level'.[137] When the

Brussels autopsy service was inaugurated in 1878, the university administrator Joseph Van Schoor hoped that the service would 'quickly lead to a
rapid progress of the medical sciences' in the city, 'after the example of a
few big university cities abroad'.[138]

Pathologists in particular regarded complete autopsies as scientific
examinations, because they reflected more general changes in their field.
As argued in the introductory chapter of this book, morbid anatomists
around 1850 were accused of being overly static, paying too much attention to the product rather than the development of disease. As a result,
they started to include physiological, microbiological and chemical
approaches in their work. Complete autopsies fitted with this trend, as
their aim was to elucidate the course of diseases. In the words of Firket,
complete autopsies echoed the wish to turn pathological anatomy into
'pathological physiology'.[139] In order to go beyond 'symptomatology',
autopsies had to examine the 'mechanism through which death had developed' rather than the direct cause of death.[140] Ultimately, lesions had to be
correlated into a 'material succession of morbid processes' that 'could
revive the struggle of our tissues'.[141]

Although this physiological approach to pathology did not rest solely
on autopsies, they remained at its basis. In several Belgian pathologists'
views, pathological knowledge could only be advanced through a combi-
nation of different methods, including clinical observation, autopsy,
microscopy, chemical analysis and experimentation.[142] Within this constel-
lation, the study of the material manifestations of disease continued to be
the most certain kind of knowledge, which laid the foundation for newer
approaches. In the words of Wehenkel, autopsies were 'the cornerstone of
positivist medicine' because they gave other approaches 'a purely empirical
basis'. In his view, pathological anatomy was both the certain basis of and
a control mechanism for newer strands of research.[143]

Apart from complete autopsies, microscopic analyses also had to guarantee the scientific value of the autopsy service. Pathologists studied tissues both macroscopically and microscopically. In Liège, for example,
Firket taught his students how to study diseases under the microscope.[144]
By 1932, Albert Dustin, then head of the Brussels autopsy service and
professor of pathological anatomy at the Free University of Brussels, stated
that he spent more time 'verifying tumours and performing microscopic
analyses' than 'making incisions by means of the scalpel'.[145] Significantly,
this emphasis on the microscope allowed pathologists to associate themselves with laboratory sciences. Their research, so they stated, formed the

link between the sick ward, the post-mortem room and the laboratory. Their knowledge of histology and bacteriology made the autopsy not only more specialised but also more profound and scientific.

In reality, however, microscopic examinations often remained descriptive, although they were represented as 'experimental' research. The argument of Alison Kraft and Samuel Alberti concerning biological research in late nineteenth-century Britain seems to apply equally well to pathological investigations in Belgium: whether in the autopsy room or in the laboratory, pathologists 'carried out descriptive morphological studies'.[146] During the meetings of the Society for Pathological Anatomy of Brussels, for example, various tumours were examined on a microscopic level. Researchers focused on the size of the cells (e.g. 'one sixty-fifth of a millimeter'), or their shape and colour (e.g. 'pear-shaped' or 'a reddish shade of brown'). Despite the fact that these studies mainly consisted of a description of form, pathologists characterised them as 'laboratory' or 'experimental' work, suggesting that the rise of the laboratory was as much about a shift in rhetoric than about an actual change in research practices.[147]

Autopsy services nevertheless appealed to universities' and hospitals' aspirations by stressing the scientific nature of their activities. Dhondt has shown that universities transformed from 'technical colleges' into scientific institutes in the late nineteenth century. From that moment onwards, universities wished to train not only medical practitioners, but also scientists. Research became a core task of medical faculties, alongside education.[148] In Brussels, the autopsy service was one of the first manifestations of this new ambition. As early as 1876, Rommelaere, the founding father of the service, had argued that the establishment of 'specialised laboratories', such as the autopsy service, would make it easier to introduce students to medical research.[149] At other Belgian universities, too, the leading figures of autopsy services were advocates of a more scientific medical training, allowing students to choose between a career as a practitioner or a scientist.[150]

Science also became increasingly important within the hospital. Several studies have shown that laboratory science started to play a part in day-to-day medical practice around the turn of the twentieth century.[151] On the one hand, these studies have highlighted instances of tension and conflict between medical science and clinical practice. Andrew Cunningham and Perry Williams, for example, have described the introduction of laboratory medicine as 'the outcome of a group struggle' between scientists and clinicians, or between universities and hospitals. In their view, clinicians resisted the appointment of scientists in hospitals, as they saw this as an intrusion of their own sphere of authority.[152] On the other hand, scholars have drawn attention to instances of peaceful cooperation. In his study on the Bergen General Hospital, for example, Morten Hammerborg has shown that clinicians were generally enthusiastic to integrate laboratory techniques and technologies, most importantly histopathology, in their everyday practices.[153] Steve Sturdy has posited that conflicts between clinicians and scientists have been overstated in historiography in general, and might have been the exception rather than the rule.[154]

In Belgium, clinicians indeed were mostly enthusiastic about the foundation and working of specialised autopsy services. This was in part due to the fact that they generally requested pathological research themselves. Clinicians mostly made use of the service when the clinical signs had been uncertain or confusing during the patient's life. The autopsy service allowed, but did not compel, them to receive certainty about a case after death, and to integrate histological microscopy into their diagnostic armamentarium. As the service's conclusions in most cases reinforced rather than undermined clinicians' judgements, it only threatened their authority exceptionally. The appointment of the professor of pathological anatomy as the head of the autopsy service in the hospitals of St. Jean and St. Pierre, which was a regulatory requirement that was seldom broken, also suggests that hospitals agreed to give professional scientists a responsible role in their institution.[155] In addition, hospital physicians' occasional participation in the meetings of the Society for Pathological Anatomy of Brussels indicates that they saw clinical and therapeutic achievements at least partly as grounded in scientific research.[156]

Yet conflicts did occur occasionally. The hierarchical division of unclaimed bodies in particular led to tensions, not only between pathologists and clinicians, but also with anatomy teachers. Although anatomists were happy to receive help from pathologists when it came to claimed bodies, unclaimed bodies were a major source of conflict. Autopsying patients who would otherwise be buried without prior examination was beneficial, because it allowed students to gain much-needed experience with internal anatomy despite a lack of cadavers in the anatomical theatre. However, anatomists deemed the preliminary autopsy of unclaimed bodies, those bodies that could in fact be used for dissection, problematic, as they resulted in 'harmful delays' and 'inevitably destroyed different regions of the body'.[157] The division of cadavers also occasionally displeased clinicians. *Chefs de service*, who came last in the division of corpses since they were not linked to the university, thought it was a shame to let dissections take priority over autopsies, especially in cases 'that were interesting from a scientific and medical point of view'.[158]

For these reasons, both anatomists and clinicians increasingly complained about pathologists' privileged position when the number of (often complete) autopsies continued to grow in the 1890s and 1900s. In 1903, for instance, Jean-Joseph Sacré, professor of descriptive anatomy, asked the commission of hospitals of Brussels to limit the number of autopsies since he received less and less intact bodies, although an ever increasing number of students attended his courses.[159] In 1906, an administrator of the Free University of Brussels reported that the anatomical theatre only received 'empty' or putrefied autopsied bodies and the corpses of foetuses and stillborn children. In the months preceding his complaint, 22 out of 50 cadavers 'had been useless for the ordinary courses for students'.[160] As aforementioned, *chefs de service* in their turn were dissatisfied when the bodies of their patients could not be autopsied because the university took priority. In 1893, for example, Jules Destrée wanted to request an autopsy on a patient he had been treating for multiple years. He was not given permission since the body had to serve for dissections.[161] In the same year, Emile Houzé was also denied access to a cadaver because he was not a *chef de clinique*.[162]

These conflicts over cadavers seem to have been a European phenomenon in the late nineteenth century, and their outcomes provide an insight into the relative prestige of autopsies and dissections, and the functions associated with them. In her research on Sweden, for instance, Eva Ahrén has shown that autopsies took priority over dissections because clinical and scientific interests were seen as more important than education. As a result, only bodies of patients who were not judged to require an autopsy could be brought to departments of anatomy.[163] Similarly, Tatjana Buklijas argued that the pathological institute of Vienna was 'the first port of call for cadavers' because the discipline of pathological anatomy was regarded as the axis of clinical research.[164] In Brussels, too, the hierarchical distribution of cadavers reflected the precedence of scientific and clinical interests over education. While autopsies were represented as examinations of both scientific and clinical value, dissections were increasingly reduced to mere didactic procedures. This evolution went hand in hand with the formation of pathology as a discipline independent from anatomy. As their field of research became more closely related to the programme of scientific medicine while anatomy was increasingly viewed as an auxiliary and descriptive science, pathologists argued that their first right to the cadaver followed from the greater importance of their discipline.

To defend their privileged position in the division of cadavers in the face of complaints from anatomists, for example, employees of the Brussels autopsy service underlined the contribution of autopsies to pathological research. In their view, anatomy remained important as the servant of surgery, but was not an innovative science anymore. Whereas they practised technologically advanced *travail de finesse*, dissections happened 'with nothing more than a knife'.[165] In addition, they said that autopsies produced both new and fundamental knowledge. By aligning themselves with laboratory sciences, pathologists restructured medical knowledge around the elaboration of general scientific principles. Whereas dissections only laid bare known structures, autopsies unlocked the mysteries of diseases on a basic, cellular level.

Pathologists in this way gradually banned the dissection from the realm of science through their rhetorical construction of a 'scientific space' limited by selective criteria.[166] In their view, science had to be innovative, structured around general principles, based on laboratory research and cell theory. Whereas pathological anatomy and the autopsy still met the requirements of this new conception of medical science, macroscopic anatomy and the dissection did not. This way of thinking was also echoed in the autopsy service's educational function: whereas anatomical theatres were represented as quintessential for the formation of medical practitioners, a traineeship at the autopsy service was seen as a stepping stone in a research career.[167]

Clinicians, too, pointed at the mere educational character of dissections in their quest for bodies. In their opinion, corpses should 'serve scientific research' before being used for dissection.[168] They also underlined the clinical importance of autopsies. In the words of Dustin, for instance, it was 'a duty of the physician to confirm the correctness of his diagnosis by means of biological and anatomical examinations'.[169] In Brussels and Liège, this demarcation between dissection and autopsy was confirmed spatially. Anatomical theatres moved to more remote locations in the late nineteenth century. The spatial division between dissections (happening in university buildings) and autopsies (happening in hospitals) enhanced the different reputations of both procedures: the one more closely tied to the clinic than the other.[170]

In the end, teaching hospitals tried to reconcile their different functions without losing sight of scientific progress. In 1898, the Brussels commission of hospitals advised against complete autopsies on unclaimed bodies unless there existed fundamental scientific reasons. The employees of the autopsy service had to restrict their work to the necessary organs, so that a dissection could still happen afterwards. Additionally, corpses had to become more durable. In order to safeguard the didactic value of unclaimed bodies, the autopsy service was urged to use preservation techniques:

> The bodies that are left by their families [...] should receive a partial injection with preservation fluid after the autopsy, following the method indicated by the teacher of the dissecting course at the university.[171]

Thanks to partial embalmment, a delay in the autopsy room no longer led to the receipt of a decomposed corpse in the anatomical theatre, and corpses could be saved during school holidays. In order to be 'suitable for dissection', preservation techniques had to respect the 'structure and texture' of tissues.[172] Finally, professors of anatomy were asked to inform the autopsy service if they needed intact bodies for their courses. A limited number of bodies could be put aside.[173]

Nevertheless, the primacy of science continued to be directly associated with the priority of autopsies over dissection. The autopsy service kept right of precedence in all 'scientifically interesting cases'.[174] Significantly, clinicians from 1898 onwards also received priority in these instances. The commission of hospitals decided that 'bodies that could be seen as scientifically rare' could be autopsied at the request of *chefs de service*, even if anatomy teachers had reserved them for the dissecting course.[175] Despite hospitals' efforts to meet anatomists' objections, dissections were increasingly subordinated to autopsies, because the interests of science and the clinic weighed more heavily than educational goals. Pathologists had not only separated themselves from, but also raised themselves above their mother science.

6   Conclusion

The autopsy was a useful alternative for the dissection in an era in which the gradual extension of democracy went hand in hand with growing criticism of harsh poverty politics. As shown in the previous chapter, the procurement of bodies for dissection had become a headache for the local authorities in the socially turbulent context of the late nineteenth century. The abolishment of the forced dissection of the poor was a symbolic measure: a political weapon meant to take the wind out of the sails of the emergent Socialist Party. Yet for medical faculties, more empathetic regulations on dissection caused glaring shortages of bodies, especially because the number of students continued to grow.

The autopsy figured as a solution for this problem. Despite the growing resistance against the forced dissection of the poor, corpses were needed for research and education. Medical faculties therefore argued that the stricter regulations were exaggerated when it came to autopsies. In the 1880s and 1890s, they increasingly distinguished dissection and autopsy from each other, even though the examinations became more similar in real terms. Firstly, the autopsy was said to be useful because its results potentially allowed families to take precaution. Although post-mortems always came too late to help the patient, autopsies could, at least rhetorically, appeal to an indirect therapeutic benefit: namely the tracing of hereditary diseases. Secondly, medical faculties argued that the autopsy did not hinder burial rituals. This was a particularly useful argument in a period in which death customs were increasingly refashioned culturally as a right. Since the body was returned after the autopsy, there was no infraction of individuals' rights on their remains.

Autopsied bodies could be given back to families because their exterior was by and large respected. Bodies might have been plundered on the inside, but remained outwardly intact. In the words of the personnel of the Brussels autopsy service, the examination therefore was 'nothing like the large and pointless mutilations of the anatomical theatre'.[176] Whereas dissections turned the body into a pile of disintegrated remains and only served students' interests, autopsies could provide important information for families and respected the individuality of the dead.

Above all, autopsies could be hidden from families' view. By promising the city council that nobody would notice the traces of the autopsy, the Brussels faculty of medicine received more lenient regulations and recuperated the bodies of the poor—at least on the inside. It is true that under the new regulations on autopsy and dissection, the bodies of paying patients technically had the same chance of ending up under the scalpel. Yet as the majority of hospital patients were poor, the scalpel in practice remained an instrument of discrimination. Moreover, the entangled histories of dissection and autopsy clearly show that paternalism continued to characterise the attitudes of both politicians and physicians in the late nineteenth century. Although political rhetorics on poverty were changing, the feelings and opinions of the hospitalised poor were not really taken into account. Patients were not informed about what would happen to their bodies if they died.

Strategical distinctions between autopsies and dissections also seem to have been introduced in other European countries. In many French cities, dissections were differentiated from autopsies in legislation, while autopsy manuals underlined the importance of complete examinations.[177] In Austria, Sweden and the United Kingdom, too, the number of autopsies augmented while regulations on dissections became more stern (that is, from the anatomist's point of view).[178] It appears that the rise of the autopsy went together with the decline of the dissection in many countries, although more research is needed to fully understand the relationship between these simultaneous developments.

In any case, this chapter has shown that studies on the number and functions of autopsies should be complemented by research on dissecting practices. Rather than distinguishing autopsies from dissections on the ground of contemporary definitions, as has been done in the existing historiography, historians should look at how these distinctions came about and evolved. This chapter suggests that, at least for the late nineteenth century, the development of the autopsy cannot be understood without looking at the simultaneous development of the dissection. The interplay between both post-mortem examinations was one of openness and secrecy: while the dissection gained visibility, the autopsy was selectively hidden from view. While consent and bequeathal gradually came to govern the dissection in an era of expanding democracy, the autopsy became a concealed way for medical faculties to keep a hold of the bodies of hospital patients without permission. Physically, legally and spatially, the autopsy was concealed because the dissection had been uncovered.

Anatomists were happy with the augmenting number of autopsies when it came to claimed bodies, as the often systematic exploration of internal anatomy provided students with opportunities to train their hands in times when anatomical material was in short supply. Unclaimed bodies, however, became a source of conflict between autopsy services and anatomists. In the face of shortages, the cultural boundaries of science were rewritten according to the programme of scientific medicine. Against the background of the disciplinary formation of pathology, the autopsy was increasingly represented as both scientifically and clinically relevant, whereas the dissection was reduced to an educational practice. The success of the autopsy dovetailed with the success of laboratory sciences in general, and shows the increasing supremacy of science over education around 1900. In addition, the successive partition of cadavers suggests that clinical interests ended up in a sort of intermediate position between scientific and educational ones.

This chapter has shown that bodies, or precisely the absence of them, deeply influenced anatomical practices. The (un)availability of cadavers not only determined if anatomical research and education was possible, but also actively shaped medical practices and disciplines. Bodies, both as object and subject, led to complex relationships between physicians, patients and families, as well as between clinicians, teachers and researchers. In a very concrete manner, in the ways in which corpses were cut apart and sewn back together, the autopsy was a compromise, shaped by scientific and didactic needs, as well as by burial rituals and social protest. As bodies were divided into different pieces and among different actors, their significances changed, whether this was from research tool to didactic object or from medical case to mourned individual.

NOTES

1. Based on: Dossier concernant la plainte par le collège au sujet de ce qui s'est passé à l'hôpital St. Pierre à la suite du décès de Henri De Ro, 1902, folder 133, Liasse divers 1896–1902, Affaires générales, ASSB.
2. Student petition sent to the faculty board, November 29, 1872, folder 107, Autopsies, Affaires générales, ASSB.
3. Rapport sur l'organisation du service des autopsies dans les hôpitaux de Bruxelles [1878], folder 107, Autopsies, Affaires générales, ASSB.
4. For example: Circular letter from Conseil des hospices de Louvain to several universities and hospitals, 1881, folder Gasthuis allerlei 1, Autopsie 1839–1901, ASSL; Administration des hospices to university administrator, May 7, 1885, folder 4A 43 (71), Ter beschikking stellen van lijken veroordeelden en ouderlingen voor onderwijs, UAG.
5. Rapport sur les autopsies et les dissections, 1924, folder 107, Autopsies, Affaires générales, ASSB.
6. Note pour le conseil, April 23, 1900, folder 105, Autopsies, Affaires générales, ASSB.

7. Pieter Dhondt, *Un double compromis: Enjeux et débats relatifs à l'enseignement universitaire en Belgique au XIXe siècle* (Ghent: Academia Press, 2011), 476.
8. Note pour le conseil, April 9, 1898, folder 107, Autopsies, Affaires générales, ASSB.
9. Tatjana Buklijas, 'Cultures of Death and Politics of Corpse Supply: Anatomy in Vienna 1848–1914', *Bulletin of the History of Medicine* 82, no. 3 (2008): 585–94.
10. The conclusions of this chapter draw on the following article: Tinne Claes and Pieter Huistra, 'Il importe d'établir une distinction entre la dissection et l'autopsie. Lijken en medische disciplinevorming in laatnegentiende-eeuws België', *BMGN: Low Countries Historical Review* 131, no. 3 (2016): 26–53.
11. Dhondt, *Un double compromis*, 476.
12. See, for example, Elizabeth T. Hurren, *Dissecting the Criminal Corpse: Staging Post-Execution Punishment in Early Modern England* (Basingstoke: Palgrave Macmillan, 2016), 33–68; Lynsey T. Cullen, 'Post-mortem in the Victorian Asylum: Practice, Purpose and Findings at the Littlemore County Lunatic Asylum, 1886–7', *History of Psychiatry* 28, no. 3 (2017): 280–96; Helen MacDonald, *Possessing the Dead: The Artful Science of Anatomy* (Melbourne: Melbourne University Publishing, 2010), 100–2; Claes and Huistra, 'Il importe d'établir', 26–53; Buklijas, 'Cultures of Death', 585–94; Fiona Hutton, *The Study of Anatomy in Britain 1700–1900* (London: Pickering and Chatto, 2013), 56–70.
13. Hurren, *Dissecting the Criminal Corpse*, 33–7.
14. Anne Carol, *Les médecins et la mort XIXe—XXe siècle* (Paris: Aubier, 2004), 242–9.
15. Sandra Menenteau, 'Stigmata of the Autopsy: Operative Liberties and Protocol in Forensic Examination of the Dead Body in Nineteenth-Century France', *Intertexts* 15, no. 1 (2011): 20–38.
16. Hurren, *Dissecting the Criminal Corpse*, 33–7; Idem, 'Other Spaces for the Dangerous Dead of Provincial England, c. 1752–1832', *History* 103, no. 354 (2018), 27–59.
17. Simon Chaplin, 'Anatomy or an Ottamy? Bodies on Show in Georgian London', in *The Morbid Anatomy Anthology*, eds. Joanna Ebenstein and Colin Dickey (New York: Morbid Anatomy Press, 2014), 254–70.
18. Cullen, 'Post-mortem in the Victorian Asylum', 280–96. Other examples are: Hutton, *The Study of Anatomy*, 56–70; Carol, *Les médecins et la mort*, 242–9.
19. Andrew Cunningham, 'The Pen and the Sword: Recovering the Disciplinary Identity of Physiology and Anatomy before 1800: I & II',

Studies in History and Philosophy of Biological and Biomedical Sciences 33, no. 4 & 34, no. 1 (2002–2003): 631–65 and 51–76.

20. On Guillaume Rommelaere: Elienne Langendries and Anne-Marie Simon-Van der Meersch, *Het Rommelaere-complex. Onderdeel van het gebouwenmasterplan voor de Gentse universiteit op het einde van de 19de eeuw* (Ghent: Archief RUG, 1999), 75–82.

21. Guillaume Rommelaere, *Des institutions médicales et hospitalières en Angleterre* (Brussels: Bols-Wittouck, 1866), 33.

22. Charles Firket, *L'éducation médicale en Angleterre, en Ecosse et en Irlande* (Paris: Armand Colin, 1893).

23. Dhondt, *Un double compromis*, 312–15; Geert Vanpaemel, 'The German Model of Laboratory Science and the European Periphery (1860–1914)', in *Sciences in the Universities of Europe, Nineteenth and Twentieth Centuries: Academic Landscapes*, eds. Ana Simões, Maria Paula Diogo and Kostas Gavroglu (Dordrecht: Springer, 2015), 211–25.

24. Rapport sur l'organisation du service des autopsies dans les hôpitaux de Bruxelles [1878], folder 107, Autopsies, Affaires générales, ASSB.

25. On autopsy services in Britain: Cay-Rüdiger Prüll, 'Pathology and Surgery in London and Berlin, 1800–1930: Pathological Theory and Clinical Practice', in *Pathology in the 19th and 20th Centuries: The Relationship between Theory and Practice*, ed. Cay-Rüdiger Prüll (Sheffield: EAHMH Publications, 1998), 71–99; Idem, *Medizin am Toten oder am Lebenden? Pathologie in Berlin und in London, 1900–1945* (Basel: Schwabe Verlag, 2004), 107–58.

26. Dossier concernant l'organisation du service des autopsies dans diverses villes du pays et de l'étranger, 1880, folder 107, Autopsies, Affaires générales, ASSB. See also: Dhondt, *Un double compromis*, 366–7.

27. Faculty of medicine to University board, December 28, 1872, folder 107, Autopsies, Affaires générales, ASSB.

28. Rapport sur l'organisation du service des autopsies dans les hôpitaux de Bruxelles [1878], folder 107, Autopsies, Affaires générales, ASSB.

29. Georges Feron to Conseil des hospices, May 28, 1903, folder 107, Autopsies, Affaires générales, ASSB.

30. Based on own calculations using information from: Dossier concernant le rapport fait par M. le Dr. Wehenkel sur le service des autopsies depuis sa création, 1879, folder 107, Autopsies, Affaires générales, ASSB; Dossier concernant l'augmentation du nombre de chefs de service d'autopsies, 1908, folder 107, Autopsies, Affaires générales, ASSB.

31. See, for example: Buklijas, 'Cultures of Death', 585–94; Carol, *Les médecins et la mort*, 256–64; Eva Ahrén, *Death, Modernity and the Body: Sweden, 1870–1940* (Rochester: University of Rochester Press, 2009), 27–30.

32. Eric J. Engstrom, *Clinical Psychiatry in Imperial Germany. A History of Psychiatric Practice* (Ithaca and London: Cornell University Press, 2003), 96–8; Cullen, 'Post-mortem in the Victorian Asylum', 280–96; Jonathan Andrews, 'Death and the Dead-House in Victorian Asylums. Necroscopy versus Mourning at the Royal Edinburgh Asylum c.1832–1901', *History of Psychiatry* 23, no. 1 (2012): 6–26; David Wright, Laurie Jacklin and Tom Themeles, 'Dying to Get Out of the Asylum: Mortality and Madness in Four Mental Hospitals in Victorian Canada, c. 1841–1891', *Bulletin of the History of Medicine* 87, no. 4 (2013): 591–621; Dolley MacKinnon, 'Bodies of Evidence: Dissecting Madness in Colonial Victoria (Australia)', in *The Body Divided: Human Beings and Human 'Material' in Modern Medical History*, eds. Sarah Ferber and Sally Wilde (Farnham: Ashgate, 2011), 75–107.

33. Andrews, 'Death and the dead-house', 17–18.

34. *BSAP* 28 (1879): 9.

35. Firket, *Du but et de l'organisation*, 19.

36. Wright, Jacklin and Themeles, 'Dying to Get Out of the Asylum', 608.

37. MacKinnon, 'Bodies of Evidence', 75–107; Cullen, 'Post-mortem in the Victorian Asylum', 280–96.

38. Dhondt, *Un double compromis*, 366–7.

39. Quote from: Rudolf Virchow, *Die Sektions-Technik im Leichenhause des Charité-Krankenhauses mit besonderer Rücksicht auf gerichtsärztliche Praxis*, 4th ed. (Berlin: August Hirschwald, 1893), 45. On the use of Virchow's method in Liège, see several students' testimonies in: *Manifestation en l'honneur du Dr. Charles Firket* (Liège: Vaillant-Carmanne, 1925).

40. Daniel Van Duyse, *Technique des autopsies* (Ghent: Annoot-Braeckman, 1896), 8.

41. Paul Héger, Eugène Mahaux and Charles Delstanche, Le Service spécial des autopsies à Vienne, October 18, 1872, folder 107, Autopsies, Affaires générales, ASSB.

42. Projet de règlement, 1878, folder 107, Autopsies, Affaires générales, ASSB.

43. Dossier concernant le rapport fait par M. le Dr. Wehenkel sur le service des autopsies depuis sa création, 1879, folder 107, Autopsies, Affaires générales, ASSB.

44. Service des autopsies à Gand [1881], Dossier concernant l'organisation du service des autopsies dans diverses villes du pays et l'étranger, folder 107, Autopsies, Affaires générales, ASSB; Hospices civils de Gand, *Règlement de l'hôpital de la Biloque* (Ghent: Imprimerie F.&R. Buyck Frères, 1908), Folder 91, BG12, Hospitaal de Bijloke, ASSG.

45. Quotes from: *BSAP* 14 (1868): 13; *BSAP* 25 (1876): 6.

46. Joris Vandendriessche, 'Anatomy and Sociability in Nineteenth-Century Belgium', in *Bodies Beyond Borders: Moving Anatomies, 1750–1950*, eds. Kaat Wils, Raf De Bont, and Sokhieng Au (Leuven: Leuven University Press, 2017), 51–72, on p. 68.
47. For example: *BSAP* 35 (1886): 1.
48. For instance: in 1868 and in 1876.
49. Bourneville and Bricon, *Manuel de technique des autopsies*, 23.
50. Réclamations de MM. Rommelaere et Stiénon au sujet du service des autopsies, 1887–1888, folder 107, Autopsies, Affaires générales, ASSB.
51. Firket, *Du but et de l'organisation*, 4.
52. Russell C. Maulitz, 'Pathology', in *The Cambridge History of Science: The Modern Biological and Earth Sciences*, vol. 6, eds. Peter J. Bowler and John V. Pickstone (Cambridge: Cambridge University Press, 2009), 367–81, on p. 375.
53. Bourneville and Bricon, *Manuel de technique des autopsies*, 23.
54. Firket, *Du but et de l'organisation*, 10.
55. Eugène Picard, *Travaux pratiques d'anatomie pathologique: Autopsie et Microscopie*, 3rd ed. (Leuven: Dewallens, s.d.), 3.
56. For example: Van Duyse, *Technique des autopsies*; Idem, *Premiers éléments de diagnostic anatomo-pathologique au cours des autopsies* (Ghent: Van Goethem, 1904).
57. Maurice Letulle, *La pratique des autopsies* (Paris: Naud, 1903), 9; Emile Goubert, *Manuel de l'art des autopsies cadavériques, surtout dans ses applications a l'anatomie pathologique* (Paris: Germer Ballière, 1867), 23.
58. Carol, *Les médecins et la mort*, 256–7.
59. Hutton, *The Study of Anatomy in Britain*, 91.
60. Charles Ledresseur, *Résumé du cours d'anatomie des régions* (Leuven: Peeters-Ruelens, 1882).
61. Félix Putzeys, 'Préface', in *Guide de dissection, résumé d'anatomie topographique*, Daniel John Cunningham, trans. Pierre Kuborn (Liège: Nierstrasz, 1890), v–ix, on p. v.
62. Rapport sur l'organisation du service des autopsies dans les hôpitaux de Bruxelles [1878], folder 107, Autopsies, Affaires générales, ASSB.
63. Conseil des hospices to Brussels city council, May 18, 1890, cited in Rapport sur les autopsies et les dissections, 1924, folder 107, Autopsies, Affaires générales, ASSB.
64. Firket, *Du but et de l'organisation*, 3.
65. Katharine Park, 'The Life of the Corpse: Division and Dissection in Late Medieval Europe', *The Journal of the History of Medicine and Allied Sciences* 50, no. 1 (1995): 111–32, on p. 129.

66. Russel C. Maulitz, *Morbid Appearances: The Anatomy of Pathology in the Early Nineteenth Century* (Cambridge: Cambridge University Press, 2002), 138–9.
67. Christoph De Spiegeleer, "Sterven, begraven en herdenken van koninklijke en politieke elites in België tussen 1830 en 1940: een culturele en politieke geschiedenis" (PhD diss., Vrije Universiteit Brussel, 2016), 140.
68. City council to Conseil des hospices, May 18, 1881, folder 107, Autopsies, Affaires générales, ASSB.
69. Charles Buls and Marcel Bots, *Het dagboek van C. Buls* (Ghent: Liberaal Archief, 1987), 103–4.
70. Conseil des hospices to city council, June 14, 1881, folder 107, Autopsies, Affaires générales, ASSB.
71. Firket, *Du but et de l'organisation*, 31.
72. Bourneville and Bricon, *Manuel de technique des autopsies*, 21.
73. Jean Wehenkel to Doucet, June 6, 1881, folder 107, Autopsies, Affaires générales, ASSB.
74. On degeneration in Belgium, see various essays in: Jo Tollebeek, Geert Vanpaemel and Kaat Wils, eds., *Degeneratie in België 1860–1940: een geschiedenis van ideeën en praktijken* (Leuven: Leuven University Press, 2003).
75. Dossier concernant l'organisation d'un service d'autopsie des cadavres des personnes décédées en dehors des établissements de l'administration 1881–1896, folder 107, Autopsies, Affaires générales, ASSB.
76. See, for example: Firket, *Du but et de l'organisation*, 46–50; Rapport sur les autopsies et les dissections, 1924, folder 107, Autopsies, Affaires générales, ASSB.
77. Firket, *Du but et de l'organisation*, 46.
78. Van Duyse, *Technique des autopsies*, 2.
79. On the importance of bodily integrity for clinicians (as opposed to forensic physicians) in France: Menenteau, 'Stigmata of the Autopsy', 20–38.
80. Letulle, *La pratique des autopsies*, 103. See also: Menenteau, 'Stigmata of the Autopsy', 35.
81. Letulle, *La pratique des autopsies*, 59.
82. Ahrén, *Death, Modernity and the Body*, 24.
83. Harris, *Manuel d'autopsies*, 84–5.
84. Firket, *Du but et de l'organisation*, 21.
85. Quotes from: Firket, *Du but et de l'organisation*, 21 and Rapport sur l'organisation du service des autopsies dans les hôpitaux de Bruxelles [1878], folder 107, Autopsies, Affaires générales, ASSB.
86. On the associations between face, hands and identity: Emmanuel Godeau, *L'esprit de corps: Sexe et mort dans la formation des internes en médecine* (Paris: Maison de science de l'homme, 2007), 31.

87. Firket, *Du but et de l'organisation*, 21.
88. Sandra Menenteau, *L'autopsie judiciaire: histoire d'une pratique ordinaire au XIXe siècle* (Rennes: Presses universitaires de Rennes, 2016), 266–88. Note that Menenteau emphasised that this 'moral protocol' did not matter in the case of forensic autopsies.
89. City council to Conseil des hospices, June 19, 1890, folder 292, Amphithéâtre, autopsies et dissections, Fonds du directeur de l'hôpital St. Pierre, ASSB.
90. Commission des hospices de Gand to Conseil des hospices de Bruxelles, December 26, 1881, folder Gasthuis Allerlei 1, Autopsies, ASSL.
91. Bourneville and Bricon, *Manuel de technique des autopsies*, 27.
92. MacDonald, *Possessing the Dead*, 100–2.
93. On Vienna: Buklijas, 'Cultures of Death', 585–94. On Berlin: Prüll, 'No Law, No Rights?' 34–5. On Oxford and Bristol: Hutton, *The Study of Anatomy*, 56–70. On the British colony Van Diemen's Land: Helen MacDonald, *Human Remains: Dissections and its Histories* (New Haven and London: Yale University Press, 2006), 136–82.
94. MacKinnon, 'Bodies of Evidence', 80–93; Cullen, 'Post-Mortem in the Victorian Asylum', 280–96; Wright, Jacklin and Themeles, 'Dying to get out of the Asylum', 608–18; Andrews, 'Death and the Dead-House', 6–26; Engstrom, *Clinical Psychiatry*, 96–8.
95. City council to Conseil des hospices, June 19, 1890, folder 292, Amphithéâtre, autopsies et dissections, Fonds du directeur de l'hôpital St. Pierre, ASSB.
96. Extrait du règlement d'ordre intérieur de l'hôpital de Bavière, August 28, 1869, folder Gasthuis Allerlei 1, Autopsies, ASSL; Bourneville and Bricon, *Manuel d'autopsies*, 27.
97. Menenteau, 'Stigmata of the Autopsy', 23.
98. Goubert, *Manuel de l'art des autopsies*, xi.
99. Richard Heschl to Jean Wehenkel, s.d. [circa 1880], folder 107, Autopsies, Affaires générales, ASSB. See also: Buklijas, 'Cultures of Death', 585–94.
100. See Chap. 3.
101. Rapport sur les autopsies et les dissections, 1924, folder 107, Autopsies, Affaires générales, ASSB.
102. Dossier concernant la plainte par le collège au sujet de ce qui s'est passé à l'hôpital St. Pierre à la suite du décès de Henri De Ro, 1902, folder 133, Liasse divers 1896–1902, Affaires générales, ASSB.
103. Constant Vanlair to Conseil des hospices, August 5, 1880, folder 107, Autopsies, Dossier concernant l'organisation du service des autopsies dans diverses villes du pays et de l'étranger, Affaires générales, ASSB.

104. André Berger, *Guide de l'étudiant à l'hôpital: examen clinique—autopsies* (Paris: Masson, 1893), 29; Firket, *Du but et de l'organisation*, 3.

105. This was similar to the architecture of autopsy rooms and morgues in asylums: Leslie Topp, 'Complexity and Coherence: The Challenge of the Asylum Mortuary in Central Europe, 1898–1908', *Journal of the Society of Architectural Historians* 71, no. 1 (2012): 9–14.

106. Firket, *Du but et de l'organisation*, 42–7.

107. Dépôts mortuaires des hôpitaux: construction, 1886, folder S107/2, ASB Bienfaisance, CAB.

108. Service d'hygiène no. 939 Amphithéâtre anatomique de l'hôpital St Jean, 1883, folder 107, Autopsies, Affaires générales, ASSB.

109. Dossier concernant extrait d'un rapport d'un médecin du bureau d'hygiène signalant la négligence qu'un garçon d'amphithéâtre apporte dans l'exécution de sa besogne, 1896, folder 107, Autopsies, Affaires générales, ASSB.

110. Louis Bertrand, 'A l'hôpital', *Le Peuple*, December 7, 1891. On similar ideas in France, see: Menenteau, *L'autopsie judiciaire*, 249.

111. Letulle, *La pratique des autopsies*, 18.

112. Firket, *Du but et de l'organisation*, 49.

113. Adolphe Samyn, Construction d'un dépôt mortuaire à l'hôpital St. Pierre, 1886, cartes et plans, ASSB.

114. Specifications concerning the construction of the morgue of St. Pierre, April 9, 1886, folder 34, Cultes 1811–1895, Affaires générales, ASSB. See also: Letulle, *La pratique des autopsies*, 17–18 and 47.

115. Firket, *Du but et de l'organisation*, 43.

116. Petition sent to city council, January 15, 1874, folder 22, Anatomisch Amfitheatre, ASSL.

117. Conseil des hospices to Infirmerie, August 18, 1886, folder 34, Cultes 1811–1895, Affaires générales, ASSB.

118. City council to Conseil des hospices, August 23, 1883, folder 107, Autopsies, Affaires générales, ASSB.

119. Conseil des hospices to city council, March 9, 1900, folder 50002 (641/8), Lijkenvervoer van en naar het anatomisch instituut van de universiteit 1844–1940, Modern Archief, CAL.

120. Topp, 'Complexity and Coherence', 8–41.

121. Dépôts mortuaires des hôpitaux: construction, 1886, folder S107/2, ASB Bienfaisance, CAB.

122. Specifications concerning the construction of the morgue of St. Pierre, April 9, 1886, folder 34, Cultes 1811–1895, Affaires générales, ASSB.

123. Thomas F. Gieryn, 'Two Faces on Science: Building Identities for Molecular Biology and Biotechnology', in *The Architecture of Science*, eds. Peter Galison and Emily Thompson (Cambridge: MIT Press, 1999),

423–55, on p. 428; Allan M. Brandt and David C. Sloane, 'Of Beds and Benches: Building the Modern American Hospital', in *The Architecture of Science*, 281–308.

124. On the (controlled) embedment of scientific buildings within the city, see: Sven Dierig, Jens Lachmund and J. Andrew Mendelsohn, 'Introduction: Toward an Urban History of Science', *Osiris* 18 (2003): 1–19, on pp. 13–16.

125. Auguste Brunin to University administration, May 23, 1900, folder 107, Autopsies, Affaires générales, ASSB.

126. For example: Demande de pouvoir autopsier les corps déposé à la disposition du Parquet, au cas où l'autopsie médico légale n'a pas lieu, 1906, folder 107, Autopsies, Affaires générales, ASSB; La demande de Mr le Bourgmestre de voir mettre un cadavre à la disposition du délégué du service d'Hygiène de la ville, 1898, folder 107, Autopsies, Affaires générales, ASSB.

127. Auguste Brunin to University administration, May 23, 1900, folder 107, Autopsies, Affaires générales, ASSB

128. Buklijas, 'Culture of Death', 586; Karin Stukenbrock, *'Der zerstückte Cörper': Zur Sozialgeschichte der anatomischen Sektionen in der frühen Neuzeit 1650–1800* (Stuttgart: Franz Steiner Verlag, 2001), 137–49; Cay-Rüdiger Prüll and Marianne Sinn, 'Problems of Consent to Surgical Procedures and Autopsies in Twentieth-Century Germany', in *Historical and Philosophical Perspectives on Biomedical Ethics: From Paternalism to Autonomy?* eds. Andreas-Holger Maehle and Johanna Geyer-Kordesch (Farnham: Ashgate, 2002), 73–93.

129. Thomas F. Gieryn, *Cultural Boundaries of Science: Credibility on the Line* (Chicago and London: Chicago University Press, 1999), 23–5.

130. Dossier concernant la préférence donnée à l'université pour les cadavres devant servir aux cliniques, 1878, folder 107, Autopsies, Affaires générales, ASSB.

131. Register of bodies brought to the Institut d'anatomie, 1920–1968, Private Archives of Stéphane Louryan, Brussels.

132. Rapport sur l'organisation du service des autopsies dans les hôpitaux de Bruxelles [1878], folder 107, Autopsies, Affaires générales, ASSB.

133. Quotes from: Firket, *Du but et de l'organisation*, 12; Paul Héger, Eugène Mahaux and Charles Delstanche, Le Service spécial des autopsies à Vienne, October 18, 1872, folder 107, Autopsies, Affaires générales, ASSB.

134. Lorraine Daston and Peter Galison, *Objectivity* (New York: Zone Books, 2007), 115–90, on p. 120.

135. For example: Réclamations de MM. Rommelaere et Stiénon au sujet du service des autopsies, 1887–1888, folder 107, Autopsies, Affaires générales, ASSB.

136. Conseil des hospices to University board, October 15, 1887, folder 107, Autopsies, Affaires générales, ASSB.

137. Paul Héger, Eugène Mahaux and Charles Delstanche, Le Service spécial des autopsies à Vienne, October 18, 1872, folder 107, Autopsies, Affaires générales, ASSB.

138. Joseph Van Schoor, *Discours d'ouverture prononcés en séance publique le 14 octobre 1878* (Brussels: Mayolez, 1878), 23.

139. Firket, *Du but et de l'organisation*, 6.

140. Bourneville and Bricon, *Manuel de technique des autopsies*, 24.

141. Firket, *Du but et de l'organisation*, 10.

142. *Manifestation en l'honneur du docteur Daniel Van Duyse* (Ghent: Vander Haeghen, 1907), 17.

143. Jean Wehenkel to Conseil des hospices, May 3, 1885, folder 107, Autopsies, Affaires générales, ASSB.

144. Giolio Bizzozero and Charles Firket, *Manuel de microscopie clinique* (Brussels: Manceaux, 1883).

145. Réunion entre les délégués de l'université et les délégués de la commission d'assistance, February 12, 1932, folder 107, Autopsies, Affaires générales, ASSB.

146. Alison Kraft and Samuel J.M.M. Alberti, 'Equal though Different. Laboratories, Museums and the Institutional Development of Biology in Late-Victorian Northern England', *Studies in History and Philosophy of Biological and Biomedical Sciences* 34, no. 1 (2003): 203–36, on p. 230.

147. Quotes from: *BSAP* 8 (1863): 12–14; *BSAP* 11 (1866): 12; *BSAP* 34 (1885): 11.

148. Dhondt, *Un double compromis*, 129–31 and 285–300.

149. Paul Héger, 'Notice sur la vie et les œuvres de Willem Rommelaere', in *Rapport sur l'Année académique, 1914–1918* (Brussels: ULB, 1919), 41–55.

150. For example: Firket, *L'éducation médicale en Angleterre*; Gustave Verriest, 'De l'organisation des études médicales', in *Révision de la loi de 1876. Les programmemes des examens de sciences naturelles et de médecine*, Jean-Baptiste Carnoy (Leuven: Fonteyn, 1889).

151. Steve Sturdy, 'The Political Economy of Scientific Medicine: Science, Education and the Transformation of Medical Practice in Sheffield, 1890–1922', *Medical History* 36, no. 2 (1992): 125–59; Joel D. Howell, *Technology in the Hospital: Transforming Patient Care in the Early Twentieth Century* (Baltimore and London: Johns Hopkins University Press, 1996); Volker Hess, 'Raum und Disziplin: Klinische Wissenschaft

im Krankenhaus', *Berichte zur Wissenschafts-Geschichte* 23, no. 3 (2000): 317–29.

152. Andrew Cunningham and Perry Williams, 'Introduction', in *The Laboratory Revolution in Medicine*, eds. Andrew Cunningham and Perry Williams (Cambridge: Cambridge University Press, 1992), 1–13, on p. 11.

153. Morten Hammerborg, 'The Laboratory and the Clinic Revisited: The Introduction of Laboratory Medicine into the Bergen General Hospital, Norway', *Social History of Medicine* 24, no. 3 (2011): 758–75.

154. Steve Sturdy, 'Looking for Trouble: Medical Science and Clinical Practice in the Historiography of Modern Medicine', *Social History of Medicine* 24, no. 3 (2011): 739–57.

155. Question de la réunion des services d'autopsies des deux hôpitaux sous la direction unique du professeur d'anatomie pathologique, 1915, folder 107, Autopsies, Affaires générales, ASSB.

156. For example: *BSAP* 28 (1879): 10; *BSAP* 34 (1885): 3.

157. Behaeghel to Conseil des hospices, August 23, 1898, folder 107, Autopsies, Affaires générales, ASSB.

158. Jules Destrée to Conseil des hospices, August 18, 1893, folder 107, Autopsies, Affaires générales, ASSB.

159. Joseph Sacré to Conseil des hospices, March 14, 1903, folder 107, Autopsies, Affaires générales, ASSB.

160. Behaeghel to Vanderlinden, January 22, 1906, folder 107, Autopsies, Affaires générales, ASSB.

161. Dossier concernant la plainte de M le Dr Destrée de n'avoir pu pratiquer l'autopsie du cadavre d'une personne décédée dans son service, ce cadavre étant réclamé par l'Université, 1893, folder 107, Autopsies, Affaires générales, ASSB.

162. Emile Houzé to Conseil des hospices, September 19, 1893, folder 107, Autopsies, Affaires générales, ASSB.

163. Ahrén, *Death, Modernity and the Body*, 28–9.

164. Buklijas, 'Cultures of Death', 588–90.

165. Réunion entre les délégués de l'université et les délégués de la commission d'assistance, February 12, 1932, folder 107, Autopsies, Affaires générales, ASSB.

166. Cf. Gieryn, *Cultural Boundaries of Science*, xii.

167. For example: Boulengier to Conseil des hospices [1887], folder 106, Autopsies, Affaires générales, ASSB.

168. Max to Conseil des hospices, July 3, 1898, folder 107, Autopsies, Affaires générales, ASSB.

169. Dustin to Conseil des hospices, January 17, 1924, folder 107, Autopsies, Affaires générales, ASSB.

170. In Brussels, dissections happened in Parc Léopold from 1896 onwards. In Liège, the anatomical institute was moved to a different location (Rue des Pitteurs) in 1885.

171. Règlement pour assurer la conservation des corps abandonnés à la dissection, 1898, folder 107, Autopsies, Affaires générales, ASSB.

172. Vandervelde to Conseil des hospices, September 10, 1898, folder 107, Autopsies, Affaires générales, ASSB.

173. Note pour le conseil, April 9, 1898, folder 107, Autopsies, Affaires générales, ASSB.

174. Feron to Conseil des hospices, May 28, 1898, folder 107, Autopsies, Affaires générales, ASSB.

175. Conseil des hospices to University board, May 13, 1898, folder 107, Autopsies, Affaires générales, ASSB.

176. Rapport sur l'organisation du service des autopsies dans les hôpitaux de Bruxelles [1878], folder 107, Autopsies, Affaires générales, ASSB.

177. For example: Roger Brocas, *Le droit d'autopsie: étude historique et juridique* (Paris: Faculté de droit, 1938), 96–122; Bourneville and Bricon, *Manuel de technique des autopsies*, 23.

178. Buklijas, 'Cultures of Death', 585–94; Ahrén, *Death, Modernity and the Body*, 27–30; MacDonald, *Possessing the Dead*, 100–2.

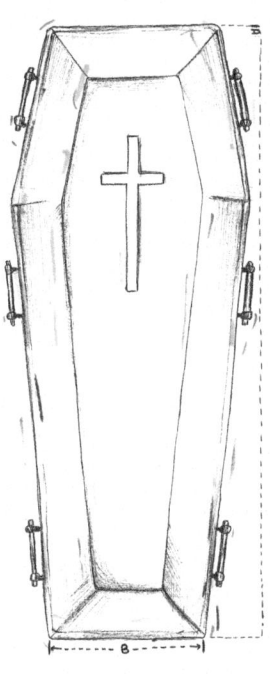

CHAPTER 5

The Jar and the Coffin: Keeping
and Disposing of the Dead

1   PREAMBLE: 'IT IS NOT PERMITTED TO PLACE TWO
CORPSES IN THE SAME COFFIN', 1886

It is a cloudy Friday in October 1886 in the maternity home of Brussels.
After hours of painful labour, Clémence Joris gives birth. Her daughter is
dead. When the doctors tell her and her husband about their loss, they
weep bitterly. In some ways, having a stillborn child is worse than losing a
child that has lived. The daughter of Clémence Joris and Jan Van
Cauwelaert will never receive a name, baptism or consecrated grave.

Overwhelmed by emotions, the mourning parents forget to claim the
remains of their child. The next day, they want to make arrangements for
the funeral. But nobody can tell them where the body of their daughter is.

*   *   *

Henri Dorrekens dies in the hospital of St. Pierre on the 30th of October.
He had not expected death: he was only 17 years old, not yet a member of
a burial society, and had not written a testament. Nobody claims his
remains. He is brought to the anatomical theatre the next day. Students
dissect him for five days straight.

On the 5th of November, an employee of the university collects all bits
and pieces of his body. Most of them are lying on a bier with his name tag

© The Author(s) 2019   209
T. Claes, *Corpses in Belgian Anatomy, 1860–1914*, Medicine and
Biomedical Sciences in Modern History,
https://doi.org/10.1007/978-3-030-20115-9_5

attached to it. A few body parts are elsewhere scattered around the dissecting room, but can still be identified because they are stamped. The coffin of Henri Dorrekens remains rather empty: approximately three fourths of his body have been wasted during dissection. Bodily fluids have leaked out, organs have been removed, soft tissue has decomposed.

The university employee looks at the empty coffin and decides he might as well fill it up with unidentified waste from the anatomical theatre. An unnamed newborn baby had just arrived. It had served for experiments in the physiological laboratory until it began to decompose. He throws it in the coffin, hence entwining the destinies of Henri Dorrekens and of Clémence Joris' daughter even though they were complete strangers.

*   *   *

The coffin is brought back to the morgue of St. Pierre for the funeral. Before closing it permanently, the *garçon d'amphithéâtre* opens the lid one last time. He sighs when he sees the contents. A few years earlier, he was reprimanded because he had buried two children together in a coffin for adults. He had been out of children's sizes, and it seemed extravagant to give them a full-size coffin each. His boss had warned him to never bury two bodies together again.

He informs his supervisor, who sends a letter to the head of the hospital. Don't worry, he replies. It is true that everybody should in principle receive a separate coffin and grave. But it occurred frequently that the university sent 'all sorts of shapeless remains' back in one coffin. The burial goes ahead as planned.

*   *   *

It has been over two weeks since that dreadful night in the hospital and the daughter of Clémence Joris and Jan Van Cauwelaert is still missing. They write a complaint to the mayor, Charles Buls. They want to find out what happened to their daughter.

Buls takes their complaint seriously. He is convinced that if one scandal after the other continues to happen in hospitals, people will only go there reluctantly. Interesting facts would be lost for science if the reputation of the hospital did not improve. He picks up the phone and starts calling.

The head of the hospital of St. Pierre picks up. He remembers the decomposed newborn in the coffin of Henri Dorrekens. He puts the receiver down, then dials the number of the head of the maternity home. They make a timeline. Yes, that probably was the child they are looking for. The head of St. Pierre writes a letter to the mayor with his apologies. He explains that he did not know anything about the provenance of the newborn child and therefore did not intervene.

Charles Buls formally replies to the commission of hospitals. Make sure that all hospital employees know that every corpse should receive a separate coffin, he instructs. The complaint of Clémence Joris and Jan Van Cauwelaert, he asserts, 'is absolutely sound'.

* * *

A few days later, the *garçon d'amphithéâtre* receives a small note from the commission of hospitals. 'It is not permitted to place two corpses in the same coffin', it reads. In addition, he is reminded that he is not allowed to send dead bodies—'not even of foetuses'—to the university without checking with the autopsy service first.

We do not know if Clémence Joris and Jan Van Cauwelaert ever learned the truth about what happened. Perhaps their daughter and Henri Dorrekens still repose somewhere together in an unknown grave.[1]

2   INTRODUCTION

An anonymous young woman's head was temporarily removed from the anatomical museum of Ghent University in 2015. Almost two centuries before, she had died in the Bijloke hospital. The anatomist Adolphe Burggraeve took her body from the morgue and decapitated her. He dissected the vessels and nerves of her head and hand, and injected colourful substances in her capillaries. He placed these dissected pieces in a jar containing preservation fluid, while the rest of the body was brought to the cemetery.[2] Now, the young woman's head was to become the showpiece of the exhibition 'Post-Mortem', which confronted anatomical specimens with works of art inspired by them. Through the interaction between science and art, the exhibition invited visitors to think about medical ethics, the beauty of the body and their own mortality. Of all the pieces on display, the woman received the most attention. Visitors responded with a variety of emotions, ranging from disgust to curiosity. Many of them wondered who she had been and how she ended up in a jar. In a way, they animated the inanimate: their imagination brought the woman back to life, as her preserved face received a new narrative.

Many histories of anatomy end with dissection. As the bulk of research has focused on the acquisition of corpses, the 'consequences' of post-mortem examinations have often been overlooked. Social historians in particular scrutinised the gender, race and class-specific traffic of dead bodies, following the ways in which corpses ended up on, rather than left, the dissecting table. Yet the journey of human remains carried on after fragmentation. Bodies or body parts could be preserved and cata-logued, after which they could be reinterpreted, re-catalogued or even re-dissected. They could be buried, partially buried or disposed of in other ways. Anatomical remains that were first preserved and put on display could still be interred or thrown away in a later stage of their post-mortem existence. In some instances, as in the young woman's case, they were lost from sight and rediscovered, and ended up on dis-play again.

What happened to bodies after they left the table? On the one hand, this chapter looks at the preservation and use of anatomical specimens in collections; on the other, it discusses the disposal of dissected and autop-sied remains. My argument is both historical and conceptual. First and foremost, this chapter shows how the developments described in the pre-vious chapters—the transformation of the medical landscape, the intro-duction of body donation and the distinctions between dissection and autopsy—influenced the final stages of the journeys of anatomical remains. Another point of attention is shifts in burial rituals and their impact on the treatment of the dead within the hospital. The emergence of a more per-sonal culture of death, in which the body took centre stage, not only transformed the middle- and upper-class burial, but also altered the face of the anatomy burial.

While drawing attention to the influence of these historical changes on the preservation and disposal of anatomised bodies, this chapter endorses the views of various scholars, who have argued that human remains change meanings dependent on the setting and the beholder.[3] Helen Lambert and Maryon McDonald, for instance, have posited that corpses are always sub-ject to (re)interpretation because their meanings are the result of a continu-ous negotiation between differing views of the body and of death.[4] Similarly, Samuel Alberti has contended that anatomical specimens are 'polysemic', inciting various associations and responses in different contexts. After their preservation, bodies continue to have different and sometimes opposite values, whether social, religious or scientific.[5] The goal of this chapter is to make clear that anatomical remains were constantly reinterpreted, whether as objects in a changing scientific environment or as subjects in a changing death culture.

Firstly, this chapter reconstructs the trajectories of anatomical specimens. Dissection, autopsy or surgery was just the beginning of a long journey for those body parts that became part of collections. Anatomical specimens shifted shapes and meanings as they moved from dissecting or autopsy table to museum shelf, in between different shelves, or from shelf back to table for reexamination. Secondly, there follows a discussion of the disposal of dissected and autopsied bodies. Whereas some anatomical remains were ritually buried in the presence of their family and friends, others were categorised as medical waste and thrown away: drained off, burned or interred without ceremony. Even for those remains that anatomists intended to keep forever, the jar was rarely a final resting place.

3   The Jar: Keeping Anatomical Specimens

In 1884, Hector Leboucq wrote a report on the anatomical collections of Ghent. The occasion was the opening of a new museum space, which brought the collections of the hospital and the university together. Leboucq praised the design of the building, which the well-known architect Adolphe Pauli had drawn after the model of prestigious foreign museums. Leboucq also made a critical assessment of the specimens at hand. When discussing the collections of descriptive anatomy, he said a few words about 'one of the prettiest pieces of Burggraeve', namely 'a young woman's head and neck, injected through the capillaries and with vessels and nerves dissected'. Although Leboucq acknowledged that the specimen was beautifully made and preserved, he ultimately decided that 'it was of no interest except for it being old'.[6]

Leboucq's judgement contrasted sharply with the opinion of both the medical community and lay visitors of the anatomical cabinet half a century before. Much like the visitors of the Post-Mortem exhibition 200 years later, they had held the specimen of the young woman's head in great admiration. The members of the Medical Society of Ghent (*Société de Médecine de Gand*) had praised the specimen for its lifelike appearance and finesse, which they saw as evidence of the anatomical skills of its maker. Journalists and lay visitors had interpreted the rosy cheeks of the dissected woman as a victory over death.[7] In the new anatomical museum, however, different specimens took priority. Whereas anatomists such as Burggraeve still made elegant preparations of the healthy body as a means to bolster their reputation in the early nineteenth century, anatomists from Leboucq's generation preferred to occupy themselves with the preservation of pathological tissues.

The following pages are about the evolutions through which the young woman's head became a historical rather than a medical artefact. By following the spatial and conceptual trajectories of anatomical remains, it will become evident that the reconfiguration of the medical landscape in the second half of the nineteenth century impacted both the preparation of new anatomical specimens and the use of older collections.

From Table to Shelf

In his research on medical museums in nineteenth-century Britain, Alberti has drawn attention to historians' neglect of the different acquisition routes of preserved human remains, although these distinctions 'had significant ramifications for the value and use of specimens'.[8] Anatomical specimens were not always gathered during dissections. On the contrary, Alberti has found that the bulk of museum collections in the second half of the nineteenth century came from autopsied patients. This was a result of anatomists' evolving interests: whereas dissections had been important for the supply of elegant preparations of healthy anatomical structures, they rarely revealed diseased body parts. The autopsy slab consequently replaced the dissecting table as the primary provider of the anatomical museum when pathological anatomy gained importance as a discipline.[9]

Yet anatomists still removed body parts during and after dissection, and these specimens were, as Alberti asserted himself, different from those that were procured in other ways. Whereas dissections were indeed seldom used to harvest museum pieces anymore, they still provided rich pickings for what Hieke Huistra has called 'handling' collections. In her masterful study on the afterlife of the anatomical collection of Leiden, Huistra put forward that these specimens, unlike specimens intended to be displayed in museums, were not meant to be watched from a distance or kept forever, but to be used by students until they were worn to shreds.[10] As primarily didactic tools of an accessory science, they were prepared, used and valued differently.

In Brussels and other Belgian university cities, healthy body parts often became handling specimens after the dissecting course. Detached organs and body parts that students had already (partially) dissected in hands-on exercises could be subsequently preserved and used for anatomical demonstrations. These handling specimens were mostly prepared by the support staff of the professor of anatomy, particularly by the *garçon d'amphithéâtre* (the factotum of the anatomical theatre, responsible for the management of dead bodies before and after dissection) and the prosector (responsible for the making and maintenance of anatomical specimens).[11]

Specimens for handling and specimens for display required different methods.[12] As early as 1818, in his bestseller *Manuel de l'anatomiste*, the Parisian anatomist Jacques-Pierre Maygrier argued that methods asking a lot of care and patience were valuable for the preparation of pieces 'destined to embellish anatomical cabinets', but 'useless' for the teaching of students.[13] Leboucq, too, stressed that different specimens asked for different techniques. Preparations of the nervous system, for instance, were better preserved in alcohol if they were to become *pièces de collection*, as this method produced 'prettier' results. Nevertheless, he preferred to prepare pieces for anatomical demonstrations with glycerin, because this required less effort and money. In addition, Leboucq noted that glycerin specimens were easier to handle, because they could be carried around (unlike heavy jars with alcohol) and hardened the muscles, which made them more robust and hence better for frequent use.[14]

Collections for handling also looked different. The late nineteenth-century anatomical museum functioned as an encyclopaedia of disease, in which students and researchers could observe pathologies that they might seldom or never come across in clinical practice. The aim of the museum was both to bring together and to perpetuate clinical cases from the past. To fulfil this purpose, anatomists in general tried to preserve the diseased body in a realistic manner.[15] For handling specimens, conversely, clarity preceded over accuracy. Since they were primarily didactic tools, anatomist left out 'useless details' and highlighted other features, for instance, by 'painting parts of osteological pieces' or by 'injecting colourful substances in the vessels'.[16]

Because they were didactic tools, handling preparations resided in lecture rooms rather than museum halls. In Ghent, for example, the auditorium of the anatomical institute contained five big armoires for the storage of frequently used demonstrative pieces of normal anatomy, whereas other specimens were displayed in the anatomical museum.[17] Handling specimens were also given a different kind of jar.[18] Smaller organs or tissues ended up in stoppered jars that could be opened easily. To save expensive preparation fluid or to give students an idea of anatomical variation, these jars sometimes housed multiple organs. Bigger body parts—for example, partially dissected torsos, arms or legs—were stored together in solid, rectangular containers. Zinc tanks the size of a coffin in particular were, in the words of Auguste Brunin, teacher of anatomy at the Free University of Brussels, 'highly useful for the conservation of anatomical pieces used in the dissecting course'.[19]

All these distinctions between handling and display specimens—different preservation techniques, styles and means of storage—were consequences of the same cause: their different use. Students and teachers actively used handling collections as anatomical teaching aids. Huistra has shown that handling specimens were meant to be taken out of their jars. When it came to these specimens, hands were as important as eyes: students passed them around, turned them around to get a look from different angles, pushed upper layers of tissue aside in order to observe the structures underneath, or paid attention to characteristics such as elasticity and texture. By not only seeing, but also touching the specimens, students were able to remember lessons better, and to train all their senses for their future career as physicians.[20]

Also within their jars, handling specimens were moved around more often than display specimens. Professors—of anatomy, but also of disciplines such as gynaecology, ophthalmology, public hygiene or surgery—frequently removed them from their shelves and took them with them to their courses. It was so difficult to keep track of handling specimens that in 1879 the conservator of the anatomical collection in Brussels, who 'had again lost seven specimens', proposed to introduce a similar system as the one already in place in the library. In his opinion, teachers should only receive permission to take a specimen with them if they signed a receipt. Much like books, anatomical specimens that were used for demonstrations seem to have been lent out, carried around and sometimes returned too late.[21]

As a result of their mobility, handling specimens were inevitably damaged. As they moved between students' hands and travelled on bumpy roads between lecture rooms, many of them eventually fell apart. Especially outside of their jars, specimens were subject to daily wear and tear.

However, anatomists did not care deeply about their deterioration. It appears that they saw specimens procured during dissections as throw-away articles: made cheaply, used for a limited period of time and discarded. If they were worn out, they were simply made anew. In Brussels, an inventory of 'decayed anatomical specimens' has been preserved (Illustration 5.1). By means of this register, prosectors were informed of pieces 'that had been lost during last year's courses' and 'that had to be replaced in due time'.[22] The replacement of handling specimens, in other words, was routine.

Late nineteenth-century preparation manuals, too, suggest that preserved pieces of the healthy body were disposables. Readers of the French

Illustration 5.1 Inventory of 'decayed anatomical specimens', 1873. © Archives of the Social Services of Brussels. From: Folder 69, Collections scientifiques, Affaires générales

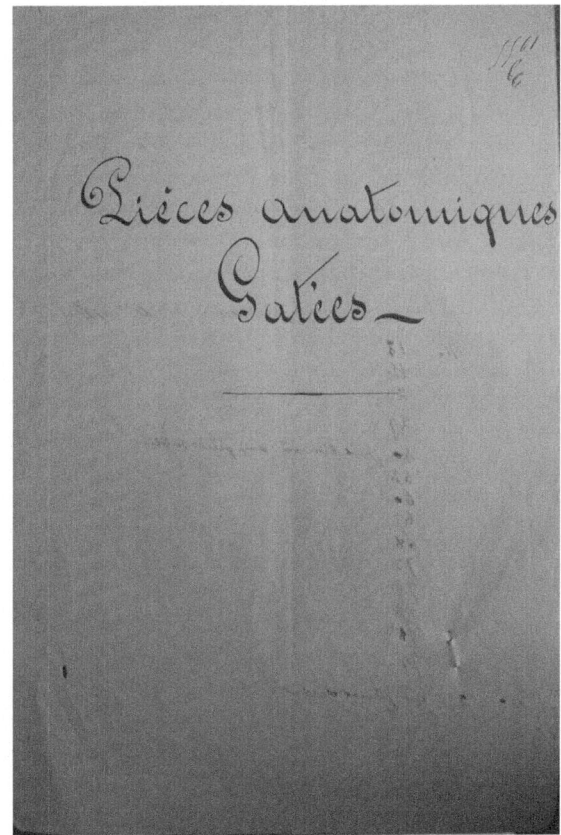

Nouveau manuel complet du naturaliste préparateur, for example, could learn how to preserve human remains, plants or animals. Remarkably, the author Pierre Boitard (who, admittedly, was a naturalist) urged them to be more careful when conserving the latter:

> It is always easy to obtain a new human cadaver in the anatomical theatre. [...] Conversely, you often only have one opportunity in your whole life to get your hands on the cadaver of a rare animal.[23]

The dissecting room, so it seems, was a bottomless pit of healthy anatomical structures that could be endlessly replaced. Yet much like rare animals were a once-in-a-lifetime opportunity for naturalists, rare anomalies were cherished by pathologists. Whereas dissected bodies only provided anatomists with disposable didactic tools, autopsied bodies, in the words of the statement of purpose of the Society for Pathological Anatomy of Brussels, enshrined 'invaluable scientific treasures'.[24]

Indeed, as Alberti has shown, most museum specimens, considered valuable enough for (expensive) permanent preservation and display, originated from autopsies in the second half of the nineteenth century.[25] In her research on the Strasbourg collection of pathological anatomy, Tricia Close-Koenig consequently has related the growth of the museum to an increase in the number of autopsies.[26] In Belgium, autopsy services from the 1880s onwards played a pivotal role in the expansion of pathological collections. Before their foundation, clinicians could in principle only retrieve pathological parts from deceased patients if their families had not claimed the body, although several sources suggest that they sometimes clandestinely removed interesting organs 'while the body was already laid in a coffin'.[27]

However, since these removals of organs could only happen covertly, pathological collections remained rather small. When the Ghent physician and professor Joseph Guislain compared Belgian collections to foreign examples in 1840, he even concluded that Belgium had no pathological museum worthy of this name.[28] Similarly, André Uytterhoeven, surgeon at the Brussels St. Jean hospital, deplored that many interesting pathological pieces 'were still left to the grave' in 1852. In his opinion, autopsies had to become both more frequent and elaborate in order to get hold of pathological specimens. Only in this way, Brussels hospitals would be able to establish impressive museums along the lines of foreign institutions.[29]

As briefly mentioned in the previous chapter, the expansion of the pathological collection indeed became a key objective of the autopsy service of

Brussels. Shortly after its foundation in 1878, its personnel promised that through their efforts, the pathological museum would soon become a 'scientific monument', comparable to the Hunterian in London or the museum of the *Pathologisch-Anatomische Institut* in Vienna.[30] From 1890 in particular, when the city council gave pathologists the right to autopsy all deceased hospital patients, pathological collections increased in size. In 1892, Jean-Hubert Thiry, then president of the Society for Pathological Anatomy, contended that Brussels finally had a collection 'that could compete with the most prestigious museums of Europe'.[31] He linked the growth of the museum to the new regulation on autopsies, which had enabled pathologists to make the most of the richness of the wards.[32]

Collection building was indeed easier under the new regulations, as anatomical specimens could be legally harvested from both claimed and unclaimed bodies without permission from the deceased or his or her family.[33] Since there were no specific laws on the partial preservation of bodies, pathologists only had to take into account the general rules on autopsies and burials.[34] This meant that they could always preserve internal body parts, unless the deceased or his or her family had lodged a spontaneous objection (which rarely happened).[35] In order to safeguard the outward appearances of the cadaver, they replaced retrieved body parts by pieces of cloth or newspapers.[36] Only anatomical specimens that could not be cut out without visible mutilations, could not be taken from claimed bodies without consent.[37]

Burial regulations also did not hinder pathologists, since they could keep smaller specimens without interfering with the timely inhumation of the remaining parts of the body. In fact, pathologists only broke the law if they preserved bodies as a whole. For example, they had to ask the city council for special permission to preserve foetuses after six months of gestation because their bodies had to be buried according to the law. In Brussels, the city council always granted these requests if the foetus was stored in a public museum (in the university or the hospital) and served the education of medical students.[38] The city council also occasionally gave permission to preserve the bodies of children or adults, mostly as skeletons. In 1890, for instance, the city council allowed the commission of hospitals 'to not inhume' the body of Georges Roberts, a 23-year-old man from Baltimore in the United States. 'For scientific purposes', his body was prepared in its entirety for the anatomical collection.[39]

Autopsy services were at the basis of the enlargement of anatomical collections not only because they gave access to more corpses, but also because they professionalised preparation practices. Whereas the preservation of pathological tissues had not been of primary importance for clinicians, it became a core task for pathologists working at autopsy services. It appears that an increasing number of body parts were kept aside when the scope of the autopsy expanded. As the autopsy became more complete, pathologists or interns working at autopsy services often preserved multiple pathological pieces from the same cadaver. In this way, collections came to reflect the ideal of 'pathological physiology', in which the dynamic course of disease, rather than its static result, was the main object of study. In a case of bronchiolitis (an inflammation of the smallest air passages of the lungs, most commonly occurring in young children), for example, a certain Dr. Roger prepared not only the lungs, but also a part of the left testicle, a bit of cellular tissue of the left arm, nerves of the same arm, the liver and the spleen. By relating the primary lesion, the lungs, to secondary lesions in different regions of the body, Roger wished to elucidate the progress of the disease.[40]

The microscopic research of autopsy services, too, led to an expansion of collections. As cellular pathology gained importance, macroscopic specimens were often supplemented with microscopic slides. Cabinets with drawers full of slides found their way to anatomical museums. In 1886, for instance, the pathological collection of Ghent University counted 1954 microscopic specimens. Two years later, this number had already risen to 2236.[41] From 1873 onwards, the Society for Pathological Anatomy of Brussels organised courses in which students learned how to interpret and preserve microscopic specimens, leading to an increase in the number of histological preparations within its collections.[42] In the same year, the president of the society even posited that no macroscopic pathological observation was complete without its microscopic counterpart.[43]

Not only microscopic studies, but also case histories, autopsy reports and biochemical analyses increasingly accompanied macroscopic specimens. In his research on the presentation of anatomical specimens in scientific societies in nineteenth-century Belgium, Joris Vandendriessche has argued that these elaborate reports attested to a changing scientific culture, in which accuracy and quantification became the shared ideals that motivated the preparation and preservation of human remains. In his view, accurate descriptions became more important than the rarity of specimens in the late nineteenth century. As a result, collections did not mainly consist of curiosities anymore. They became, as Vandendriessche eloquently pointed out, 'collections of ordinary facts'.[44]

Much like the autopsy itself, the anatomical museum revealed the changing boundaries of the scientific enterprise. The mere acquisition and preservation of rare specimens no longer counted as a true contribution to medical science. In order to be scientific, presentations of anatomical specimens had to go further than the mere description of form. In 1885, for instance, the jury of the student competition of the Society for Pathological Anatomy of Brussels decided that the prize could not be rewarded because the specimens were not accompanied by clinical information and microscopic research. Without this information, the jury report stated, 'most of these pieces are of no interest'.[45] These stringent standards were a response to the broader scientific evolutions that have been discussed in the introductory chapter of this book: by combining macroscopic specimens with both microscopic analyses and clinical case studies, the society's members refuted the image of (pathological) anatomy as an accessory science. Whereas microscopic research related specimens to the science of cellular pathology, case histories linked specimens to clinical practice.

Remarkably, these detailed clinical case studies often contained personal information on patients. Various historians have claimed that patients disappeared in collections. In their view, anatomical specimens were fungible (meaning that they were interchangeable with other specimens that represented the same condition) and anonymous. If specimens were identified with a person at all, this was mostly with the maker of the specimen rather than with the deceased to whom the body part once belonged.[46] In the words of historians Lisa O'Sullivan and Ross L. Jones, for instance, the creation of anatomical specimens went hand in hand with 'the stripping of individual identity'.[47] Jonathan Reinarz, too, has linked the 'disappearance of the sick man' in the second half of the nineteenth century to the objectification of human remains in anatomical collections. He has contended that anatomical museums transformed the individual patient into a universal object of investigation.[48]

The Brussels case, however, suggests that the patient became more present in collections from the 1870s onwards. Inventories, for example, contained case histories or references to medical publications that had discussed the patient's condition during life.[49] In the proceedings of scientific societies, too, specimens were linked to clinical information, often mentioning the name of the patient alongside other personal characteristics, such as age, occupation, marital status, physical appearances and 'temperament' (mostly captured in humoural terms, even as late as the 1890s). In these patient histories, specimens clearly related to individuals, such as Marie Moerenhout, Adrienne Vervleeck, Catherine Vanderhouten, Henri Van den Plas or Anne-Marguerite Spilleboul 'who for a year had slept in room 15' of the St. Pierre hospital.[50] In a few instances, pathological specimens were even presented side by side a drawing or picture of the living patient or next to a post-mortem photograph of the deceased patient.[51]

The relation with individual patients was most clear when specimens were taken during surgery. In meetings of scientific societies, surgeons occasionally presented removed body parts—mostly tumours and amputated members—together with surviving patients. In 1890, for instance, a certain Dr. Bastin brought a living child to a meeting of the Society for Pathological Anatomy of Brussels. Two weeks before, he had given the child surgery in order to cure a strangulated hernia, a medical condition in which a portion of the small intestine pushes through an area within the lower abdominal muscles. By presenting pieces of removed intestines next to the pain-free patient, the surgeon proved that his intervention had been successful, both from a medical and humane perspective.[52]

The most important reason for the continued individuality of pathological specimens was their link with the clinic. Students learning anatomy were taught to see the body as a universal entity that could be generalised. The dissecting knife divided the body into fungible, healthy body parts that stood for human anatomy as a whole. For pathologists working in hospital services, however, individual case histories were more important because they promised to link their work with therapeutics and prognosis. Pathological specimens were more particular not only because they were harder to come by than healthy parts, but also because they remained related to a patient's history. As Lorraine Daston and Peter Galison have asserted with regard to pathological atlases, representations of disease were mostly 'characteristic': not the 'pure phenomenon' or the ideal, but an individual object that at the same time represented similar objects.[53] Embedded in textual records, pathological specimens simultaneously stood for an individual patient and for a whole class of similar patients in the future.

As such, pathological specimens and their textual reflections bridged different epistemological ideals. Daston and Galison have convincingly shown that what constituted scientific or genuine research changed profoundly throughout the nineteenth century (much like it did before, and still is doing today). Whereas early nineteenth-century scientists mostly sought 'true-to-nature' knowledge, a process that required their mediation as they had to interpret nature in order to uncover its underlying types and principles, the ideal of 'mechanical objectivity' grew in significance as the century progressed. Idealising intervention became a sin. The scientist, who had once been quintessential for the construction of knowledge, now had to restrain the impulse to intervene, interpret or perfect, so that nature could speak for itself.[54]

Significantly, these different epistemic ideals required different objects of study. Whereas eighteenth- and early nineteenth-century scientists were after types or ideals that showed the essence of nature, late nineteenth-century scientists shifted their attention to individual items that were unaltered by the observer and therefore 'objective'. In the words of Daston and Galison, the 'object-as-particular' came to replace the 'object-as-type'.[55] Against this background, the pathological specimen became a hybrid of idealising and naturalising modes. As a representation of an individual patient that had to inform physicians about similar patients in the future, the pathological specimen marked a transition between successive scientific traditions.[56]

The persistent link between the pathological specimen and the individual patient followed not only from the rise of pathology or the ideal of mechanical objectivity, but also from the increased attention for accounting practices within the hospital. Specimens mostly had a number rather than a name, but these numbers in many instances were retraceable to a name through a complex ensemble of paper records. Close-Koenig has recently drawn attention to the importance of recordkeeping for the history of anatomical collections, stating that paper records not only accompanied collections of preparations, but also were a collection in themselves. Inventories and catalogues organised, stored and *produced* knowledge.[57] In a similar vein, Volker Hess and Sophie Ledebur have paralleled the filing of clinical records to collecting.[58] In Brussels, volumes of documents that were linked together—patient files, autopsy records, inventories of collections, articles in the medical press and proceedings of scientific societies—enabled and shaped medical research. By perpetuating older medical cases both materially and textually, newer cases could be put in a comparative framework, even if they were rare. For instance, if a new specimen that resembled older specimens became part of the museum collection, researchers traced back the patient histories of the older specimens in order to gain a more profound insight into the development of the disease.[59]

It appears, then, that different routes of acquisition impacted not only the value and use of anatomical specimens, but also their identity. Because dissected parts resided in the realm of anatomy (the dissected part stood for human anatomy as a whole) and autopsied parts belonged to clinical pathology (the autopsied part remained related to a patient), the first were generalised and lost their singularity, whereas the latter remained exact observations of particulars.

To sum up, body parts that originated from dissected bodies mostly were handling specimens in the second half of the nineteenth century. They were disposable and objectified didactic tools: as these specimens did not refer to one specific individual but to a universal healthy body, they could be easily replaced. Therefore they were mostly not inventoried for further preservation but simply exhausted and made anew. Yet body parts taken during autopsies or surgeries received a different treatment. As the material manifestations of particular diseases, they were both more valuable and more individual. Anatomists mostly could not replace specimens of the morbid body because they referred to an individual case history, and therefore, they had to preserve them perpetually. Whereas dissections brought forth anonymous and replaceable didactic tools, autopsies and surgeries provided pathologists with scientifically valuable reminders of the patients to whom they once belonged.

In the remainder of this chapter, it will become clear that the ways in which (parts of) patients made it from table to shelf also deeply influenced their further post-mortem journey. Whereas dissected parts were exhausted and disposed of (often as medical waste), autopsied remains were mostly used and reinterpreted for a long time before they were thrown away.

From One Shelf to Another

For the specimens intended for display, the museum was not a static home either. Historians have characterised the nineteenth century as a period in which both art and scientific collections were institutionalised. Against the background of patriotism, private collections across Europe transformed into public museums.[60] Both these broader cultural evolutions and the reconfiguration of medicine had an impact on anatomical collections. Formerly privately owned anatomical specimens, which were often part of larger cabinets of curiosity, gradually became the property of institutions. As private collections turned into institutional museums—first in hospitals, later in universities—their content was delineated more markedly. They became more strictly medical, both by excluding objects from other disciplines and by labelling and arranging the specimens according to medical conditions and theories.[61] As a result, they were increasingly unintelligible for general audiences. Anatomical collections, as opposed to many other collections, became less accessible in the second half of the nineteenth century, and were mostly visited by researchers and students in search of specialised medical knowledge.[62]

Anatomical museums in Belgium by and large developed along these lines. Whereas in the beginning of the nineteenth century, many anatomical collections had been scattered over the country, there were only four major museums left at the end of the century, which were all owned and curated by universities. Formerly privately owned collections of anatomical specimens—held by medical schools, hospitals, scientific societies, private practitioners or students—had often purchased by or donated to universities.[63] The growing number of autopsies was hence not the only reason for the expansion of the anatomical museum. Equally important was the number of donations, which peaked in the late nineteenth and early twentieth centuries. In this period, many anatomical specimens moved from small cabinets in private homes to the increasingly overloaded shelves of the university museum.[64]

In Brussels, the Society for Pathological Anatomy figured as a hub for the transfer of anatomical specimens to the university. Immediately after the society was founded under the patronage of the Free University of Brussels in 1857, the central hospital administration allowed its members to collect specimens at all institutions under its supervision, most importantly, the hospitals of St. Jean and St. Pierre.[65] Originally, these specimens remained the property of the hospitals, which had established their own cabinets in 1836.[66] In the 1870s, however, the creation of a catalogue of the society's collection prompted the donation of hundreds of specimens from the hospital to the university cabinet.[67] The reasons for this transfer were twofold. First and foremost, the members of the commission of hospitals believed that the specimens would be more frequently used at the university. In addition, they took in precious space that could otherwise be used for the storage of new medical instruments, objects that did not breathe harmful 'miasmas' as anatomical specimens allegedly did.[68]

Another important collection was integrated into the university cabinet of Brussels in 1890. On the initiative of Thiry, the Royal Academy of Medicine of Belgium (*Académie royale de médecine de Belgique*) gifted its collection of anatomical specimens.[69] In so doing, the Academy recognised its position within the new scientific landscape, in which universities were the scientific centres par excellence. Whereas the Academy had once been the backbone of a movement for the national organisation of medicine, including the wish to establish a national museum of anatomy, the transfer reflected the supremacy of urban universities in the late nineteenth century.[70] Other scientific societies, too, gifted their collections to university cabinets. The Medical Society of Ghent, for instance, donated multiple specimens from its cabinet of pathological anatomy to the new anatomical museum of the university in the 1880s.[71]

As a result of the growing size, visibility and prestige of the university museum, the number of individual donations increased as well. The proceedings of the Society for Pathological Anatomy are telling in this respect, because they provide an insight into the personal networks through which anatomical specimens became part of the museum of the Free University of Brussels. These networks were both socially and geographically diverse. Established researchers from other universities, for instance, Charles Poelman and Gustave Boddaert, both based at Ghent University, sometimes sent specimens to the Brussels society for discussion. Yet also private practitioners, often from rural areas, gave interesting pieces to their former professors in the city. In 1890, for example, a certain Léon Belière, based in Moustier-sur-Sambre, a small town in the province of Namur, donated a 'monstrous foetus' along with a report of its delivery.[72] Midwives, too, donated embryos, foetuses, umbilical cords and placentas to the society.[73] These donations were occasionally sent by post. In 1884, for example, two physicians urged each other to make it look like their package with human remains contained meat by labelling it as 'game'.[74]

The most important donors, however, were students. In the opening statement of the academic year of 1892, the president of the Society for Pathological Anatomy urged hospital interns to make the most of their working place, 'a vast field of exploration where so many pathological treasures are waiting to be found'.[75] As shown in the previous chapter, students did respond to these encouragements: in several years, they accounted for half, exceptionally even for two thirds, of the specimens presented at the society. Collection building, in short, cut across the divisions of the medical profession, bringing medical practitioners from different social classes and regions together.

Networks of supply nevertheless attested to internal hierarchies within the medical community. Vandendriessche has argued that they reflected the growing division between medical practice and scientific inquiry.

Private practitioners mostly only delivered specimens and left their analyses and the publication of articles to established professors. In the end, their contribution to science remained quite minimal: it was the professor who presented the specimen, clarified its scientific value and participated in debate. This was in part due to the strict scientific requirements that came to govern the presentation of anatomical specimens. As the more gentlemanly tradition of collecting, in which curiosity had been quintessential, was disappearing against the background of specialisation and 'scientification', private practitioners gradually had to withdraw themselves from the active scientific community.[76]

Yet they did continue to receive some credit, because their donation made interesting cases available to scientists. To compensate for their loss of private ownership, they were praised through various 'mechanisms of recognition'.[77] By making a contribution to the anatomical museum, donors could enhance their scientific prestige in various ways: they could strengthen their network, receive an honourable mention in a scientific journal or in a museum catalogue or be acknowledged as a corresponding member of a scientific society. This scientific recognition was most important for students and young researchers working within hospitals. For them, private ownership was not a possibility since regulations across Belgium stated that anatomical remains belonged to the hospital or the university.[78] Only by donating specimens, they could receive some credit for their collecting efforts—though not financial, but scientific.

In fact, it appears that the institutionalisation of anatomical collections coincided with the development of a notion of scientific ownership. Veronique Deblon has argued that Belgian anatomists in the early nineteenth century gained individual prestige from collections, even if they became the property of the state or the university. Anatomists continued to be seen as the 'intellectual' owners of specimens because they had skilfully prepared them.[79] The hospital archives of Brussels, too, show that a form of intellectual property rights *avant la lettre* governed anatomical collections. It appears that research, rather than preparation, became the first criterion in the assignment of scientific ownership in the second half of the nineteenth century. After the establishment of the autopsy service, for instance, Brussels clinicians asserted that they remained the 'owners' of their patients' case histories, even if specialised pathologists prepared their body parts. In the words of the pathologist and surgeon Emile Spehl, 'it went without saying that the scientific ownership of observations resided with the medical attendant'.[80] Although many names filled the proceedings of the Society for Pathological Anatomy of Brussels—from patients, private practitioners, prosectors etcetera—the name of the professor or student who had examined the patient always came first. Whereas specimens were originally owned by their maker (in particular in an age of private collections), they came to belong to their interpreters in an institutional setting. Scientific ownership gradually became more important than possessing the actual material object.

As collections expanded through the increased number of donations and autopsies, the need for proper categorisation grew. While anatomical museums were centralised—as multiple collections were integrated into university cabinets—they in the same process were divided into a growing number of subcollections. Historians such as Reinarz and Alberti have contended that disciplinarity and materiality went hand in hand. In their view, museums both mirrored and moulded the growth of specialisms in the medical field and curriculum. Consequently, once unified anatomical collections fell apart into a growing number of departments in the second half of the nineteenth century.[81]

In Brussels, too, inventories attest to the impact of medical specialisation on anatomical collections. A catalogue from 1834, for example, discussed anatomical specimens alongside the cabinet of mineralogy and the library collection. In a four-page register of anatomical specimens, foetuses (with or without pathology), 'different parts of the skeleton (35 pieces)' and 'the head of a cat' were listed underneath each other without further categorisation.[82] In 1852, the anatomical cabinet of St. Jean housed not only approximately 500 anatomical specimens, but also 'other remarkable objects, foreign to anatomy', such as 'objects belonging to natural history' and a collection of torture instruments.[83] When the Society for Pathological Anatomy of Brussels composed a new catalogue in 1872, however, the collections from the hospitals and the university were both brought together and reorganised. The new inventory listed specimens according to physiological function (e.g. blood circulation), body part (e.g. heart or arteries) and pathology (all cases of hypertrophy were, for example, listed underneath each other). In addition, the catalogue shows that the collection had become more strictly medical. Specimens belonging to the fields of mineralogy or botany, as well as the instruments of torture, disappeared completely, and most animal specimens became part of the collections of veterinary medicine or comparative anatomy.[84]

The differentiation of the disciplinary landscape was also reflected in the establishment of distinct university museums around 1860. In Ghent, for instance, separate museums existed for normal anatomy, pathological anatomy and comparative anatomy.[85] In Liège, too, specimens belonging to different disciplines—physiology, descriptive anatomy, pathological anatomy and comparative anatomy—were housed in different museums. Healthy, morbid and animal specimens were demarcated both by space and by staff, as curators and prosectors were responsible for but one collection.[86] As the century progressed, collections were divided further, since scientists building new disciplines often did so materially. In Brussels, for instance, the professionalisation of gynaecology, a field that was institutionalised around 1890, went hand in hand with efforts to establish a 'Special Museum for the Anatomy, Physiology and Pathology of Women'.[87] Around the same time, Brussels anthropologists also established their own collection.[88]

In fact, disciplinary identities were not only materially manifested in collections. In a few cases, collections also actively shaped them. The Society for Pathological Anatomy of Brussels, for instance, was at the basis of the construction of a professional community of pathologists, and was a key factor for the strengthening of the position of pathological anatomy in the medical curriculum. From its foundation in 1857, the society's leading figures, most importantly Gottlieb Gluge, had been driven by the wish to turn pathological anatomy into an obligatory course, a goal that they partially established in 1876.[89] The society also figured as a vehicle for newer disciplines. The presentation of specimens from the fields of ophthalmology, forensic medicine, bacteriology, embryology or physical anthropology, for example, led to discussions on the importance of these fields of research. Ultimately, however, these specimens were simply integrated into the existing collection. Although the reconfiguration of the disciplinary landscape clearly influenced the organisation of anatomical museums in Brussels and other Belgian university cities, the precipitation of specialism did not map directly onto collections. The impact of specialisation on collections should not be overstated: many specimens remained part of general anatomical collections, even after their related disciplines had established their own journals, societies and professorial chairs.[90]

As a result of both the institutionalisation and specialisation of medical collections in the second half of the nineteenth century, anatomical specimens moved between different shelves: from private cabinets to university museums, from unified to specialised collections. Along the way, their original identity was sometimes lost from sight or transformed. When reorganising collections, curators often replaced the inventory numbers, which made it harder to link anatomical specimens to their paper trail, including patients' records. In 1897, for instance, Louis Stiénon, former professor of pathological anatomy, investigated the history of the Canzius collection. The commission of hospitals and the university had bought this collection for educational purposes from the Dutch anatomist Jan Lubbertus Onderdenwijngaart Canzius in 1837.[91] Sixty years later, when Stiénon retraced the specimens, they had moved multiple times: from Canzius' home to the cabinet of the St. Jean hospital, from the hospital to the museum of the Society for Pathological Anatomy at the university and ultimately to the new anatomical institute in Parc Léopold. On their journey, the specimens had become part of different subcollections. The originally unified collection fell apart into specialisms:

> 36, interesting pieces for embryology and the physiology and pathology of pregnancies
> 4, comparative anatomy
> 4, normal and pathological anatomy of the skin
> 12, normal and pathological anatomy of the bones and joints
> 6, normal and pathological anatomy of the sensory organs
> 1, pathology of the nervous system
> 15, normal and pathological anatomy of organs of the respiratory and circulatory system
> 10, pathology of the intestines
> 10, gynaecology[92]

By his own account, Stiénon was unable to retrace the bulk of the collection, 'maybe because the specimens had been destroyed, [...] or because the indication of their origins had not been recopied on the new labels'.[93] Like most specimens for display, the specimens of the Canzius collection had changed together with the medical landscape: they had received new labels, had moved to new places and had been utilised in new disciplinary frameworks. As they were re-catalogued and reinterpreted, their older meanings faded and were replaced by new ones. Canzius himself had not made a 'gynaecological' specimen: it was through late nineteenth-century shifts in the medical landscape that a specimen of a woman's body part became classified as such.

In the view of Stiénon, who wished to reconstruct the history of the collection, the disconnection of the specimens from their past was the result of the 'neglect' of the cabinet's curator.[94] The following paragraphs, however, argue that he was wrong. Curators consciously changed specimens' labels to safeguard their relevance and use in a changing medical environment. In order to remain up-to-date, anatomical specimens had to be reinvented when new disciplines, theories or technologies developed.

From Shelf to Table

In recent years, historians have asserted that museums remained at the heart of medical education and research in the second half of the nineteenth century.[95] Unlike earlier histories, these studies have posited that museums and laboratories did not develop in opposition to, but next to, each other.[96] In the much-cited words of Alison Kraft and Samuel Alberti, 'the laboratory supplemented, rather than eclipsed, the museum'.[97] Brussels anatomists held similar views. As late as 1890, for instance, Thiry called pathological collections 'the corollary of the laboratory', with which he meant that museum work was often indispensable for experimental research. In the view of Thiry, museums provided pathologists not only with a certain, closed body of knowledge to build on, but also with the necessary pathological materials to experiment on.[98]

Anatomical specimens remained relevant because of their adaptability. Huistra has shown that old specimens could become the raw material for new facts. Specimens contained facts other than those they were made to display, since they were 'made of what they represent'. For example, early modern specimens consisted of cells, although they were made before the cell was discovered, and nineteenth-century preparations contained DNA, although these structures were unknown to their makers. Precisely because they were composed of 'original' body parts, anatomical specimens could transform from representations of knowledge to 'raw' research materials when new sciences developed.[99] As Leboucq already put forward in 1884, old anatomical collections that seemed to show 'not much that is new' could nevertheless contain 'details that appeared insignificant before'.[100]

The reuse of a collection of skulls at Ghent University may illustrate this point.[101] In the 1820s, physicians at the main prison of Ghent began to preserve the skulls of inmates, a practice that had been customary since the early modern period. Twenty years later, Daniel Mareska, head physician at the prison, organised the collection of roughly 170 skulls from a phrenological point of view. He linked the development of areas in the brain not only to personality characteristics (the typical research agenda of phrenologists), but also to living conditions. His research mostly focused on the shape and size of the skulls, which he compared to the heads of other social groups in society. In a study on the living conditions of textile workers, for example, Mareska used the prisoners' skulls as a point of comparison to make claims about the impact of factory work on the intellect and morality of the workers. He compared factory workers' skulls not only with those of prisoners, who 'had to be seen as expressing the results for the rural population' because they lived in the country, but also with those of orphans, who worked in small industries. In other words, Mareska did not study the prisoners as criminals, but as humans engaged in a certain kind of work.[102]

In 1844, Belgian physicians started lobbying to transfer the collection to an institution that would allow for its proper display, in order to enable more physicians to view and study the skulls. They regarded the phrenological collection as particularly valuable because it contained a large number of 'authentic skulls' instead of the plaster copies of famous heads that were found in most phrenological museums.[103] The collection became part of the anatomical museum of Ghent University one year later. Even though the scientific credibility of phrenology was waning, the curators of the anatomical cabinet enthusiastically welcomed the transfer of the skulls because they could still serve as teaching tools on the structure of head and brain.[104] In the 1860s, curators complemented the skulls with the heads of decapitated convicts and moved them to the cabinet of comparative anatomy. There, the anatomical preparations became part of a display that compared the anatomical properties of the heads of criminals, 'foreign' skulls and even animal skeletal remains.[105]

By the end of the nineteenth century, the collection returned to the anatomical institute. In the care of Leboucq, the skulls gained further international fame and caught the interest of a number of foreign researchers. The French anatomist Charles Marie Debierre, for example, performed extensive research on the craniological specimens, calling the collection 'unique in its genre' (Illustration 5.2).[106] Researchers who were then developing criminal anthropology as a discipline in Belgium used Debierre's data in order to nuance Cesare Lombroso's idea of the born criminal. In their view, the anatomical deviations of the born criminal (which they did not deny) had to be nuanced by looking at external causes, in order to gain real insight into the dynamics between nature and nurture with regard to criminal behaviour.[107]

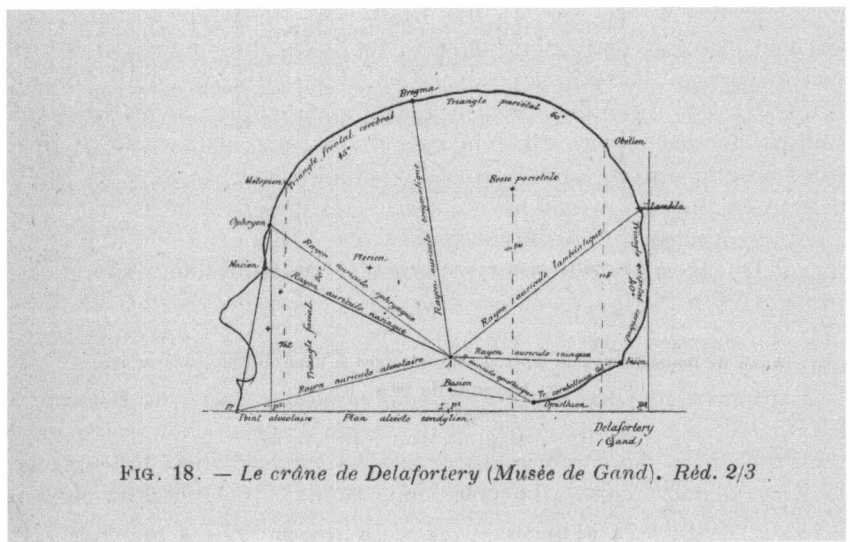

FIG. 18. — *Le crâne de Delafortery (Musée de Gand). Réd. 2/3*

Illustration 5.2   Criminological research on 'phrenological' skulls in the anatomical museum of Ghent. From: Charles Marie Debierre, *Le crâne des criminels* (Lyon: Masson, 1895)

This long history of reallocation and reinterpretation came to an end in 1940, when the collection of skulls was caught up in the fervour of the Second World War. Roger Soenen, professor of anatomy during the German occupation of Ghent University, integrated the skulls in a racist ideology based on white supremacy, anti-Semitism and eugenics.[108] Although Soenen never explicitly referred to the prisoners' skulls in his publications, it is likely that he placed them in a racist framework: either to study the 'Northern' origins of the Flemish race (an important focus of his research) or to teach students how to measure facial bones in his course on race and ethnology.[109] Where once they had been material sources of phrenological or anthropological research and anatomical teaching aids, now the skulls became associated with a racial ideology, one that, in the postwar context, had to be repudiated. Shortly after the war, the new professor of anatomy Julien Fautrez made sure the skulls were destroyed.

Until they were disposed of, the early nineteenth-century skulls were used to develop new approaches or disciplines that evolved: whether this was phrenology, criminal anthropology or eugenics. Other specimens experienced a similar fate. In the 1880s and 1890s, for instance, the German anatomist and anthropologist Paul Albrecht reinterpreted various specimens from the collection of the Society for Pathological Anatomy of Brussels under the influence of degeneration theories. In his view, the study of 'monsters (with which he meant congenital malformations) provided insights into the hereditariness of pathological characteristics rather than in the normal evolution of a species (as these specimens had been interpreted before).[110] Evolution theory, too, had a profound impact on the interpretation of collections, in particular of embryological specimens. In the words of Leboucq, embryology was 'entirely governed by the laws of evolution', and consequently, specimens could not be understood without this framework.[111]

Anatomists revisited older specimens not only with new theoretical knowledge, but also by means of new technologies. They occasionally took specimens out of their jars to study them with new instruments. Perhaps the microscope is the most telling example. As Huistra has shown, it was not unusual to reinterpret macroscopic pathological preparations on a microscopic level in the mid-nineteenth century, when the instrument improved through the use of multiple lenses that were optically corrected.[112] Another example is the X-raying of anatomical specimens. In 1908, the forensic doctor Fernand Héger-Gilbert X-rayed both recently deceased babies (obtained through the autopsy services of Brussels and Paris) and foetal specimens (from the collection of the Society for Pathological Anatomy of Brussels and the Parisian Orfila Museum).[113] By means of the new technique, Héger-Gilbert wanted to solve an old medical problem: distinguishing stillbirth from infanticide. Héger-Gilbert hence reinterpreted anatomical specimens in a new disciplinary framework (forensic medicine) and by means of a new technology (X-rays). As raw material, the specimens produced new research results: Héger-Gilbert found that X-rays proved if infants had breathed or not (Illustration 5.3).

This process of reexamination could be at odds with the aim of preservation. The proceedings of the Society for Pathological Anatomy of Brussels show that anatomists were well aware of this problem. They sometimes decided that the preservation of a specimen was more important than its analysis. In 1861, for instance, Jean-Joseph Sacré presented a foetus with agenosomia (absence of genital organs) in its integrity as a more complete study of the pelvis 'would have destroyed a teratological piece that is too interesting to not be conserved in the collections'.[114] Tensions between conservation and use continued to influence the man-

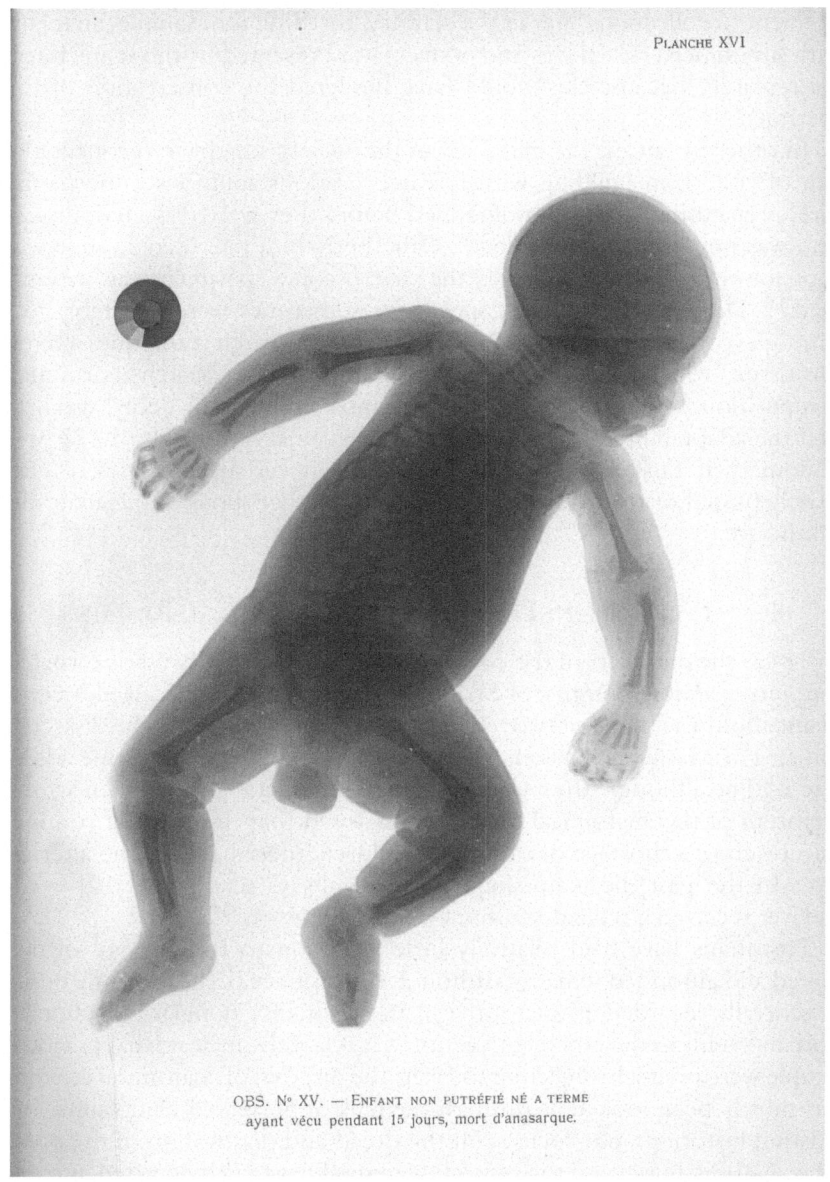

OBS. N° XV. — ENFANT NON PUTRÉFIÉ NÉ A TERME
ayant vécu pendant 15 jours, mort d'anasarque.

Illustration 5.3   X-ray image from Fernand Héger-Gilbert, *La radiographie fœtale envisagée au point de vue médico-légal* (Brussels: Piette, 1908). In the view of Héger-Gilbert, infants who had breathed were more transparent on X-ray images than stillborn babies

agement of collections later in the century. In 1889, for example, an infant with sirenomelia (fused legs and no feet) was presented 'without any internal research' because this would have hindered the conservation of the specimen.[115]

In other instances, the members of the society sought to reconcile the aim of collection building with research. Various solutions came to the fore: preparations were photographed before they were dissected, dissection was limited to a certain part of the body or a microscopic slide was kept for the collection whereas the macroscopic specimen was reexamined.[116] Despite these efforts, specimens sometimes were inevitably lost through (re)use. Injection fluids or destructive research techniques altered the tissue, which made it harder for anatomists to research its original composition. Other material conditions, most importantly decay, also limited the adaptability of preserved human remains. As a result, the jar was seldom their final resting place. Most anatomical specimens were ultimately buried or thrown away, much like the other spoils of the anatomical theatre.

4   The Coffin: Disposing of Anatomical Remains

In 1882, the members of the commission of hospitals of Brussels carpeted the *garçon d'amphithéâtre* of St. Jean because they had received a complaint about the burial of two 'child cadavers'. He had placed the dissected remains of Marie Schoekaerts and Charles Gilbert together in one adult-size coffin, although they were complete strangers. In his defence, the factotum of the anatomical theatre emphasised that this was not routine. As a result of a shortage of coffins for children, there had been no alternative. In the past, he maintained, 'he had always made sure that every cadaver received a separate coffin, even the foetuses'.[117]

Historians have paid relatively little attention to the disposal of dissected and autopsied remains. Although the disgrace of the anatomy burial is generally regarded as one of the main reasons for popular objection to post-mortems in the nineteenth century, it is largely unclear what practices people were precisely objecting to.[118] As the disposal of anatomical remains has mostly been looked at through the lens of preserved complaints, the existing historiography focuses on the dread and deprivation of the poor. The anatomy burial and the respectable burial have been depicted as complete opposites, irrespective of the exact time period.[119] For the British context, however, historians of death have argued that the second half of the nineteenth century was a period of profound change. To name but a few examples: sanitary principles began to influence the organisation of cemeteries, the poor increasingly refashioned death customs as civil rights, and burials became increasingly individual rituals.[120] Only a handful of historians of anatomy have interacted with this elaborate historiography of death, drawing attention to the ways in which these nineteenth-century changes influenced the inhumation of anatomical remains.[121]

The next pages take a step towards a more comprehensive study of the ways in which the spoils of the anatomical theatre were handled. Instead of looking at the popular perception of the disposal of anatomised bodies, the next pages shed light on actual practices and their evolution over time. Firstly, attention is paid to the burial of hospital patients. Although hitherto largely unnoted in historical research, the burial of dissected and autopsied patients changed profoundly under the influence of both an evolving culture of death and the development of body donation. As the complaint from 1882 shows, the individual burial of anatomical remains gradually became usual, or at least became considered as the decent thing to do. Then follows a discussion of the other ways in which the detritus that was created in post-mortem rooms was disposed of. The classification of certain remains as medical waste was not a neutral activity, but a cultural decision on what counted as 'human'. The plea of the *garçon d'amphithéâtre*, for example, suggests that some body parts or types of bodies, such as foetuses, were perhaps less entitled to a decent burial than others. Lastly, the two seemingly opposite topics of this chapter—preservation and disposal—are brought together by looking at the reasons for, and meanings of, the disposal of previously preserved anatomical remains.

From Table to Coffin

In one of the few studies on the burial of anatomical remains, Helen MacDonald has drawn attention to the indecent burial of the devastated 'bodies' left behind following dissection in nineteenth-century Britain. Although the burial clause of the Anatomy Act described that remains had to be 'decently interred', the end products of dissection were often disposed of in deplorable ways: in poorly made coffins, boxes or bags, mixed together with other human or non-human remains and rubbish, and on cheap dumping grounds rather than consecrated plots. In the view of MacDonald, highlighting these abuses was one of the most effective ways for critics to mobilise popular opposition against the Anatomy Act.[122] In a

similar vein, Elizabeth Hurren has highlighted that the poor were fright-
ened by the prospect of an anatomy burial by means of letters and complaints preserved in Poor Law archives.[123]

In Belgium, too, the archives reveal traces of the inhumane circumstances in which the bodies of the dissected and destitute were sometimes
buried. Surviving complaints sketch the dreadful treatment of dissected
and autopsied remains in graphic terms: there were coffins leaking blood,
hearses followed by a trail of bodily fluids trickling from the trunk, coffins
so cheaply made that nails split the wood and gaps exposed putrefied
remains, coffins that were too narrow 'for slightly obese people' or too
short for tall people.[124] Strangers occasionally ended up together after
death. In 1886, for instance, three legs and three arms were discovered
together in one coffin in Brussels, and in 1889, five corpses were found
scattered over two coffins in Ghent.[125] Then there was the state of the
cadaver: eyes wide open (in the words of one complainant, looking as if
the deceased had been 'surprised by death'), putrefied or shaved bald.[126]
Most often though, families complained that they had not been informed
about the burial, which had left their deceased relative (hopefully) alone in
his or her coffin when being consigned to the earth.[127]

In studies on Britain, historians have shown that these complaints considered breaches of deeply felt cultural sensitivities. Even after the Anatomy
Act was established in 1832, bereaved indigent families applied for welfare
assistance to bury their deceased 'decently'.[128] It appears that poor law
officials mostly recognised their requests until the last third of the nineteenth century, when policy changed radically under the influence of the
so-called 'crusade against outdoor relief', an initiative by the central government that was welcomed by poor law unions wishing to cut expendi-
ture. During the crusade, applicants for relief were less likely to obtain
outdoor assistance. Many of them were forced to enter the workhouse
instead.[129] Customary funeral payments also were increasingly refused.
The poor were forced to beg for burials or to give the bodies of their loved
ones up for dissection.[130]

Several studies have shown that the poor had always resented these
more stringent relief policies; yet it appears that they only became more
vocal when democracy extended. More people became prepared to speak
out in the more liberal political climate of the 1890s and 1900s. This went
hand in hand with rhetorical shifts in the discourse about relief and poverty. Even if paupers and pauper advocates had no absolute rights to welfare in the discretionary poor law, they tried to claim and maintain

entitlement to relief, for example, by stressing their contribution to the community. What was before seen as a custom, gradually became refashioned culturally as a right. As the idea arose that all citizens were entitled to a decent funeral, the number of complaints sent to local authorities grew.[131]

In Belgium, too, the balance of archival evidence suggests that the growing number of complaints were tied up with the widening of democracy. In 1886, for instance, a certain Ferdinand Dupont urged the mayor of Brussels to improve the burial conditions in public hospitals because this was 'both a democratic and a humane task'.[132] Several pauper letters from the 1880s and 1890s portrayed death customs as rights, for instance, by claiming that every pauper was 'entitled to a decent and dignified burial' or by stressing their contribution to the community during life. In these complaints, citizenship came to encompass respect for the dead.[133]

These findings fit with a recent study on pauper burials in late nineteenth-century Brussels, in which Jolien Gijbels has found that indigent hospital patients were buried under better conditions from the 1880s onwards. Against a background of ideological tensions and gradual democratisation, so Gijbels argued, the Brussels city council became a powerful actor pressing for improvements in burial conditions, both from a hygienic and humane point of view. Putrefied or disintegrated bodies or body parts, for example, had to be soaked with chemical substances, and mortuaries were constructed within hospitals, allowing relatives from different religions to pay their last respects to the deceased.[134]

Municipal regulations were a much-needed amendment to an inadequate legal framework. There was only one general law on the burial of hospital patients (many of whom were autopsied or dissected) in Belgium, which stated that 'the costs for the inhumation of the poor who die in hospitals [...] are included in the general costs of these institutions'.[135] In real terms, this meant that hospitals saved money by cutting funeral costs. The use of plain coffins, measured to *just* fit the corpse in order to save expensive wood, or the choice for a collective transport of bodies to the cemetery was in a sense logical, as it had an immediate impact on hospitals' budget. Towards the end of the nineteenth century, however, municipal regulations put some restraints on these savings.

The Brussels city council imposed more stringent rules not only as a result of the growing number of complaints, but also because of heightened ideological and social tensions. As has been discussed in more detail in Chap. 3, conflicts on burials had a profound impact on the treatment of the dead in hospitals in the late nineteenth century. Catholic burial societies tried to mobilise opposition to the secularisation of burials by pointing at the deplorable ways in which the poor were inhumed. For the Socialist Party, too, the indecent burial of the poor was an important political weapon. In this context, Liberal city councils rose municipal funeral standards in order to steal a march on their political opponents. However, also anatomists benefitted from better burial conditions, since they allowed them to improve their reputation in an era in which patients' consent grew in significance. Much like the construction of neo-Gothic anatomical institutes or meticulously designed post-mortem rooms, the improvement of the conditions of the burial of dissected and autopsied patients was a way to procure more corpses for education and research.[136]

New standards came to govern the burial of hospital patients in Brussels. Since the completion of the construction of new mortuaries in the St. Pierre and St. Jean hospitals in 1886, for instance, families received more privacy during their last visits. Bodies before burial lay in small separate chambers from approximately five square metres, separated from each other by heavy curtains.[137] The fact that every corpse laid in its own room was an improvement of the previous situation, in which, according to a complaint, each body rested on a bier 'placed among other biers without symmetry', so that visitors saw not only the corpse of their own loved one, but also those of other patients who had died around the same time.[138]

Bodies not only waited for burial in more decent rooms; they also received a more respectable coffin. Around 1900, pentagonal coffins with full lining replaced the rectangular, plain coffins that had been used before.[139] Before this measure, families had occasionally 'pimped' plain pauper coffins themselves. They, for instance, added ornaments and flowers, or paid an additional fee to the carpenter for 'a more luxurious coffin'.[140] In a few instances, they delivered a coffin of their own choice to the hospital, even though they did not have the means to pay for the rest of the funeral.[141] In 1905, the hospitals of St. Jean and St. Pierre started to use screws instead of nails to close coffins, as it was judged insensitive and indecent to disturb the dead with the knocking of hammers. In 1910, the commission of hospitals even debated if perhaps a soft pillow should support the head of the deceased.[142]

The transport of cadavers, too, was reorganised. In 1904, the Brussels city council abolished collective transports from hospitals to the burial ground because they 'lacked decency'.[143] In Ghent and Liège, this measure had already been taken a few years before.[144] Thanks to the introduction of individual transports, poor families could accompany their deceased loved one to his or her final resting place in a funeral procession, a ritual that was commonplace in the middle and upper classes.[145] Within medical institutions, too, conditions improved. The commission of hospitals bought expensive black cloth to assure the decent transport of dead bodies to and from the mortuary, and at the anatomical institute of the Free University of Brussels, the handcart that was for a long time used to carry dead bodies from the hearse to the building was replaced by a carriage.[146] Those carrying coffins had to make sure they were 'dressed properly' at all times.[147]

Hospital patients were increasingly allowed to die as they had lived: whether this was as a Roman Catholic, Protestant, Jew or freethinker. As a result of the presence of burial societies within hospitals, deceased Roman Catholic patients more often received an individual service for their salvation.[148] Individual services were also frequently held for freethinkers or for those of different faiths.[149] As burial societies democratised the idea that the funeral should reflect the ideological convictions of the deceased, the funeral gradually transformed into a final expression of identity.[150] In the face of a complaint by a Liberal burial society, for instance, the commission of hospitals of Brussels in 1911 declared that they in the future would always recognise 'the last wishes of the deceased regarding the organisation of the burial'.[151]

Moreover, corpses were for the first time transported to the chapel for burial services in the late nineteenth century. Whereas before the body of the deceased stayed in the morgue while relatives celebrated his or her life, the corpse from 1896 onwards could be brought to the chapel on request in Brussels hospitals. In so doing, the commission of hospitals gave in to a decades-old wish of families. Roman Catholics in particular thought the presence of the coffin (preferably an open coffin) was important, as they believed that they could improve the fate of the dead by praying over their remains, and by letting a priest bless the body. Tellingly, the chaplain noted that services with the corpse present 'had become the rule rather than the exception' just one month after the change of regulations.[152]

All these changes in the burial conditions of hospital patients—privacy in the morgue, pentagonal coffins, individual transports and open-casket burial services—reflected broader changes in the culture of death. Historians have characterised the nineteenth century as a period in which death was both 'beautified' and 'individualised'. Rituals such as the laying out of the dead and the viewing of the corpse aestheticised death. Thanks to clothing, make-up or embalming, dead bodies became objects suitable for visual display.[153] Furthermore, funerals were increasingly sentimental and personal rituals, in which each individual received a separate grave and coffin. The establishment of large, communal burial grounds outside the city allowed for the replacement of temporary, anonymous burial places by marked graves devoted to an individual or a family, which could be visited for a long time.[154] At the basis of this new culture of death were fundamental changes in Western society, such as the development of public health policy (individual graves, for instance, were also a sanitary requirement because they fastened decomposition), the professionalisation of death practices (think about the emergence of mortuaries and funeral parlours), the growth of consumer markets (leading to the manufacture and exchange of death goods such as elaborate coffins with lacquered finishes or finely made mourning clothes) and the rise of bourgeois 'respectability' (burials became an expression of class identity).[155]

Whereas historians have mostly studied these evolutions for middle- and upper-class burials, Julie-Marie Strange has reached similar findings in her excellent study on the late nineteenth-century mourning practices of the British working classes. Strange asserted that the corpse figured as the material focus of grief despite the growth of a sanitised culture. As the 'nucleus for personal and sentimental reflection', the dead body was inseparable from notions of memory, decency, dignity and respect. In rendering the corpse as lifelike as possible—for example, by dressing the body in a shroud resembling a night dress, as if the deceased was merely asleep—the bereaved could continue to associate the body with the known personality of their lost one for a short time. The laying out of the corpse could be a final moment of intimacy with the deceased for relatives, or an opportunity to reassess relationships between the dead and the living. The last visit and the wake were important mourning practices, which allowed the bereaved and those who sympathised with them to communicate in a symbolic way, and to express care towards the deceased by safely guiding his or her remains to the grave.[156] Much like the middle and upper classes, the poor continued personal bounds with their deceased after the body was laid in a coffin, for example, by organising burial services for them, by visiting their graves or by sharing memories.[157]

Most significantly, Strange has interpreted the institutional pauper burial—burials for individuals whose families could not pay the interment fees themselves, as was mostly the case with hospital patients—as a 'contested site for notions of respectable burial'.[158] Unlike many other historians, who have argued that the pauper burial was the exact opposite of respectability—in the words of Ruth Richardson 'a carefully constructed negative image of the respectable burial'—Strange has put forward that pauper and respectable burials were not mutually exclusive.[159] On the one hand, the poor were not without agency: they took elements of the middle- and upper-class burial and applied them, where possible, to the inhumation of the dead in a pauper grave. On the other hand, public authorities tapped into popular attitudes towards the corpse and accommodated the desire of the poor to perform certain rites in order to break resistance against public health measures or post-mortem examinations.[160]

This interpretation applies equally well to the burial of hospital patients in Brussels. The adornment of plain coffins, for instance, can be seen as a personal gesture of grief that salvaged respectability and decency for the deceased despite a further lack of financial means. Poor families facilitated their own understandings of a respectable burial: by placing floral wreaths, nameplates or other decorations on the coffin, they expressed their personal sorrow and affirmed the identity of the deceased. Much like in Britain, these simple rites, which were first met with hostility as they allegedly proved that the poor took advantage of 'free' burials, were perceived more positively towards the end of the nineteenth century.[161] In the words of the Brussels city council, for instance, they were 'loving gestures one had to respect'.[162]

Requests to view the corpse before burial, too, can be seen as an attempt to maintain death customs. As aforementioned, relatives and friends from 1896 onwards could pay their respects to the coffin and pay their condolences to close mourners—the traditional double function of the last visit—during burial services within Brussels hospitals.[163] In addition, families visited their dead in the morgue. In 1902, for instance, the head of the St. Jean hospital explained to the commission of hospitals that he could not meet a new sanitary requirement, according to which the coffin had to be closed one hour before burial, because mourners 'nearly always' requested to view the corpse before funeral processions. These demands in part followed from suspicions that bodies were mutilated or shaved bald, and were a way to assert the identity of the dead; the head of St. Jean noted that relatives in this way 'made sure that they guided the "right" deceased to the cemetery'.[164] Yet the wish to be present when the coffin was closed can also be interpreted differently. In a domestic context, it would have been usual for Roman Catholic families to wake the body:

they would have stayed with the body until burial in order to ensure its safe passage to the next world.[165] Families' presence in the hospital morgue might have fulfilled a similar purpose: by being there, they expressed their care for the passage of their loved one to his or her final resting place, both in a literal and symbolical sense.

The introduction of pentagonal coffins, individual transports and 'decent' burial services for destitute patients, conversely, can be seen as top-down initiatives meant to enhance the reputation of hospitals. Gijbels has rightly pointed at the leading role of the mayor of Brussels, Charles Buls, who, as mentioned in the previous chapter, wanted to increase the impact of science upon society.[166] Buls was aware that sympathising with working-class attitudes was a good way of gaining acceptance for his progressive policy. When he constructed a mortuary in the Rue Saint-André, for example, he made sure that the building was designed with care:

> It concerns a public building destined to receive the dead. Therefore it is important that the respect, which we feel for the unlucky ones who are brought there, is reflected in the decency of the institution's architecture.[167]

By turning the mortuary into a stately building, Buls made an unpopular hygienic measure, namely the removal of bodies from family homes, easier to swallow.

In a similar vein, the improvement of burial conditions was an important way to bolster the image of public hospitals. In order to turn hospitals into 'magnificent' institutions, central to the spread of medical knowledge (one of Buls' ambitions), the hospital had to be disassociated from the dishonourable burial.[168] The improvements discussed above had to ensure that dying in hospital no longer implied the anonymity of the grave, the inability to claim ownership of the body or the denial of mourning rites. In addition, these measures were related to the admittance of paying patients and the generalisation of the autopsy: as different people entered the scene, the burial had to be accommodated to their standards.

When it came to autopsied bodies, the measures taken were quite successful. By the turn of the century, autopsied patients received a burial ritual that hardly deviated from usual standards: families could claim ownership of their bodies, pay them their last respects, pray in their presence during a burial service and escort them to a marked grave in a funeral procession. Yet these rituals were more difficult when it came to dissected 'bodies', which, as a result of their complete disintegration, could not be visited or moved.

Still, anatomists took more care when it came to these remains, too. Apart from parts that were kept for anatomical collections, all pieces of the cadaver had to be placed back together in one coffin. As early as 1875, the head of St. Jean urged his employees to make sure that 'the corpse in all its parts is disposed of in the coffin destined for this purpose' because 'inhumations have to happen with respect for the human remains and [...] the identity of the deceased has to be confirmed according to burial regulations'.[169] In 1886, the head of St. Pierre reassured the hospital board that the employees of his institution 'always make sure that all parts of the body that belonged to the deceased are placed in the same coffin'.[170] There was only one instance in which the remains of several persons could be buried together: if a mother-to-be had died together with her unborn or newborn child during or shortly after labour.[171]

Keeping the parts of the deceased together during dissection, however, was not easy. In order to be able to identify the remains, bodies received tags when they entered the morgue.[172] Instructions from the Brussels St. Jean hospital show that this card moved along with the, increasingly unrecognisable, remains:

> The card of the bodies given to anatomical studies is detached from the neck and fixed to the legs of the table on which the subject is placed, and if it [the body] is moved to the coffin, this card is pinned on the linen that covers the remains of the cadaver.[173]

In order for this system to work, the cadaveric remains could 'in no instance be moved from the table'.[174] To overcome this difficulty, stamps were sometimes used to identify different parts; a technique that had been introduced by public health officials who in this way marked that bodies had been certified as dead.[175] Ghent University introduced identity certificates for dissected bodies in 1889.[176]

Despite these instructions, employees of the anatomical theatre remained puzzled about how to rejoin the remains after dissection. First and foremost, time was not in their favour: as the dissection progressed, more and more body parts putrefied. Before students arrived at the bones, muscular and other soft tissues were often in a state of decomposition. In order to solve this problem, hospitals in Brussels bought special (temporary and reusable) coffins, allowing anatomists to saturate 'finished' remains with carbonic acid while other body parts were still on the table (Illustration 5.4). There were also dissecting tables with built-in storage tanks for 'detached intestines' (Illustration 5.5). In addition, *garçons d'amphithéâtre* sometimes found it hard to keep track of the whereabouts of different parts, as bodies could be divided between different disciplines, the thorax, for example, being used in a course of surgery while the leg was being prepared for the anatomical museum. As a result of these diffi-

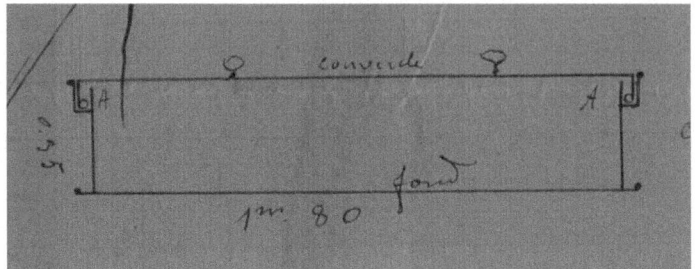

Illustration 5.4 Design of a coffin in zinc with lid, meant for the storage of dissected body parts. On the inside, the coffin was dressed with 'cotton soaked with disinfectant liquids'. © Archives of the Social Services of Brussels. From: Laurent Anciaux ('plumber and gasman') to St. Jean Hospital, August 26, 1880, folder 76, Divers, Affaires générales

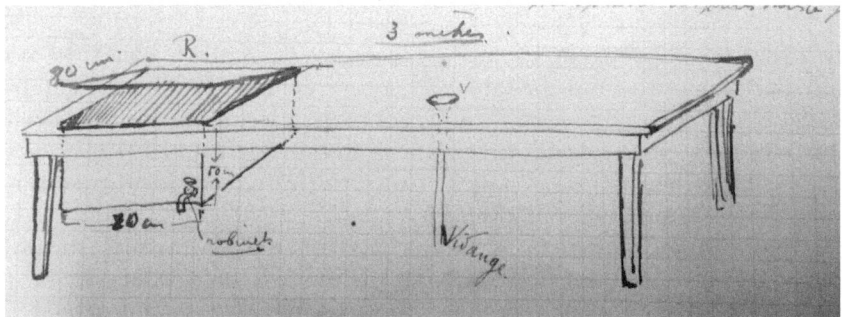

Illustration 5.5 Design of a table with built-in storage tank for 'detached intestines', s.d. [circa 1900]. © Archives of the Social Services of Brussels. From: Construction d'un nouveau dépôt mortuaire et amphithéâtre 1904–1909, folder 21, Travaux, établissements hospitaliers

culties, quite a deal of bodily material was not buried but thrown away, in spite of good intentions.

The burial of dissected patients, unlike the burial of autopsied patients, therefore, continued to differ markedly from popular customs. Families could not view a corpse after dissection, and as a result of their extensive use, dissected remains were often only buried weeks after death: in Brussels an average of three weeks later in 1900 and even an average of five months after death in 1920, as a result of improved preservation techniques.[177] The identity of the deceased might have become more important from the end of the nineteenth century onwards—dissected remains did not end up in an anonymous grave anymore—but dissection still implied yielding ownership of the body and a denial of certain burial rituals.

Perhaps this is the main reason why dissections continued to be negatively perceived. In her research on Vienna, Tatjana Buklijas has argued that protest against dissection in this city might have been less outspoken than in Protestant Britain because Roman Catholic citizens found 'burial in a coffin and in consecrated ground' more important than 'whether the body had been dissected'.[178] The Brussels case conflicts with this view. Even when dissected bodies were being inhumed in separate coffins and in consecrated ground, Roman Catholic families continued to claim their deceased to keep them out of the hands of anatomists. Perhaps, this difference might be explained by Buklijas' lack of attention for the distinctions between the burial of dissected and autopsied remains.[179] When taking into account the findings of both this chapter and the previous chapter it is plausible that there was little objection to the autopsy, a procedure that was more common than the dissection in Viennese hospitals. For dissections, however, the story might have been different.

In fact, the Brussels case fits better with the interpretation of Elizabeth Hurren, who, in her study on Britain, suggested that the poor above all feared dissection because of its 'corporeal reality'.[180] Regular pauper burials were not necessarily stark or abhorred. On the contrary, Hurren and King have argued that 'we must be aware of generalising scandals' as many paupers were laid to rest in decorated coffins, with signification and a ceremony.[181] Anatomy burials, conversely, dealt with perishing human waste. Even when care was taken to label the material, various steps of the funeral rite could not be maintained because the cadaver had effectively vanished.[182]

For Belgium, the archival evidence suggests that different standards and levels of acceptance with regard to the burial of autopsied and dissected remains were tied up with the material reality of the corpse. Whereas the autopsy did not hinder families to take possession of the body and ritu-ally care for the deceased, though with a slight delay, the dissection was difficult to reconcile with a culture of death in which the corpse took centre stage. Dissected remains could possess a name and a grave, but not a body and a face. Even when coffins and ceremonies were provided for, dissection implied that relatives had to give up the material locus of their grief.

This problem continued to bother anatomists when body donation gained importance in the first half of the twentieth century. From the 1930s onwards, in order to promote anatomical bequeathal, they stressed the possibility of holding funeral services for the deceased before or after dissection, and of the return of the coffin to families for final disposi-tion.[183] In Brussels, the anatomical institute reassured potential donors that giving your body to science 'did not exclude cremation or religious services' and that remains were always buried 'in an individual grave' after-wards, even if the deceased or his or her descendants had not made any arrangements themselves.[184] In most instances, a public funeral service was held for relatives, friends and acquaintances of the donor shortly after his or her death. The body was then buried post-dissection several months (up to a year) later, in the circle of one's most intimate friends (Illustration 5.6).[185]

Anatomical institutions also took various initiatives to honour and commemorate individuals who had donated their bodies to science. In a case study on mid-twentieth-century anatomical bequeathal at the univer-sity of Aberdeen in Scotland, Elizabeth Hallam has shown that the grow-ing importance of bequests for anatomical institutions went hand in hand with initiatives to honour donors. On the one hand, the anatomical insti-tute of Aberdeen, much like in Brussels, respected the observance of indi-viduals' requirements for their burial after dissection; on the other hand, donors were memorialised publicly, by means of a memorial book (inscribed with their names), a yearly memorial service and a cemetery memorial.[186]

The anatomical institute of Brussels took similar initiatives to publicly express gratitude towards those who had donated their bodies to science, and to hence promote anatomical bequeathal. By means of 'a discrete campaign', they 'wished to increase the number of voluntary donations of human remains'.[187] In 1949, for instance, Jean La Barre, then dean of the faculty of medicine, researched the attitudes and intentions of donors, for he saw donation as the best solution for the shortage of anatomical mate-

Illustration 5.6 The obituary card of Louise Strubbe, born Valentin, who died on January 2, 1951 and donated her body to the Free University of Brussels. This kind of card was typically distributed during a public service before dissection. The body was then buried amongst close relatives and friends after the dissection 'later in the year' (the date could not be specified as anatomists could not tell in advance how long they would use the body). © Archives of the Free University of Brussels, folder 1 FD 703

rial that was impeding medical education at the time. He found that people who bequeathed their body to science were often driven by 'philanthropic intentions', and by 'the concern—which may seem stingy but which exists nevertheless—of avoiding the costs related to a normal funeral'. Others were afraid of being buried alive or wanted a financial compensation in exchange for their donation, yet this was never granted. With these findings in mind, La Barre proposed to publish an announcement in a 'big newspaper':

> The education of physicians and surgeons and the progress of science is impossible without a profound knowledge of the human body. To donate your body to a medical faculty is an act of the purest philanthropy. No costs, no recompense.[188]

In addition, the anatomical institute of Brussels decided to honour 'the voluntary donors whose bodies [they] were able to use in the last few years by inscribing their names on a commemorative plaque'.[189] Much like memorial services, these kinds of inscriptions, as Hallam has shown, worked in a number of ways: they expressed gratitude for individual persons, linked their donation to a wider culturally valued process (the progress of science) and increased public awareness of the need for bodies for dissection and the laudability of anatomical bequeathal.[190]

Despite the care taken with burial requests and commemorative practices, however, the prospect of dissection remained painful for relatives, who sometimes objected to the anatomical donation of their loved ones. La Barre, who did not donate his body himself, understood their concerns. In his opinion, the anatomical institute should not 'hurt the very respectable feelings of a family or enter in conflicts for the ownership of a cadaver', and only accept donations from persons without close relatives or with progressive relatives who agreed with their decision. The anatomical institute of Brussels ultimately decided to specifically target two groups in their promotion of body donation: medical alumni and advocates of cremation. Despite the fact that cremation, which had only been legal since 1931, was very unusual, the anatomical institute contacted the Belgian Society for Cremation (*Société belge pour la crémation*) in the hope that they would spread information on anatomical donation. The best way to gain more bodies, so it seems, was to target a minority that did not place the intact corpse at the centre of the burial ritual.[191] Only when the body became a less important site of ritual, could anatomical donation become truly dignified.

Down the Drain

Although hospitals increasingly cared about burial rituals for autopsied and dissected remains from the end of the nineteenth century onwards, they remained the privilege of certain body parts. A considerable amount of human material was (and is until today) classified as medical waste, and thrown away without further ado. MacDonald has pointed out that in the case of dissection specifically, all that was left to bury was a pile of bones denuded of the flesh, which weighed no more than one fourth of the intact body and fitted in a small box. The remaining three quarters of the corpse was consigned to troughs and pits rather than the grave.[192]

It is important to note that nineteenth-century anatomists themselves did not talk about 'medical waste', a concept with a rich history that has unfortunately not been subject to sufficient research.[193] They rather distinguished 'bodies' (*corps* or *cadavres*) from 'intestines' (*viscères*), 'remains' (*débris*), 'pieces' (*pièces* or *morceaux*) or 'jars' (*bocaux*). Their choice of words was important: the use of 'body' or 'cadaver' mostly signified that the remains had to be inhumed in a significant way, whereas the 'debris' could be wasted. This does not mean that the 'body' always was an intact corpse. With this word, anatomists rather referred to those parts that would receive a proper burial. In real terms, 'bodies' often signified skeletons or body-shells robbed of internal organs. Those parts that would be wasted were also described in euphemisms. By using vague terms, anatomists obscured the fact that they were talking about human remains. By referring to the inhumation of 'jars', for example, they disguised the fact that they were actually talking about jars containing organs that had once belonged to a person.[194]

Distinct ways of disposal could, and mostly did, coexist. Classifying certain parts of the body as 'remains' or 'intestines' did not mean that the body as a whole was denied a funeral: a part of the body could be wasted while another part was carefully inhumed. In fact, preservation, disposal and burial could go together: anatomists could remove and preserve a pathological organ for the museum during an autopsy, after which they removed and threw away parts of flesh and intestines during a dissection, after which the bones and other remaining tissues could ultimately be buried. On a more abstract level, this implies that certain remains of the same body could be preserved or disposed of as objects, whereas other parts continued to be treated as a subject, buried in the presence of family and friends.

The partial wasting of the body during post-mortems was in fact inevitable. There was the problem of blood: gushing from incisions, staining scalpel blades, gloves and surgical gowns. With the advancement of putrefaction, ever more liquids and mucus started dripping from the body, as organs burst and tissues softened and eventually liquefied. For reasons of hygiene and health—contact between these liquids and small wounds in dissectors' hands was a common cause of dangerous infections—anatomists drained of these bodily fluids as efficiently as possible. Autopsy and dissection tables were designed with care: increasingly made from solid stones resistant to antiseptics rather than from wood, in a rectangular shape with rounded angles (much like a bath) and with a border of minimum four centimetres that avoided 'that liquids ran over'.[195] Around 1900, tables with an elaborate drainage system increasingly replaced tables with only one wastepipe at the foot, with the aim of removing liquids faster. In order to assure a swift flow off, tables were often slightly tilted, and to avoid blockages, drains had a diameter of approximately five centimetres (Illustration 5.7). Once removed from the table, liquids came together in a subterranean reservoir for sterilisation, together with the water that had served to wash the room as well as anatomists' clothes and hands. Ultimately the content of this reservoir ended up in the regular sewage system.[196] Bodily fluids and softened tissues were never kept for burial because coffins would leak and nauseating scents would erupt from hearses.[197]

Yet medical waste was not only a practical matter. Susan Lawrence has argued that ongoing conflicts about the medical use of human material not only stem from divergent conceptions of the body in different areas of life—for example, religious worship, criminal law or medicine—but are also dependent on the kind of body part involved. Whereas we mostly believe that human remains clearly embodying a sense of identity or personhood should be disposed of in meaningful ways, we are largely indifferent towards the treatment of other bodily materials, such as tissue samples, hair or vials of blood.[198] Similarly, in a study on the disposal of embryos, Lynn Morgan has posited that the propensity to classify certain human materials as 'waste' follows from historically rooted assumptions about what kind of entities different types of bodies or body parts are.[199]

Medical waste is not a neutral category, but a decision made on the basis of the significance of certain bodily tissues, and in turn a justification

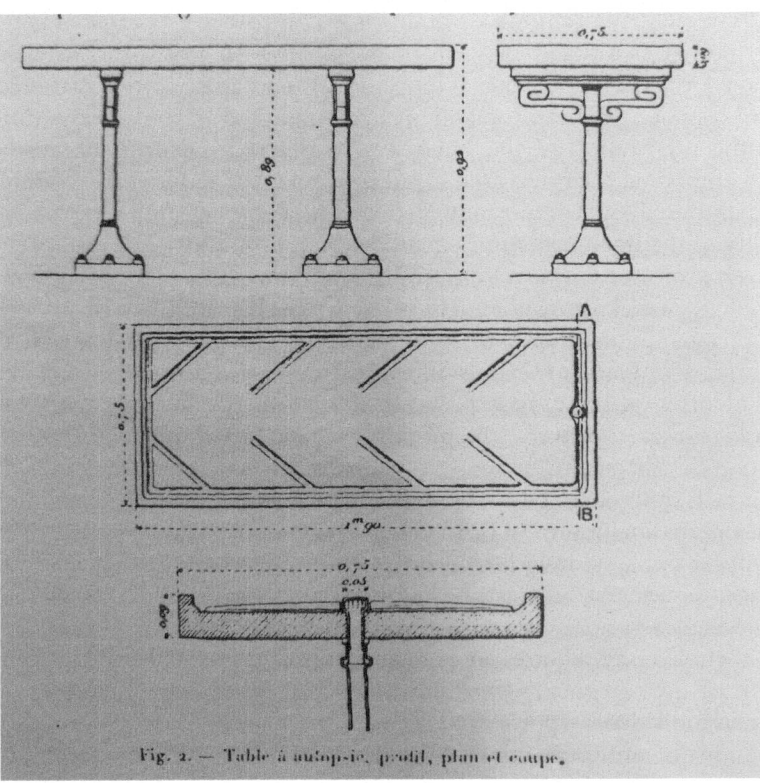

Fig. 2. — Table à autopsie, profil, plan et coupe.

Illustration 5.7 An 'ingenious' table design from 1903, with drains and gutters to drain 'liquids and other cadaveric pieces' easily. From: Letulle, *La pratique des autopsies*, 21–2. © University of Amsterdam, Bijzondere Collecties, UBM: 1320 B 24

for the exemption of other human materials from burial. Studying different ways of disposal hence casts light on the social status and emotional value of different bodies and their parts. For instance, pathological alterations of the body, such as tumours, are mostly seen as 'foreign' and therefore not imbued with personal identity. It appears ridiculous to worry about a proper burial for 'renewable' body parts, such as the hair and skin cells we lose every day, or blood and saliva.[200] In a similar vein, one could argue that the drainage of blood and other liquids from the anatomical theatre was not only a hygienic or practical issue, but also an indication that these fluids were not significantly 'human'.

Conversely, the removal of brains was controversial because this body part figured as a materialisation of personality. In 1923, for instance, a scandal over the clandestine removal of a brain arose in the Brussels St. Jean hospital. Although the patient had consented to the autopsy during life, the preparation of the brain was considered indecent: because the brain embodied their deceased loved one, relatives and friends thought its 'proper burial' was important.[201] The other way around, the personal quality of the brain sometimes incited research. In 1901, for instance, the Leuven professor Gustave Verriest dissected the brains of his friend Guido Gezelle, a well-known poet. He related the characteristics of his brains with his genius and linguistic competence.[202] The importance attached to the integrity of face and hands during autopsies also says something about the 'identity' of these body parts. In the words of the Liège pathologist Charles Firket, they were 'the physical traits by which relatives remember the one they had lost'.[203]

In fact, differences between whole bodies, internal organs and 'significant' organs are present in Belgian legislation until today. Whereas there is an opting-out system with regard to organ transplantation (meaning that organs, tissues and cells can be removed after death unless the deceased has lodged a formal objection during life), the anatomical donation of one's whole body requires an explicit permission. Tellingly, the transplantation of eyes (cornea)—an important facial feature, strongly identified with an individual—appears to remain more painful for relatives than the donation of internal organs.[204] The opting-out system reminds of nineteenth-century authorisations by tacit agreement. Even though the secret use of the internal parts of the dead body seems like something from the past, one could wonder if the removal of organs for transplantation today is really that different. As many people do not know that Belgium has an opting-out system, the absence of an objection does not necessarily mean that the person would have agreed with the procedure, much like not all patients who ended up on the autopsy slab in the late nineteenth century would have consented to this examination if they had been asked. The abstract knowledge that all our cells contain unique versions of DNA (our biological 'identity') has changed the meanings and value of tiny pieces of human material, leading to legislation on biobanks. This new importance attached to our smallest parts, however, does not apply for the dead: although regulations on biobanks attest to a growing importance of consent when it comes to living patients, only the use of entire bodies requires an explicit agreement when it comes to the dead.

Whereas the history of the wasting of 'insignificant' parts is mostly characterised by continuity, there have been shifts in what we consider as 'significant' human tissue. The fate of foetal remains is a telling example. In her research on the United States, Morgan has argued that evolutions in the treatment of embryonic and foetal remains echoed changes in their 'social value'. Morgan has posited that whereas anatomists casted embryos and foetuses as 'inert, impersonal entities' 100 years ago, they have become 'active, personified, anthropomorphised agents' today. In her view, this is clear from an evolution in the ways in which embryonic and foetal remains are disposed of: whereas around 1900, embryos and foetuses came to reside within the exclusive jurisdiction of scientific medicine and often ended up in teaching and research collections, the advent of the personi-fied public foetus—think, for example, about the use of foetal images in anti-abortion campaigns has led to calls for more respectable and humane burials.[205]

It is telling in this respect that embryonic and foetal remains received a different treatment than infant remains in late nineteenth-century Brussels. Whereas anatomists often preserved less developed foetuses in collections, stillborn and newborn children were mostly buried individually in a sepa-rate coffin or alongside their mother if she had died during childbirth. Apart from legal requirements—foetuses from six months of gestation fell under burial law—this criterion of age was tied up with the shape of the remains. As stillborn foetuses already looked like children, it was consid-ered important that they received a proper burial ritual.[206] Conversely, foetuses which were gestated for only a few months often ended up in scientific collections because they were not recognised (in both senses of the word) as humans.[207] The active participation of parents, who in many cases deliberately gave their deceased foetus to physicians, too, suggests that miscarried or abnormal foetuses hardly counted as individuals worthy of a funeral, although the economic benefit of saving burial costs might have influenced their decision as well.[208]

Foetuses with malformations, back then tellingly called *monstruosités*, were often preserved even if they had been gestated longer. This can be partly explained by the scientific value of their rare pathologies; yet in addition, their preservation followed from their different social value. In 1889, for example, the 24-year-old woman Ida Conard gave birth to two children after nine months of pregnancy: the first was a 'well-formed' girl, the other 'a monster child'. The intern who assisted her during the diffi-cult delivery, a certain Hector Nicaise, immediately planned to keep the latter for the scientific collections, even though it was still alive. The child died after 27 hours. In his request 'to not bury' the remains and in his report for the Society for Pathological Anatomy, Nicaise explained that this child had to be classified 'outside of the human type'.[209] The city council endorsed Nicaise's request. However, the child ultimately did not end up in the anatomical collection, because it decomposed too fast.[210]

This example again clarifies that the treatment of human remains as waste was-and is-both dependent on their social value based on enduring associations with the deceased, and on their materiality. The transformation of human materials into medical waste followed not only from the natural or scientific qualities of the flesh, but also from social considerations. Whether before, during or after dissection, the treatment of anatomical bodies remained a negotiation. Studying different ways of disposal together therefore provides an insight into changing sensitivities with regard to the relations between body and personhood, and into what it means to be (dis)qualified as human.

From Jar to Coffin

In 1878, Sacré informed the commission of hospitals about the ongoing transfer of the anatomical collections to the university. Regarding the pathological specimens preserved in the St. Jean hospital, he noted that 'he had those that had not deteriorated with time transferred to the university', while 'the others were sent to the cemetery'. Dozens of specimens ended up together in a grave.[211] This short update offers a rare insight into the ultimate fate of most human remains residing in collections. In many instances, anatomical specimens ended their post-mortem journey at the cemetery, after moving between different shelves, classrooms and examination tables. A jar was rarely held in perpetuity.

Even though anatomical collections, in the words of Alberti, were dynamic organisms 'growing and shrinking, changing over space and time', many studies have focused on what is left of collections today.[212] Whereas a few remaining specimens have received object biographies, showcasing the dynamic roles and meanings of human remains in collections, the 'missing majority' has not been subject to a great deal of scrutiny.[213] However, when one compares the dozens or hundreds of pages of nineteenth-century inventories with the scarce remainders of collections, it becomes painfully clear that the few surviving specimens, which were mostly intended for display, are not at all representative of the collections to which they once belonged.

Historians of anatomy therefore should try to include absent objects—such as those that Sacré sent to the cemetery—in their analysis. The study of missing specimens could be revealing for various reasons. In his well-known essay on the 'missing footstool', Glenn Adamson has argued that the appearance and disappearance of things allow for the explanation of historical (meta)narratives.[214] In the same way that the deficiency of surviving British domestic footstools before 1800 may tell us something about larger historical issues (like class, luxury, sexuality and the exotic), the disposal of anatomical specimens may cast light on changing standards with regard to the preservation and use of human remains in collections.

In her research on North American medical museums, for instance, Erin McLeary has related the partial dismantling of anatomical collections in the 1920s to changing research and didactic practices. In her view, anatomical specimens were increasingly neglected in this period, since students and researchers failed to utilise them as tools for learning or for generating knowledge.[215] In more recent decades, specimens have turned into redundant or inferior teaching tools in comparison to their three-dimensional digital representations. The University of Liège, for example, provides medical students with online photographs of their anatomical collection, while the actual specimens are locked away from view.

Both the disposal and repurposing of anatomical specimens reflect reconfigurations of the medical landscape. The Brussels hospital archives reveal that research interests shaped anatomical collections and vice versa. Changes in the scientific landscape often went hand in hand with the removal of dozens of anatomical specimens 'that no longer offered any scientific insights'.[216] The same goes for the relegation of specimens for display. In 1884, for instance, Leboucq explained that museum specimens that had lost their relevance 'could still be used for demonstrations and the teaching of students' until they were worn to shreds. Changing scientific standards could turn former display specimens into (finite) handling specimens.[217] In other instances, teaching or research tools gradually transformed into historical heritage, as happened with the young woman's head with which this chapter began.

Anatomical specimens were discarded not only for scientific, but also for material reasons. As aforementioned, especially anatomical specimens that were used for teaching did not last long. Curators regularly reported that they threw them away 'as they had perished through intensive use'.[218]

If pathological specimens representing conditions that appeared fairly frequently decomposed, they were replaced. If rarer specimens were in decay, anatomists tried to save what they could: they sometimes re-dissected specimens in order to remove bits that were in decomposition.[219] Perhaps ironically, also limited use was a threat to anatomical specimens. If collections remained untouched, jars gathered dust and their contents deteriorated. For example, as early as 1864, the curator of the hospital of St. Jean complained about the deplorable state of older, little used anatomical specimens. If no one took care with them, 'the alcohol evaporates or becomes cloudy'. Inadequately sealed bottles, unsuitable temperatures and broken jars further endangered his collection. In order to remain viable for preservation, the curator concluded, he needed more time for 'maintenance work': jars needed to be regularly opened, filled with preservation fluid and relabelled.[220]

Most significantly, the removal of anatomical specimens sometimes attested to changing ethical considerations. In her pioneering anthropological research, Mary Douglas connected disposal to issues of boundary and order. In her account, disposal was about getting rid of all that threatens to pollute a community: by placing unwanted things 'outside', members of a community create a sense of shared identity and belonging.[221] More recently, Kevin Hetherington has stressed that disposal is 'thoroughly constitutive of social and indeed ethical activity'. In his view, getting rid of unwanted things is important for the maintenance of a recognisable state of social order.[222] In this sense, the disappearance of anatomical specimens is *performative*: disposal can be an expression of changing considerations, whether scientific or ethical.[223]

In the case of the aforementioned phrenological skull collection from Ghent University, for instance, the immediate cause for destruction was an attempt to distance anatomy from an atrocious recent past. Following Lynn Meskell's definition of 'negative heritage', they were conflictual objects that became 'the repository of negative memory in the (collective) imaginary'.[224] After the Second World War, the new professor of anatomy Julien Fautrez decided that the skulls were no longer viable for recycling and discarded them to wipe out their most recent significance. Although archival traces connect the collection to the emergence of phrenology and the subspecialisation of anatomical collections, the skulls were eventually abandoned to dismantle any link between anatomy and Nazi ideologies. The destruction of the specimens was meant to dissolve a problematic past with as little controversy as possible.

The other way around, the few remaining specimens might also tell us something about ethical considerations and sensitivities. To give one concrete example: it is estimated that the Ghent anatomist Adolphe Burggraeve and his aide Edouard Meulewaeter made over 200 preparations per year. Only three remain today. The most well-known is a newborn child dressed in a white christening gown, lauded for its rosy cheeks and natural skin tone. The others are the young woman's head (a specimen that continues to evoke emotions until today), and a piece of skin injected with quicksilver. In my view, it is no coincidence that two out of three remaining specimens are made of body parts that are strongly invested with a sense of identity, whereas the other is shiny like a piece of jewellery.

How did the other specimens disappear? Little is known about what ultimately happened with the less emotionally charged or aesthetically pleasing preparations of Burggraeve and his nineteenth-century colleagues. In the Brussels city archives, requests to bury 'jars with organs' have been preserved.[225] At least until the 1940s, discarded anatomical specimens were brought to the cemetery, where they were buried anonymously, together in a grave. A few sources suggest that deteriorated specimens occasionally ended up in coffins with other dissected remains.[226] The phrenological skull collection presumably found its final resting place in the central heating boilers of the university, where the skulls were sent up in flames along with the university's history of collaboration.[227]

Interestingly, these anonymous burials or cremations have become subject to criticism in recent decades. Historians and public figures have put forward that anatomical remains, in particular if they had been acquired without consent, deserve named commemoration and full documentation. In Ghent, the phrenological skull collection has been linked to a register with the prisoners' names. Although the specimens have been destroyed, their original identity is known again today.[228] More generally, Elizabeth Hurren has argued that historians of anatomy should 'seek to make new human connections about our collective medical past' by giving every pauper who was compelled to dissection a name, an address, a place of birth, a sex, an age, family connections and other personal characteristics.[229]

This new importance attached to the identity of the 'victims' of anatomy was in part influenced by recent scandals around the retention and display of human tissue in hospital settings and museums.[230] In 1999, for instance, an organ retention scandal at the Alder Hey Hospital in Liverpool caused public outrage, which ultimately led to the Human Tissue Act, which regulates the removal, storage, use and disposal of human remains in England, Northern Ireland and Wales since 2004.[231] Another controversy arose over the display of the skeleton of Charles Byrne, 'the Irish giant', at the Hunterian Museum of the Royal College of Surgeons in London. The skeleton, which is still on display today, was acquired by the anatomist John Hunter in the eighteenth century, in contravention of the last wishes of Byrne, who wanted to be buried at sea in order to avoid dissection. In 2011, ethicists Thomas Muinzer and Len Doyal posited that the Hunterian Museum should 'morally rectify' this injustice by burying the skeleton at sea at last.[232] In addition, there has been activism from source communities, leading to the repatriation of anatomical specimens and skeletons from various colonial contexts.[233]

Yet perhaps these scandals were only the direct cause of a change that had been in the making for a long time. The recent care for individual commemoration in fact attests to much slower developments: the developments that have been the subject of this book. On the one hand, it reflects the increased importance of consent: whereas anatomical material in the early nineteenth century was mainly acquired from members of the lower classes of society without permission, the unauthorised collection of human material has gradually become ethically unacceptable. On the other hand, criticism on the anonymous disposal of specimens shows that our relationship to the dead has changed. Not only has the funeral become an expression of the identity of the deceased, we also commemorate the dead in more personal terms. As a result, the journey from person to thing has been reversed. Even if the matter has ceased to exist, a few anatomical specimens have regained an identity, and are commemorated as a subject.

5   Conclusion

Registers of the anatomical institute in Brussels show that bodies received a number when entering the dissecting room. Louise Veekman, for instance, became no. 85, and Augustin Eggermont became no. 92. The students who dissected them did not know their name. For them, they were not the remains of specific persons, but teaching tools that represented general human anatomy. However, remains regained their name when moving out of the dissecting room for final disposal. On their burial day, 85 and 92 transformed into individuals again, commemorated by a marked grave.[234] Body parts also received a number when they became part of the anatomical museum. The skull of Marie-Therese van Helmont, for instance, turned into no. 59. However, the inventory also mentioned the number of her patient file, enabling students and researchers to retrace her name and case history.[235]

Whether the dead ended up as autopsied bodies, dissected remains or anatomical specimens, their humanness lingered on. After they left the post-mortem room, anatomical remains could transform into deceased persons again. Anatomical specimens might have been physically turned into objects through preservation—in the sense that they could be carried around and handled like other things—but they nevertheless often continued to refer to a deceased subject and his or her illness. In general, anatomical bodies were never fixed but constantly *converted*: materially, scientifically and socially. Materially, as the form of the body changed through fragmentation, preservation or deterioration. Scientifically, as the changing disciplinary landscape, new medical theories or technologies imbued remains with new meanings. Socially, because the treatment of body parts, whether as significant remainders of a deceased individual or as medical waste, was always in part determined by their social value.

The individuality of human remains lived on in various ways. From a scientific point of view, the patient gained importance as an individual case history with the rise of pathology. Whereas early nineteenth-century anatomists had been interested in the rules of nature (the 'normal' structures of the healthy body), pathologists studied its exceptions (the myriad variations of the pathological body). As the sick body was a lot more difficult to generalise than the healthy body, pathologists promoted exact observations of individual patients. In the late nineteenth century, this preference for the individual case was enhanced by the new epistemological ideal of mechanical objectivity and facilitated by the development of accounting practices within hospitals.

It is important to note, however, that this increased attention for the individual case did not imply that there was more attention for the patient as a person. Objectification in a Foucauldian sense—signifying the detachment from body and person, the notion of medicine based on the analysable body undisturbed by the patient's discourse—did not require the stripping of the singularity of the patient. The descriptions of pathological specimens within the Society for Pathological Anatomy of Brussels clearly show that a named individual could become a research object without the need for anonymising generalisation. The body could be detached from the person and nevertheless remain an individual body. A number might be impersonal, but is still unique.

The individual person *did* become more important when it came to the burial of autopsied and dissected remains. The fate of certain body parts, namely those associated with the known personality of the deceased, changed as a result of the new emphasis on the corpse in sentimental burial rituals. The individualisation of death altered the face of the anatomy burial: from the end of the nineteenth century onwards, anatomical remains were no longer buried anonymously, but were, to the extent possible, placed together in an individual coffin and in a marked grave. In this way, significant parts of autopsied and dissected bodies were 'reunited' with the person they once were. The rise of body donation further facilitated this development. Whereas unclaimed bodies had been detached from their social relations, donated bodies were still enmeshed in ongoing contacts with relatives and friends. After serving as scientific objects, these bodies reverted to being social subjects.

By studying distinct ways of disposal together, this chapter has sketched a multifaceted view of the fate of anatomical remains that was not necessarily disgraceful or abhorrent. The final disposal of anatomical remains was a compromise, in which divergent meanings of the dead body in different areas of life came together. The division of the body into different categories—scientifically valuable specimens, significant parts associated with the deceased or medical waste—allowed anatomists to reconcile the social requirement of a decent burial with the medical use and exhaustion of other parts. Much like the autopsy was a useful alternative for the (socially more sensitive) dissection, the distinction between significant remains and medical waste allowed for the reconciliation of social and religious customs with didactic and scientific needs. From acquisition to disposal, the treatment of anatomical bodies was a politically, scientifically and socially negotiated compromise, determined by different actors, laws, beliefs, relationships and material realities.

NOTES

1. Based on: Dossier concernant la remise préalable à l'un des hôpitaux de tout cadavre destiné à l'université, 1886, folder 66, Université, maternité, amphithéâtre 1834–1922, Affaires générales, ASSB; Conseil des hospices to head of St. Pierre, 4 December 1886, folder 292, décès-autopsies 1852–1922, Fonds du directeur de l'hôpital St. Pierre, ASSB.
2. On the history of the young woman's head: Veronique Deblon and Pieter Huistra, 'Het geheim van de anatoom: Adolphe Burggraeve en de ontwikkeling van de Belgische anatomie in de negentiende eeuw',

Studium: Tijdschrift voor Wetenschaps—en Universiteitsgeschiedenis 9, no. 4 (2016): 202–16; Veronique Deblon, 'Het zachte gedruis van het leven. De preparaten van Adolphe Burggraeve', in *Post Mortem. Vesalius tussen kunst en wetenschap*, ed. Marjan Doom (Ghent: Universiteit Gent, 2015), 63–5.

3. Cara Krmpotich, Joost Fontein and John Harries, 'The Substance of Bones: the Emotive Materiality and Affective Presence of Human Remains', *Journal of Material Culture* 15, no. 4 (2010): 371–84; Elizabeth Hallam, 'Articulating Bones: An Epilogue', *Journal of Material Culture* 15, no. 4 (2010): 465–92; Elizabeth Hallam, Jenny Hockey and Glennys Howarth, *Beyond the Body: Death and Social Identity* (London and New York: Routledge, 1999).

4. Helen Lambert and Maryon McDonald, 'Introduction', in *Social Bodies*, eds. Helen Lambert and Maryon McDonald (Oxford and New York: Berghahn, 2011), 1–15.

5. Samuel J.M.M. Alberti, *Morbid Curiosities: Medical Museums in Nineteenth-Century Britain* (Oxford: Oxford University Press, 2011), 164.

6. Leboucq, *Le musée anatomique de l'Université de Gand* (Ghent: Vanderhaeghen, 1884), 12–13.

7. Deblon and Huistra, 'Het geheim van de anatoom', 202–16.

8. Alberti, *Morbid Curiosities*, 75.

9. Ibid., 53–66.

10. Hieke Huistra, *The Afterlife of the Leiden Anatomical Collections: Hands On, Hands Off* (London and New York: Routledge, 2019), 18–56.

11. For example: Règlement d'ordre intérieur pour les élèves de l'université de Bruxelles, admis aux dissections qui ont lieu à l'amphithéâtre de l'hôpital de St Jean, 1872, folder 8, Minutes, Affaires générales, ASSB; Charles Firket, *Du but et de l'organisation du service des* autopsies (Liège: Vaillant-Carmanne, 1883), 12–16.

12. Huistra, *The Afterlife*, 29–31.

13. Jacques-Pierre Maygrier, *Manuel de l'anatomiste*, 4th ed. (Paris: Gabon, 1818), xxxvii.

14. Leboucq, *Le musée anatomique*, 17–18.

15. Carin Berkowitz, 'Systems of Display: The Making of Anatomical Knowledge in Enlightenment Britain', *The British Journal for the History of Science* 46, no. 3 (2012): 359–87, on p. 370.

16. Quotes from: Leboucq, *Le musée anatomique*, 9. Chloé Pirson has discussed differences between realistic and didactic representations in her study on anatomical models: Chloé Pirson, *Corps à corps. Les modèles anatomiques entre art et médecine* (Paris: Mare et Martin, 2009), 114–42.

17. Leboucq, *Le musée anatomique*, 6.
18. Huistra, *The Afterlife*, 29–31.
19. Auguste Brunin to Conseil des hospices, August 26, 1901, folder 66, Université. Maternité, amphithéâtre, etc. 1834–1922, Affaires générales, ASSB.
20. Huistra, *The Afterlife*, 18–56.
21. Bralion to Conseil des hospices, October 16, 1879, folder 69, Collections scientifiques, Affaires générales, ASSB.
22. Rapport au conseil, October 21, 1875, folder 69, Collections scientifiques, Affaires générales, ASSB.
23. Pierre Boitard, *Nouveau manuel complet du naturaliste préparateur. Deuxième partie. Taxidermie, préparation des pièces anatomiques* (Paris: Roret, 1890), viii.
24. *BSAP* 1 (1859): 2.
25. Alberti, *Morbid Curiosities*, 75.
26. Tricia Close-Koenig, 'Cataloguing Collections: The Importance of Paper Records of Strasbourg's Medical School Pathological Anatomy Collection', in *The Fate of Anatomical Collections*, eds. Rina Knoeff and Rob Zwijnenberg (New York: Ashgate, 2015), 211–27, on p. 221.
27. *BSAP* 11 (1866): 4.
28. Joseph Guislain, *Lettres médicales sur l'Italie, avec quelques renseignements sur la Suisse* (Ghent: Gyselynck, 1840), 313.
29. André Uytterhoeven, *Notice sur l'hôpital Saint-Jean de Bruxelles ou l'étude sur la meilleure manière de construire et d'organiser un hôpital de malades* (Brussels: N.J. Gregoir, 1852), 202–7, on p. 204.
30. Rapport sur l'organisation du service des autopsies dans les hôpitaux de Bruxelles [1878], folder 107, Autopsies, Affaires générales, ASSB.
31. *BSAP* 41 (1892): 4.
32. *BSAP* 40 (1891): 3.
33. See Chap. 4.
34. Maurice De Laet, 'Le droit à l'autopsie et à prélèvements post-mortem', *Bulletin de l'Académie royale de médecine de Belgique* 107 (1948): 84–92.
35. In the hospital archives of Brussels, only one objection of this kind has been preserved: Parmentier Fernand to head of St. Pierre, February 5, 1905, folder 292, Autopsies-décès 1852–1922, Fonds du directeur de l'hôpital St. Pierre, ASSB.
36. Thomas Harris, *Manuel d'autopsies ou méthode de pratiquer les examens cadavériques au point de vue clinique et médico-légal*, trans. H. Surmont (Brussels: Manceaux, 1888), 84–5.
37. City council to Conseil des hospices, June 19, 1890, folder 292, Amphithéâtre, autopsies et dissections, Fonds du directeur de l'hôpital St. Pierre, ASSB.

38. Dozens of requests for the preservation of bodies are held by the Brussels City Archives: Corps conservés dans des collections scientifiques, 1878–1986, folder ASB Cultes 2843, CAB.

39. Dossier concernant la demande au collège de ne pas inhumer le cadavre du nommé Roberts, 1890, folder 131, Liasse divers, Affaires générales, ASSB.

40. *BSAP* 1 (1879): 9.

41. Récolement de la collection d'anatomie pathologique, December 31, 1886, folder 4A 43 (82), UAG; Récolement de la collection d'anatomie pathologique, December 31, 1888, 4A 45 (82), UAG.

42. *BSAP* 22 (1873): 11.

43. Ibid., 18.

44. Joris Vandendriessche, 'Anatomy and Sociability in Nineteenth-Century Belgium', in *Bodies Beyond Borders: Moving Anatomies, 1750–1950*, eds. Kaat Wils, Raf De Bont and Sokhieng Au (Leuven: Leuven University Press, 2017), 51–72, on pp. 63–8.

45. *BSAP* 34 (1885): 5.

46. See, for example: Alberti, *Morbid Curiosities*, 95–8; Anita Guerrini, 'Anatomists and Entrepreneurs in Early Eighteenth-Century London', *Journal of the History of Medicine and Allied Sciences* 59, no. 2 (2004): 219–39, on pp. 231–2; Ruth Richardson, 'Human Remains', in *Medicine Man: The Forgotten Museum of Henry Wellcome*, eds. Ken Arnold and Danielle Olsen (London: British Museum Press, 2003), 319–45, on p. 342.

47. Lisa O' Sullivan and Ross L. Jones, 'Two Australian Fetuses: Frederic Wood Jones and the Work of an Anatomical Specimen', *Bulletin of the History of Medicine* 89, no. 2 (2015): 243–66, on p. 245.

48. Jonathan Reinarz, 'The Age of Museum Medicine: The Rise and Fall of the Medical Museum at Birmingham's School of Medicine', *Social History of Medicine* 18, no. 3 (2005): 419–37, on p. 437.

49. For example: *BSAP* 8 (1863): 21; *BSAP* 25 (1876): 189; M. Hanaut, 'Catarrhe chronique de la vessie', *Presse médicale belge*, January 10, 1875, 41–2.

50. Examples from: *BSAP* 8 (1863): 21; *BSAP* 16 (1869): 28; *BSAP* 18 (1870): 79; *BSAP* 23 (1874): 23; *BSAP* 26 (1877): 23.

51. For example: *BSAP* 26 (1877): 10; *BSAP* 29 (1880): 189; *BSAP* 35 (1886): 26.

52. *BSAP* 39 (1890): 75. Another example is: Lucien Thiriar, 'Fracture exposée de la jambe', *La clinique: organe officiel des hôpitaux de Bruxelles* 7 (1903): 41–4.

53. Lorraine Daston and Peter Galison, *Objectivity* (New York: Zone Books, 2007), 82.

54. Ibid., 55–190.
55. Ibid., 161.
56. Daston and Galison make a similar argument with regard to pathological atlases: Ibid., 82.
57. Close-Koenig, 'Cataloguing Collections', 211. For a more general argument on paper technologies and the production of knowledge, see: Volker Hess and Andrew Mendelsohn, 'Case and Series: Medical Knowledge and Paper Technology, 1600–1900', *History of Science* 48, no. 3–4 (2010): 287–314.
58. Volker Hess and Sophie Ledebur, 'Taking and Keeping: A Note on the Emergence and Function of Hospital Patient Records', *Journal of the Society of Archivists* 32, no. 1 (2001): 21–32.
59. For example: Vandervelde, 'Lésions cutanées dans le sclérème des nouveau-nés', *BSAP* 44 (1879): 67; Albert Delcourt, 'Le rachitisme. Sa pathogénie' (PhD diss., Université Libre de Bruxelles, 1899).
60. Tony Bennett, *The Birth of the Museum: History, Theory, Politics* (London: Routledge, 1995), 17–88. On Belgium: Liesbet Nys, *De intrede van het publiek: museumbezoek in België, 1830–1914* (Leuven: Leuven University Press, 2012), 129–56.
61. Alberti, *Morbid Curiosities*, 46–60.
62. Huistra, *The Afterlife*, 90–124. Idem, 'Weg met pottenkijkers! Hoe het publiek verdween uit het Leids anatomisch kabinet', *Negentiende Eeuw* 34, no. 3 (2010): 193–208.
63. Veronique Deblon, 'Commercialized Bodies: The Trade in Anatomical Preparations'. Paper presented at the EAHMH Conference, Cologne, September 2015.
64. This was also the case in Britain: Alberti, *Morbid Curiosities*, 85.
65. Renaud Bardez, 'La Faculté de médecine de l'Université Libre de Bruxelles: entre création, circulation et enseignement des savoirs, 1795–1914' (PhD diss., Université Libre de Bruxelles, 2016), 181–2.
66. Ibid., 184–7.
67. Dossier concernant le catalogue de la collection anatomo-pathologique dressé par M. le docteur Carpentier, 1872, folder 69, Collections scientifiques, Affaires générales, ASSB.
68. Head of St. Jean to Conseil des hospices, February 29, 1878, folder 69, Collections scientifiques, Affaires générales, ASSB.
69. *BSAP* 39 (1890): 1.
70. Vandendriessche, 'Anatomy and Sociability', 51–72.
71. Ibid., 56–7.
72. *BSAP* 39 (1890): 31.
73. For example: *BSAP* 27 (1878): 27. See also various requests in: Corps conservés dans des collections scientifiques, 1878–1986, folder ASB Cultes 2843, CAB.

74. Dossier concernant les pièces anatomiques demandées par M le docteur Casse, 1884, folder 261, matériel médical/pharmacie, Affaires générales, ASSB.

75. *BSAP* 41 (1892): 3.

76. Vandendriessche, 'Anatomy and Sociability', 62.

77. On these 'mechanisms of recognition': Ibid., 51–72; Lynn M. Morgan, *Icons of Life: A Cultural History of Human Embryos* (Berkeley: University of California Press, 2009), 76.

78. For example: Conseil des hospices to Université Libre de Bruxelles, December 9, 1870, folder 69, Collections scientifiques, Affaires générales, ASSB.

79. Deblon, 'Commercialized Bodies'.

80. Emile Spehl to Conseil des hospices, May 5, 1892, folder 107, Autopsies, Affaires générales, ASSB.

81. Reinarz, 'The Age of Museum Medicine', 429–32; Alberti, *Morbid Curiosities*, 54–66.

82. M. Guiette, Inventaire générale des collections scientifiques mises à la disposition de l'Université de Bruxelles, 1834, folder 69, Collections scientifiques, Affaires générales, ASSB.

83. Uytterhoeven, *Notice sur l'hôpital Saint-Jean*, 203–4.

84. Catalogue de la collection anatomo-pathologique dressé par M. le docteur Carpentier, 1872, folder 69, Collections scientifiques, Affaires générales, ASSB.

85. Procès-verbaux de récolement, 1874, folder 4A 25 (82), UAG.

86. Alphonse Le Roy, *Liber memorialis. L'Université de Liège depuis sa fondation* (Liège: Vaillant-Carmanne 1869), 1142–56.

87. Julie De Ganck, 'Cultiver la différence: Histoire du développement de la gynécologie à Bruxelles, 1870–1935' (PhD diss., Université Libre de Bruxelles, 2016), 385–8 and 400.

88. Paul Héger, 'Création d'un musée anthropologique à Bruxelles', *Bulletin de la société d'anthropologie de Bruxelles* 7 (1888–1889): 284–6.

89. Bardez, 'La faculté de médecine', 180–2. On the change of the curriculum in 1876: *Recueil de documents concernant la révision de la loi du 20 Mai 1876 sur la collation des grades académiques et le programmeme des examens universitaires* (Brussels: Adolphe Mertens, 1883), 5.

90. For example: *BSAP* 44 (1897): 10; Achat de sept crânes indiens, 1888, folder 45 (71), UAG.

91. Bardez, 'La faculté de médecine', 186.

92. *BSAP* 44 (1897): 81.

93. Ibid., 82.

94. Ibid., 81.

95. Alison Kraft and Samuel J.M.M. Alberti, 'Equal though Different. Laboratories, Museums and the Institutional Development of Biology in

Late-Victorian Northern England', *Studies in History and Philosophy of Biological and Biomedical Sciences* 34, no. 2 (2003): 203–36; Reinarz, 'The Age of Museum Medicine', 419–37.

96. Examples of this more oppositional view are: Steven Conn, *Museums and American Intellectual Life, 1876–1926* (Chicago: University of Chicago Press, 1998), 15–18; Andrew Cunningham and Perry Williams, eds., *The Laboratory Revolution in Medicine* (Cambridge: Cambridge University Press, 1992).

97. Kraft and Alberti, 'Equal though Different', 207.

98. *BSAP* 39 (1890): 2.

99. Huistra, *The Afterlife*, 58–60.

100. Leboucq, *Le musée anatomique*, 6.

101. This section draws on: Tinne Claes and Veronique Deblon, 'When Nothing Remains: Anatomical Collections, the Ethics of Stewardship and the Meanings of Absence', *Journal of the History of Collections* 30, no. 2 (2018): 351–62.

102. Daniel Mareska and J. Heyman, *Enquête sur le travail et la condition physique et morale des ouvriers employés dans les manufactures de coton à Gand* (Ghent: Gyselinck, 1845), 135.

103. Dossier concernant la collection phrénologique à la Maison de Force à Gand, 1844, folder T015/583, SAB.

104. *Annales des Universités de Belgique: contenant des mémoires de professeurs et d'agrégés des universités, les mémoires couronnés aux concours universitaires, des mémoires de docteurs spéciaux, et d'autres documents et pièces académiques* (Brussels: Lesigne, 1850), 828–9.

105. Charles Poelman, *Catalogue des collections d'anatomie comparée y compris les ossements fossiles de l'université de Gand* (Ghent: Vanderhaeghen, 1868), 8.

106. Charles Marie Debierre, *Le crâne des criminels* (Lyon: Masson, 1895), 4; *Actes du troisième congrès international d'anthropologie criminelle tenu à Bruxelles* (Brussels: Hayez, 1893), 238.

107. For example: Jules Dallemagne, *Les stigmates anatomiques de la criminalité* (Paris: Masson, 1894), 36–8. On late nineteenth-century criminal anthropology in Belgium, see: Leen Beyers, 'Rasdenken tussen geneeskunde en natuurwetenschap: Emile Houzé en de Société d'Anthropologie de Bruxelles 1882–1921', in *Degeneratie in België 1860–1940: een geschiedenis van ideeën en praktijken*, eds. Jo Tollebeek, Geert Vanpaemel en Kaat Wils (Leuven: Leuven University Press, 2003): 43–78.

108. Wouter De Raes, 'Roger Soenen: een Vlaams-nationalistische rassentheoreticus (1902–1977)', *Wetenschappelijke Tijdingen* 61 (2002): 79–86.

109. Roger Soenen, 'Over de rassenkundige samenstelling van het Vlaamsche volk', *DeVlag* 4 (1942): 392–3; Idem, *Enkele begrippen over ras—en rassenkunde* (Brussels: De Burcht, 1943).

110. For example: *BSAP* 32 (1883): 145–52; *BSAP* 34 (1885): 14 and Alfred Stocquart, 'La théorie d'Albrecht concernant la signification morphologique de bec-de-lièvre compliqué de fissure palatine', *Bulletin de la société d'anthropologie de Bruxelles* 11 (1892): 185–93.

111. Hector Leboucq, *L'anatomie humaine et les tendances modernes de la morphologie* (Ghent: Annoot-Braeckman, 1907), 9.

112. Huistra, *The Afterlife*, 70.

113. Fernand Héger-Gilbert, *La radiographie foetale envisagée au point de vue médico-légal* (Brussels: Piette, 1908), 4.

114. *BSAP* 4 (1861): 31.

115. *BSAP* 38 (1889): 95.

116. For example: *BSAP* 34 (1885): 3; *BSAP* 20 (1880): 188; *BSAP* 44 (1897): 67.

117. Dossier concernant le placement de 2 cadavres d'enfant dans un même cercueil à St Jean, 1882, folder 38, Inhumations, Affaires générales, ASSB.

118. For exceptions, see note 120.

119. For example: Richardson, *Death, Dissection and the Destitute*, 275; Thomas Laqueur, 'Bodies, Death, and Pauper Funerals', *Representations* 1, no. 1 (1983): 109–31; Michael Sappol, *A Traffic of Dead Bodies: Anatomy and Embodied Social Identity in Nineteenth-Century America* (Princeton and Oxford: Princeton University Press, 2002), 36.

120. See, for example: Julie Rugg, 'Constructing the Grave: Competing Burial Ideals in Nineteenth-Century England', *Social History* 38, no. 3 (2013): 328–45; Julie-Marie Strange, *Death, Grief and Poverty in Britain, 1870–1914* (Cambridge: Cambridge University Press, 2005); Steven King, *Sickness, Medical Welfare and the English Poor, 1750–1834* (Manchester: Manchester University Press, 2018), 220–48; Elizabeth T. Hurren and Steve King, 'Begging for a Burial: Form, Function and Conflict in Nineteenth-Century Pauper Burial', *Social History* 30, no. 3 (2005): 321–41.

121. Helen MacDonald, 'A Body Buried is a Body Wasted: The Spoils of Human Dissection', in *The Body Divided: Human Beings and Human 'Material' in Modern Medical History*, eds. Sarah Ferber and Sally Wilde (Farnham: Ashgate, 2012), 9–27; Elizabeth Hallam, 'Anatomical Bodies and Materials of Memory', in *Death Rites and Rights*, eds. Belinda Brooks-Gordon et al (Oxford: Hart Publishing, 2007), 279–98; Elizabeth T. Hurren, *Dying for Victorian Medicine: English Anatomy and its Trade in the Dead Poor, ca. 1834–1929* (Basingstoke and New York: Palgrave Macmillan, 2012), 41–73; Jolien Gijbels, 'Reassessing the Pauper Burial:

The Disposal of Corpses in Nineteenth-Century Brussels', *Mortality: Promoting the Interdisciplinary Study of Death and Dying* 23, no. 2 (2018): 184–98.

122. MacDonald, 'A Body Buried is a Body Wasted', 9–27.
123. Hurren, *Dying for Victorian Medicine*, 41–73.
124. Examples from: Conseil des hospices to the head of St. Jean, June 17, 1864, folder 38, Inhumations, Affaires générales, ASSB; city council of Brussels to Conseil des hospices, December 20, 1878, folder 38, Inhumations, Affaires générales, ASSB; Comité de charité to Conseil des hospices, November 5, 1885, folder 38, Inhumations, Affaires générales, ASSB.
125. Conseil des hospices to the head of St. Pierre, December 4, 1886, folder 292, Décès/Autopsies, Fonds du directeur, ASSB; Hippolyte Lippens to University board, July 27, 1889, folder 46 (100), UAG.
126. Ferdinand Dupont to Charles Buls, January 12, 1886, folder ASB Bienfaisance S107/2, Dépôts mortuaires des hôpitaux, CAB.
127. Dossier concernant la plainte de la veuve Pouillon relative à l'enterrement de son mari décédé à l'hôpital de St. Jean, sans avoir été prévenue à temps, 1885, folder 113, Hôpital St. Jean: généralités, Affaires générales, ASSB; Dossier concernant un article des 'nouvelles du jour' à charge de l'hôpital St. Pierre, 1880, folder 130, Hôpital St. Pierre: divers, Affaires générales, ASSB.
128. For example: King, *Sickness*, 220–48; Julie-Marie Strange, 'Only a Pauper Whom Nobody Owns: Reassessing the Pauper Grave c.1880–1914', *Past & Present* 178, no. 1 (2003): 148–75; Anne Digby, *The Poor Law in Nineteenth-Century England and Wales* (London: Historical Association, 1982), 19–26.
129. Mary MacKinnon, 'English Poor Law Policy and the Crusade Against Outrelief', *The Journal of Economic History* 47, no. 3 (1987): 603–25; Karel Williams, *From Pauperism to Poverty* (London: Routledge, 1981).
130. Hurren and King, 'Begging for a Burial', 321–41; Elizabeth T. Hurren, 'A Pauper Dead-House: The Expansion of the Cambridge Anatomical Teaching School Under the Late-Victorian Poor Law, 1870–1914', *Medical History* 48, no. 1 (2004): 69–94.
131. See, for example: Elizabeth T. Hurren, *Protesting about Pauperism: Poverty, Politics and Poor Relief in Late-Victorian England, 1870–1900* (Woodbridge: Boydell Press, 2007), 191–241; Peter Jones and Steven King, 'Obligation, Entitlement and Dispute: Navigating the English Poor Laws 1600–1900', in *Obligation, Entitlement and Dispute under the English Poor Laws*, eds. Peter Jones and Steven King (Newcastle upon Tyne: Cambridge Scholars, 2015), 1–19; Elizabeth T. Hurren, 'World

Without Welfare. Pauper Perspectives on Medical Care under the Late Victorian Poor Law 1870–1900', in *Obligation, Entitlement and Dispute*, 292–320.

132. Ferdinand Dupont to Charles Buls, January 12, 1886, folder ASB Bienfaisance S107/2, Dépôts mortuaires des hôpitaux, CAB.

133. Examples in notes 122–125.

134. Gijbels, 'Reassessing the Pauper Burial', 184–98.

135. Charles de Gronckel, *Hospices civils et bureaux de bienfaisance. Précis du régime légal de l'assistance public* (Brussels: Bourlard and Havaux, 1884), 509.

136. See Chap. 3.

137. Adolphe Samyn, Construction d'un dépôt mortuaire à l'hôpital St. Pierre, 1886, cartes et plans, ASSB.

138. Ferdinand Dupont to Charles Buls, January 12, 1886, folder ASB Bienfaisance S107/2, Dépôts mortuaires des hôpitaux, CAB.

139. Head of St. Pierre to Conseil des hospices, May 14, 1902, folder 39, Inhumations, Affaires générales, ASSB.

140. Conseil des hospices to City council, February 18, 1887, folder 38, Inhumations, Affaires générales, ASSB.

141. Head of St. Pierre to M. Vanderton, December 20, 1886, folder 130, Hôpital St. Pierre: divers, Affaires générales, ASSB.

142. Gijbels, 'Reassessing the Pauper Burial', 184–98.

143. 'Service des inhumations', *Le Peuple*, June 11, 1904.

144. 'Inhumations et transports funèbres: suppression des transports collectifs de corps d'indigents adultes non décédés à domicile ainsi que des corps de tous les enfants pauvres', *Bulletin communal de Bruxelles* 55 (1904): 921.

145. 'Amélioration des transports funèbres des indigents', *Bulletin communal de Bruxelles* 54 (1903): 1650; 'Inhumations et transports funèbres', 894.

146. Dossier concernant l'achat d'un drap mortuaire pour recouvrir les cercueils des personnes venant à décéder dans les hôpitaux en dehors de toute communion religieuse, 1876, folder 76, Instructions concernant les hôpitaux, Affaires générales, ASSB; 'Inhumations et transports funèbres', 890.

147. Conseil des hospices to head of St. Pierre, May 29, 1886, folder 292, Autopsies-décès 1852–1922, Fonds du directeur de l'hôpital St. Pierre, ASSB.

148. See Chap. 3.

149. Dossier concernant les locaux à affecter aux cérémonies des cultes dissidents 1885–1886, folder 34, Cultes, Affaires générales, ASSB.

150. Gijbels, 'Reassessing the Pauper Burial', 184–98.

151. Dossier concernant la protestation de La Libre Pensée de St. Gilles au sujet des conditions dans lesquelles a lieu à l'hôpital St. Pierre l'enterrement de sieur Pardaens, 1911, folder 35, Cultes, Affaires générales, ASSB.

152. Dossier concernant l'autorisation générale de célébrer des services corps présent, 1896, folder 35, Cultes, Affaires générales, ASSB.

153. Jenny Nyberg, 'A Peaceful Sleep and Heavenly Celebration for the Pure and Innocent: The Sensory Experience of Death during the Long Eighteenth Century', in *Making Sense of Things: Archaeologies of Sensory Perception*, eds. Fredrik Fahlander and Anna Kjellström (Stockholm: Stockholm University Press, 2010), 15–33; John Troyer, 'Embalmed Vision', *Mortality: Promoting the Interdisciplinary Research of Death and Dying* 12, no. 1 (2007): 22–47.

154. Régis Bertrand, *Mort et mémoire. Provence, XVIIIe-XXe siècles, Une approche d'historien* (Marseille: La Thune, 2011), 21–56; Thomas Laqueur, *The Work of the Dead: A Cultural History of Mortal Remains* (Princeton and Oxford: Princeton University Press, 2015), 388–412.

155. The standard work on these evolutions is: Philippe Ariès, *L'homme devant la mort* (Paris: Seuil, 1977). See also: Sappol, *A Traffic of Dead Bodies*, 29–30; Rugg, 'Constructing the Grave', 328–45.

156. Strange, *Death, Grief and Poverty*, 66–97. Ruth Richardson, too, posited that 'the centrality of the corpse was popularly crucial': Richardson, *Death, Dissection and the Destitute*, 15.

157. Strange, *Death, Grief and Poverty*, 98–130 and 163–229.

158. Ibid., 133.

159. Richardson, *Death, Dissection and the Destitute*, 275. Other examples of this view are: Laqueur, 'Bodies, Death, and Pauper Funerals', *Representations* 1 (1983): 109–31; Sappol, *A Traffic of Dead Bodies*, 36.

160. Strange, *Death, Grief and Poverty*, 131–62 and 91–7.

161. Ibid., 148–51.

162. City council of Brussels to Conseil des hospices, March 1, 1887, folder 38, Inhumations, Affaires générales, ASSB.

163. On the ritual of viewing the corpse: Strange, *Death, Grief and Poverty*, 80–7; Richardson, *Death, Dissection and the Destitute*, 23–6.

164. Head of St. Jean to Conseil des hospices, May 12, 1902, folder 39, Inhumations, Affaires générales, ASSB.

165. On the wake: Strange, *Death, Grief and Poverty*, 87–91.

166. Gijbels, 'Reassessing the Pauper Burial', 184–98.

167. 'Dépôt mortuaire', *Bulletin communal de Bruxelles* 46 (1895): 708.

168. Charles Buls and Marcel Bots, *Het dagboek van C. Buls* (Ghent: Gent Liberaal Archief, 1987), 104.

169. Instructions concernant les cadavres—amphithéâtre, 1875, folder 107, Autopsies, Affaires générales, ASSB.

170. Head of St. Pierre to Conseil des hospices, July 15, 1886, folder 292, Décès—Autopsies, Fonds du directeur de l'hôpital St. Pierre, ASSB.
171. City council of Brussels to Conseil des hospices, November 27, 1886, folder 66, Université—Maternité—Amphithéâtre, Affaires générales, ASSB.
172. Note that Hurren linked the use of identifying tags to the trade in body parts. In her view, bodies were monitored to prevent students from taking away body parts to sell on. However, there is no evidence for such a trade in the Belgian context. Hurren, *Dying for Victorian Medicine*, 74.
173. Instructions concernant les cadavres—amphithéâtre, 1875, folder 107, Autopsies, Affaires générales, ASSB.
174. Ibidem.
175. City council of Brussels to Conseil des hospices, October 25, 1883, folder 76, Généralités, Affaires générales, ASSB.
176. Hippolyte Lippens to University administrator, July 8, 1889, folder 46 (100), Autopsies, UAG.
177. Répertoire des macchabées remis à l'institut d'anatomie 1899–1907, folder 1 FD 701, UAB; Register of bodies brought to the Institut d'anatomie, 1920–1968, Private Archives of Stéphane Louryan, Brussels.
178. Tatjana Buklijas, 'Cultures of Death and Politics of Corpse Supply: Anatomy in Vienna, 1848–1914', *Bulletin of the History of Medicine* 82, no. 3 (2008): 570–607, on p. 606.
179. Ibid., 588. Note that Buklijas did draw attention to differences between dissection and autopsy, or rather between 'whole bodies' and autopsied bodies, when discussing disciplinary competitions concerning the distribution of cadavers.
180. Hurren, *Dying for Victorian Medicine*, 67–8.
181. Hurren and King, 'Begging for a Burial', 321–41.
182. Hurren, *Dying for Victorian Medicine*, 68.
183. See Chap. 3.
184. Albert Dalcq to M. Barzin [1951], folder 1 FD 703, Corps légués ou mis à la disposition du service d'anatomie 1945–1968, UAB.
185. Jean La Barre, Rapport concernant la pénurie de matériel anatomique, 1949, folder 1 FD 703, Corps légués ou mis à la disposition du service d'anatomie 1945–1968, UAB.
186. Hallam, 'Anatomical Bodies and Materials of Memory', 279–98.
187. Jean La Barre to Albert Jottrand, December 14, 1949, folder 1 FD 703, Corps légués ou mis à la disposition du service d'anatomie 1945–68, UAB.
188. Jean La Barre, Rapport concernant la pénurie de matériel anatomique, 1949, folder 1 FD 703, Corps légués ou mis à la disposition du service d'anatomie 1945–68, UAB.

189. Ibidem.
190. Hallam, 'Anatomical Bodies and Materials of Memory', 292.
191. Jean La Barre, Rapport concernant la pénurie de matériel anatomique, 1949, folder 1 FD 703, Corps légués ou mis à la disposition du service d'anatomie 1945–68, UAB.
192. MacDonald, 'A Body Buried', 15 and 22.
193. Morgan shows the richness of the history of the concept of 'medical waste' in: Lynn M. Morgan, 'Properly Disposed Of. A History of Embryo Disposal and the Changing Claims on Foetal Remains', *Medical Anthropology* 21, no. 3–4 (2010): 247–74.
194. Demandes d'inhumer les viscères, folder ASB Cultes 2843, Corps conservés dans des collections scientifiques, 1878–1986, CAB.
195. Maurice Letulle, *La pratique des autopsies* (Paris: Nand, 1903), 20; Uytterhoeven, *Notice sur l'hôpital Saint-Jean*, 122–3.
196. Letulle, *La pratique des autopsies*, 21; Construction d'un nouveau dépôt mortuaire et amphithéâtre, 1904/1909, folder 21, Travaux—établissements hospitaliers, ASSB.
197. For instance: Dossier concernant les mesures prises pour la désinfection des amphithéâtres, 1883, folder 107, Autopsies, ASSB.
198. Susan C. Lawrence, 'Beyond the Grave: The Use and Meaning of Human Body Parts: A Historical Introduction', in *Stored Tissue Samples: Ethical, Legal and Public Policy Implications*, ed. Robert F. Weir (Iowa: University of Iowa Press, 1998), 113–14.
199. Morgan, 'Properly disposed of', 249.
200. Lawrence, 'Beyond the Grave', 113–14.
201. Dossier concernant l'autopsie pratiquée clandestinement du cadavre de l'infirmière Lankester, 1923–1924, folder 107, Autopsies, ASSB.
202. Raf De Bont, 'Onbeschaamde geleerden hebben zijn naaktheid betast: genialiteit, waanzin en degeneratie omstreeks 1900', in *Degeneratie in België*, 121–53, on p. 130.
203. Firket, *Du but et de l'organisation*, 21.
204. Belgisch raadgevend comité voor bio-ethiek. 2011. Advies betreffende bepaalde ethische aspecten van de wijzigingen door de wet van 25 februari 2007 aangebracht aan de wet van 13 juni 1986 betreffende het wegnemen en transplanteren van organen. https://www.health.belgium.be/sites/default/files/uploads/fields/fpshealth_theme_file/19104555/Advies%2050.pdf. Accessed July 17, 2017.
205. Morgan, 'Properly disposed of', 247–74.
206. Dossier concernant la remise préalable à l'un des hôpitaux de tout cadavre destiné à l'université, 1886, folder 66, Université, maternité, amphithéâtre 1834–1922, Affaires générales, ASSB.
207. Demandes de pouvoir conserver des fœtus 1889–1894, folder 246, Maternité: service de santé 1855–1923, Affaires générales, ASSB.

208. See various letters in: Corps conservés dans des collections scientifiques, 1878–1986, folder ASB Cultes 2843, CAB. Researchers working on the United States came to similar conclusions: Shannon K. Withycombe, 'From Women's Expectations to Scientific Specimens: The Fate of Miscarriage Materials in Nineteenth-Century America', *Social History of Medicine* 28, no. 2 (2015): 245–62.

209. 'Monstre pseudo-encéphalien', *BSAP* 38 (1889): 52–61, on p. 56.

210. Dossier concernant demande par M. le Dr. de la Maternité de ne pas inhumer le corps de l'enfant Conard Ida 1889, folder 246, Maternité: service de santé 1855–1923, Affaires générales, ASSB.

211. Joseph Sacré to Conseil des hospices, March 24, 1878, folder 69, Collections scientifiques, Affaires générales, ASSB.

212. Alberti, *Morbid Curiosities*, 80.

213. Cf. Linda Hurcombe, *Perishable Material Culture in Prehistory: Investigating the Missing Majority* (London: Routledge, 2014). Examples of object biographies of anatomical specimens are: Henrik B. Lindskoug and Anne Gustavsson, 'Stories from Below: Human Remains at the Gothenburg Museum of Natural History and the Museum of World Culture', *Journal of the History of Collections* 27, no. 1 (2015): 97–109; Lynn M. Morgan, 'A Social Biography of Carnegie Embryo no. 836', *The Anatomical Record* 276, no. 1 (2004): 3–7.

214. Glenn Adamson, 'The case of the missing footstool. Reading the absent object', in *History and Material Culture: A student's Guide to Approaching Alternative Sources*, ed. Karen Harvey (London: Routledge, 2009), 192–207.

215. Erin McLeary, 'Science in a Bottle: The Medical Museum in North-America, 1860–1940' (Unpublished PhD diss., University of Pennsylvania, 2001): 156–212.

216. Rapport sur les inventaires des collections scientifiques confiées à l'université, December 28, 1861, folder 69, Collections scientifiques, Affaires générales, ASSB.

217. Leboucq, *Le musée anatomique*, 24.

218. University board to Conseil des hospices, January 9, 1861, folder 69, Collections scientifiques, Affaires générales, ASSB.

219. For example: Van Volxem to Conseil des hospices, October 16, 1861, folder 69, Collections scientifiques, Affaires générales, ASSB.

220. Roger to Conseil des hospices, July 11, 1864, folder 69, Collections scientifiques, Affaires générales, ASSB.

221. Mary Douglas, *Purity and Danger*, 2nd ed. (London: Routledge, 1984), 35 et seq.

222. Kevin Hetherington, 'Secondhandness: Consumption, Disposal and Absent Presence', *Environment and Planning D: Society and Space* 22, no. 1 (2004): 157–73, on p. 158.

223. Morgan Meyer has discussed the performativity of absence in: Morgan Meyer, 'Placing and Tracing Absence: A Material Culture of the Immaterial', *Journal of Material Culture* 17, no. 1 (2012): 103–10, on p. 103.

224. Lynn Meskell, 'Negative Heritage and Past Mastering in Archaeology', *Anthropological Quarterly* 75, no. 3 (2002): 557–74, on p. 558.

225. Demandes d'inhumer les viscères, folder ASB Cultes 2843, Corps conservés dans des collections scientifiques, 1878–1986, CAB.

226. Conseil des hospices to head of St. Pierre, December 4, 1886, folder 292, décès-autopsies 1852–1922, Fonds du directeur de l'hôpital St. Pierre, ASSB.

227. Interview with Alexander Evrard, medical student under Roger Soenen, April 14, 2016.

228. Catalogue des crânes recueillis dans la maison de force de Gand, Museum of the history of medicine, Ghent.

229. Hurren, *Dying for Victorian Medicine*, xviii.

230. Tiffany Jenkins, *Contesting Human Remains in Museum Collections: The Crisis of Cultural Authority* (New York: Routledge, 2011); Gareth D. Jones and Maja Whitaker, 'The Contested Realm of Displaying Dead Bodies', *Journal of Medical Ethics* 39, no. 10 (2013): 652–3.

231. Michael Redfern, Jean W. Keeling and Elizabeth Powell, *The Royal Liverpool Children's Inquiry: Report* (London: The Stationary Office, 2001).

232. Len Doyal and Thomas Muinzer, 'Should the skeleton of 'the Irish giant' be buried at sea?' *British Medical Journal* 343 (2011), http://www.bmj.com/content/343/bmj.d7597. Accessed March 29, 2017.

233. Laurajane Smith, 'The Repatriation of Human Remains: Problem or Opportunity?' *Antiquity* 78, no. 300 (2003): 404–13.

234. Register of bodies brought to the Institut d'anatomie, 1920–1968, Private Archives of Stéphane Louryan, Brussels.

235. Pièces anatomiques sèches, Dossier concernant le transfert d'objets du cabinet d'anatomie pathologique de l'hôpital St Pierre à l'université, 1873, folder 69, Collections scientifiques, Affaires générales, ASSB.

CHAPTER 6

Conclusion

'The social disadvantages of dismemberment were far outweighed [...] by the cultural authority of the fragment', wrote Samuel Alberti in his research on nineteenth-century anatomical museums in Britain.[1] In his view, corpses could be divided into pieces and collected, despite the troubling associations of dismemberment, since fragments were regarded as the key to understanding nature around 1800. The small number of eighteenth-century anatomical collections could give rise to a range of museums containing thousands upon thousands of body parts because the scientific enterprise revolved around fragmentation. Anatomy thrived in a context in which men of science sought to know natural entities by breaking them into parts. In the view of Alberti, it was this belief in the fragment that legitimised the dismissal of the inviolability of the corpse. The disadvantages of anatomy were inextricably linked to its merits: the various meanings of the dead body—religious, cultural, emotional—could be set aside because the dissected cadaver held the promise of scientific progress.[2]

Alberti was not the first to point at the social cost of modern medicine. Authors such as Michel Foucault and Nicholas Jewson related physicians' shift of focus from the narrative of the patient to the empirical body to the changing power relations between relatively high-status doctors and poor patients within the hospital. In the eighteenth century, the dominance of the patient, who hired the physician, was reflected in a client-based kind of

© The Author(s) 2019
T. Claes, *Corpses in Belgian Anatomy, 1860–1914,* Medicine and Biomedical Sciences in Modern History,
https://doi.org/10.1007/978-3-030-20115-9_6

medicine, in which the doctor tried to make sense of the illness through the patient's account. However, the inversed power dynamics of the hospital went hand in hand with the emergence of a new kind of medicine. Post-mortems, alongside physical examinations and statistics, allowed physicians to construct an analysable body that they could understand without the mediation of the individual patient. The body became a book that could be read without the person. In fact, it became a book that only the physician could read, as knowledge of the anatomical body was inaccessible to the patient without medical interpretation.[3]

Inspired by the research of Ruth Richardson, historians of anatomy made clear that this distorted balance of power also determined the treatment of poor hospital patients when they died. Clinical detachment continued to be the heart of the doctor-patient relationship: much like during life, poor hospital patients after death were considered research or teaching material.[4] Anatomists could use their corpses for scientific and didactic purposes, irrespective of their wishes. Their bodies could be subsequently autopsied and dissected; partly preserved, partly wasted, and partly buried in a pauper grave. It is no wonder that the poor under these conditions only entered the hospital reluctantly: disconnected from their social networks and dissected into unrecognisable pieces, they suffered the most from the belief in the fragment as a source of knowledge.

However, this book has revealed that the balance between the fragment and the individual started to tip in favour of the latter in the second half of the nineteenth century. On the one hand, anatomists' access to dead bodies was impeded by their loss of scientific prestige and the expansion of the disciplinary landscape. On the other hand, social tensions and gradual democratisation resulted in, and attested to, a general softening of attitudes towards the poor. In addition, the individualisation of burial rituals and the development of medical ethics based on consent came to influence the workings of the anatomical theatre. To turn back to the words of Alberti, the social disadvantages of dismemberment began to outweigh the scientific value of the fragment. Around 1900, cultural, religious and personal sensibilities affected the treatment of the corpse in anatomy from acquisition to disposal: anatomical donation gradually replaced the involuntary dissection of the poor, the outward appearances of the corpse became more important during autopsies, and anatomical remains no longer received an anonymous inhumation, but an individual grave.

1   THE DECLINE OF THE ANATOMICAL PART

First of all, the scales tipped in favour of the individual because the discipline of anatomy suffered a loss of prestige. The idea arose that anatomy's part in the medical sciences was played out because there was not a single fibre left to discover. At the same time, anatomy was redefined as a descriptive science through the adjustment and expansion of the disciplinary landscape, although anatomists in practice integrated new approaches into their work. To put it simply: when anatomists in the second half of the nineteenth century occupied themselves with something else than the mere description of macroscopic structures—which they did—it was no longer called anatomy. Reduced to observation and taxonomy, the discipline of anatomy was no longer part of the scientific enterprise, which increasingly relied on experimentation rather than fragmentation.

Dissections no longer counted as scientifically valuable since they only laid bare known structures. Yet as the dissected body came to be seen as a closed and certain source of knowledge, its importance as an educational tool increased. Because the discipline was 'done' from a scientific point of view, anatomy could become the foundation for other, seceded branches of medicine, and accordingly, its position in the curriculum was strengthened. The Belgian law on higher education of 1876 compelled medical students to participate in macroscopic and microscopic anatomical exercises. This in part confirmed an existing practice: dissection already had acquired a central position within teaching hospitals around 1800. In Brussels especially, dissection was taught as the basis of clinical diagnosing and surgery. Microscopic exercises, conversely, were above all meant to fill students with scientific ambition, as they introduced them to the context of the laboratory. Due to this dual purpose of anatomy, the discipline was considered as a 'stepping stone' and a 'connecting thread', helping students to bridge the growing gap between clinic-based education and laboratory-based research.[5]

The ambiguous status of anatomy—of minor importance as a science, but quintessential for education—had seemingly irreconcilable consequences for the procurement of corpses. The use of the bodies of the poor became harder to legitimise because dissection no longer held the promise of scientific progress, but more cadaveric material was needed as every student was required to dissect. As enrolment increased dramatically during the nineteenth century, the demand for cadavers rose even more. Body shortages were not uncommon in the second half of the nineteenth century. Students in Brussels complained that they could not train their hands, their professors grudgingly illustrated their courses with shrivelled preparations. The medical faculty struggled to reestablish a steady supply of cadavers in a society in which the wishes of the deceased and his or her relatives had grown in significance.

The autopsy was the most effective solution for this scarcity of anatomical material. The medical faculty of Brussels regained access to the bodies of hospital patients by distinguishing the autopsy from the dissection. As a result of this distinction, bodies could still be used for medical purposes—at least on the inside—without consent. Significantly, one of the reasons why the city council passed more lenient regulations for the autopsy in 1890 was the perceived scientific and clinical merit of this examination. When it came to the autopsy, it was still possible to make a claim on the dead body based on the progress of knowledge. Although the autopsy in real terms often served educational purposes, it was represented as a scientifically and clinically relevant procedure. As such, the autopsy took priority over the (didactic) dissection, as was evident from the hierarchical division of bodies between different medical practitioners and disciplines.

Its possible scientific gains were not the only explanation for the success of the autopsy. The argument that the autopsy did not render popular burial rituals impossible was even more important. Since the outward appearances of the corpse were by and large respected during the procedure and the body was returned to the bereaved afterwards, there was no infraction of individuals' or families' rights on the remains. The triumph of the autopsy hence also sheds light on the other side of the scale: while the scientific authority of the dissected part diminished, the social and personal value of the corpse gained importance.

2   The Rise of the Anatomical Subject

In fact, the social disadvantages of anatomy began to outweigh the scientific authority of the fragment not only as a result of the reconfiguration of the medical landscape, but also because of broader changes in nineteenth-century society: most importantly the coming of democracy and shifting attitudes towards poverty, the increased importance of consent and self-determination, and the individualisation of burial rituals. As poverty ceased to be regarded as one's own fault, living in destitution no longer had to be repaid by an afterlife on the dissecting table. When the consent of hospital patients grew in significance, anatomists could not claim corpses unilaterally anymore. And as the culture of death became more personal, autopsied and dissected bodies, or at least those parts that were seen as significant, were restored as persons after they left the post-mortem room.

Within this changing context, anatomists in Brussels started to treat the body as a subject rather than an object. To a certain extent, they reconciled their need for corpses with the meanings that the dead body accrued as the material remainder of a person. After their passage through the spaces of anatomy, the deceased could transform from a didactic or research 'object' into a social 'subject' again. The link between body and person was seldom severed entirely or in perpetuity: on their way from deathbed to grave, anatomical remains shifted from subject to object, and the other way around again.

Firstly, the shift from unclaimed to donated bodies had a profound influence on the social status of the anatomical subject, who transformed from a socially disconnected body—stripped of its autonomy, name and social network—to a deceased person with ongoing relationships and convictions. Secondly, the requirement of bodily integrity changed the institution of the autopsy. Since pathologists left face and hands—the bastions of identity—intact, bodies could be restored as persons after the procedure. As the corpse was clothed and stitched back together, it transformed from a medical case to a mourned individual, who could be cared for and buried by his or her friends and family. Lastly, the growing importance of the corpse in burial rituals and the transition from involuntary dissection to body donation impacted the disposal of anatomical remains. Autopsied and dissected bodies were no longer buried anonymously, but were, to the extent possible, placed together in an individual, pentagonal coffin and in a named grave.

By drawing attention to the ways in which corpses could retain or regain their subjectivity and individuality, this book has questioned explicit and implicit readings of nineteenth-century anatomical practices as practices of objectification, turning persons into objects. For example, Chap. 3 has shown that dead bodies ceased to be seen as commodities. When the body in the late nineteenth century shifted from a thing that could change hands to an inalienable extension of the living being—a person who had the right to defend oneself from unauthorised intrusions—anatomical material was no longer subject to commercial dealing, and gradually came to rely on the principle of donation instead. In an era of anatomical bequeathal, dissection no longer equalled the denial of the subjectivity of the deceased. Anatomists increasingly took into account the feelings and wishes of the person whose body was on the table, as is for instance evident from their observance of individuals' requirements for their burial. Anatomical bequeathal could even become a final expression of identity. The conservation of human remains did not necessarily go together with the stripping of individuality either: embedded in textual records ranging from autopsy reports to patient records, pathological specimens became increasingly particular objects of research that continued to refer to specific clinical cases.

To conclude, objectification is not the most appropriate concept to discuss the journeys of anatomical remains through the spaces of anatomy. Although the idea of objectification has proven its value, allowing historians to elucidate the social implications of anatomy in the early nineteenth century, it has largely obscured our knowledge of late nineteenth-century evolutions. To get a grasp on these changes, recent insights from interdisciplinary death scholarship and material culture studies are more fruitful. Helen Lambert and Maryon McDonald's interpretation of the dead body as a 'continuous negotiation' in particular is useful to gain an understanding of the treatment of the corpse in anatomy in the late nineteenth and early twentieth century.[6] This idea better describes anatomists' struggles than the concept of 'polysemy', because it takes into account that other stakeholders' beliefs and intentions determined anatomical practices even if they were not present. Whereas the concept of polysemy has mainly been used to explain that dead bodies' meanings changed dependent on the context, the notion of a negotiation or compromise allows us to see how different values and actors affected the treatment of the corpse, even if the setting was medical.[7] For example, relatives' wish to view the body before burial influenced the autopsy, even if they mostly did not know about the procedure.

3   NEGOTIATIONS AND COMPROMISES

This book has shown that the discipline of anatomy was as much influenced by scientific evolutions as by social relations, and that the trajectories of anatomical remains were entangled with those of the persons who interacted with them. Anatomical practices were compromises, which accommodated various goals and sensitivities. The treatment of the body was determined not only by its material conditions (decomposition was especially important in this respect) but also by different actors, most importantly anatomists and families, but also politicians, clergy, jurists and the dead themselves. From acquisition to disposal, the anatomical encounter with the corpse was the result of its physical state on the one hand, and of ongoing negotiations between opposing beliefs about, and interests in, the dead body on the other.

On the first page of this book, I told the story of the body of a certain Paul, who died in the hospital of St. Pierre in 1866. Against his family's wishes, the physician Edouard Van den Corput removed his spleen while he was already lying in his coffin. This limited autopsy was a compromise: were it not for the family's objection, he would have performed a more elaborate examination. Van den Corput got his hands on a certain bit of the body that he wanted, while the unsuspecting family could carry on with the burial ritual. This book has shown that this kind of compromise became more usual as the nineteenth century progressed, because previously silenced actors—including poor hospital patients and their families—received a voice. The power dynamics between anatomists and other stakeholders shifted due to broader evolutions in society. For example, when the city council of Brussels stripped the claiming of dead bodies of financial implications in 1889, poor families were for the first time in a position to save their deceased loved ones from the dissecting table.

The resulting shortage of bodies, which the medical faculty solved by increasing the number of autopsies, might be the most telling example of negotiations between different actors, their goals and sensitivities. The late nineteenth-century entangled histories of the autopsy and the dissection laid bare complex relationships between physicians, patients and families, as well as between clinicians, teachers and researchers. First and foremost, religious and personal sensitivities had a profound impact on the incisions that pathologists used when handling the corpse. Much like in the case of Van den Corput, the autopsy was at least nominally reconciled with the rights and wishes of relatives, who wanted to pay their last respects to the corpse before burial. By scrutinising the inside of the body while respecting its outward appearances, pathologists detached the medical value of the dead body from the social disadvantages of the dissection. In addition, the embalmment of unclaimed bodies solved tensions between scientific, clinical and didactic needs. Thanks to preservation methods, the unclaimed cadaver could still serve medical purposes after having been autopsied.

The trajectories of autopsied and dissected remains continued to be influenced by different actors and contexts after their examination. Late nineteenth-century anatomists operated in a social and cultural setting where conventions and regulations guided the decent treatment of dead bodies, their ritualised disposal and commemoration. The division of the corpse into different categories—scientifically valuable specimens, parts associated with the known personality of the deceased and bodily material that could be wasted—allowed anatomists in Brussels to reconcile their activities with increasingly personal and sentimental burial rituals. By distinguishing 'bodies' from 'intestines', for instance, the social requirement of a decent funeral could go hand in hand with the exhaustion of other parts for scientific purposes. Much like the autopsy was a useful compromise for the socially more sensitive dissection, distinguishing 'significant' from 'insignificant' parts of the body allowed anatomists to reconcile research on certain bits and pieces with the proper burial of other parts. From acquisition to disposal, anatomical practices reflected and shaped relations between different actors and their views of the dead body.

4   SPECIFIC CONTEXT, BROADER EVOLUTIONS

As different actors in different situations influenced anatomical practices, they were always context specific. In her research on Vienna, Tatjana Buklijas discovered that her findings were remarkably different from the dominantly Anglo-Saxon historiography, and therefore drew attention to the importance of national and regional differences for the study of death and dissection.[8] This book endorses her view: as anatomy took shape in interaction with social, cultural and political settings, historians should be careful when projecting Anglo-American narratives on other contexts. For example, unlike in Britain or the United States, grave robbing was not an issue in Brussels, and social protest against dissection did not surge before the late nineteenth century.

Anatomy did not take place in a vacuum: anatomical practices changed under the influence of broader evolutions happening in the city. Heightened ideological tensions in particular were a catalyst for change. The involuntary dissection of the poor, for example, first came under question as a result of the burial conflict in the 1860s. The Roman Catholic Church used the image of the dissected pauper to criticise the (increasingly anticlerical) burial policy of the Liberal city council. Twenty years later, social protest against dissection also served a hidden agenda. The dissected pauper became part of Socialist political propaganda because it was a means to encourage class consciousness.

Context was everything. To name but a few examples, the image of the dissected pauper was only a successful political strategy for Socialists because the unequal treatment of indigent and propertied patients had become visible. The social composition of the hospital had changed due to the introduction of health insurance, which in its turn was the result of new attitudes towards poverty. In a similar vein, changes in the burial of anatomical remains followed not only from the centrality of the corpse in the evolving culture of death, but also from the Liberal mayor Charles Buls' ambition to improve the reputation of the hospital. The autopsy could become an effective solution for body shortages because teaching hospitals held a central position in Brussels medical education. The principle of consent became the foundation for body donation under the influence of simultaneous legal discussions on the status of the corpse (resulting from the burial conflict), and so on.

And yet, this book raises questions about the social history of anatomy elsewhere. In spite of all these peculiarities, the Brussels case might be representative for broader changes in late nineteenth-century anatomy. After all, many of the contextual factors that impacted anatomists' treatment of the corpse in Brussels were part of broader, European evolutions. 'Culture wars', for example, broke out from Northern to Southern Europe and in the United Kingdom. In many countries, these ideological conflicts revolved around the burial and led to an increased emphasis on the last wishes of the deceased.[9] The face of the hospital, too, changed profoundly across Europe. With the development of insurance policies and the welfare state, healthcare came to be seen as a right rather than a favour. In various countries, the hospital gradually transformed from a charitable institution for the poor into a prestigious medical centre for all.[10] The principle of consent was also codified across Europe, with foreign examples and scandals deeply influencing the interpretation of the legal principle across national borders.[11] Historians have described the sentimental, personal culture of death as a widespread phenomenon, tied up with the rise of individualism in Western society as a whole.[12]

Of special importance was the gradual extension of democracy. In Brussels, progressive politicians associated the dissection of the poor with the ongoing quest for universal suffrage. They argued that the discrimination of the poor, both before and after death, would only end if every citizen received political representation. The coming of democracy went hand in hand with rhetorical shifts in the discourse about poverty and relief. There was a growing awareness that poverty was not the fault of an individual but a structural problem that required state intervention, and accordingly, relief was refashioned from a privilege to a right that citizens were entitled to. Similar developments seem to have problematised the distribution of pauper bodies to anatomists in other European countries, including in Germany and Britain. For example, Elizabeth Hurren has argued that democracy was the 'death-knell' for harsh poverty politics in England, of which the forced dissection of the poor was the ultimate symbol.[13]

When it came to science, foreign medical centres were a source of inspiration for Brussels anatomists, who travelled around and brought ideas back home. The autopsy services that were established in Belgian teaching hospitals in the 1880s, for instance, were modelled after British autopsy services that offered applied courses in pathological anatomy to students. Medical centres in German-speaking Europe (most importantly Vienna) and in France (Paris in particular) were equally influential. Not only pathological institutes, but also anatomical museums and course schedules were inspired by foreign examples.

In addition, this book has placed the position of Belgium's main religious institution, the Roman Catholic Church, into perspective. The findings of this book in part fit with those of other researchers studying continental European centres, as they reveal that the use of the bodies of the poor was not necessarily accompanied by protests. Whereas authors working on German-speaking Europe, most importantly Buklijas, have linked the absence of popular resistance against dissection to the permissive stance of the Roman Catholic Church and the 'decent' inhumation of dissected remains, the evidence presented in this book points to a different conclusion.[14] Much like in Vienna, the Roman Catholic Church in Brussels indeed only protested against dissection when burial in consecrated ground was in doubt. Yet, the poor themselves—whether they were Roman Catholics or not—continued to abhor the scalpel when the burial of dissected remains improved. Popular resistance against dissection was not always grounded in religion: it was also part of broader, socially motivated protests from the lower classes, centred around the emergent democratic principles of equality and class justice.

Religious denomination was not the only factor that determined attitudes towards the medical use of the dead. Equally important were more general beliefs about the timing of death and the status of the dead body, local politics, popular burial customs and the scientific prestige of dissection. In addition, this book has revealed that the acceptance or rejection of post-mortems was tied up with the materiality of the cadaver. Historians therefore should pay sustained attention to differences between autopsy and dissection in order to improve our understanding of post-mortem examinations and their contestation. A more systematic study of the autopsy within teaching hospitals—a topic that, as Fiona Hutton has rightly noticed, remains poorly researched for the British context[15]— would allow for more comprehensive comparisons between the history of anatomy in the Anglo-Saxon world and continental Europe. Was protest more vivid in Britain and the United States because dissections were more usual than autopsies? Or was the relative lack of protest in continental Europe influenced by other contextual factors?

In short, there is a need for more profound comparative research in order to establish if the developments discussed in this book were part of a wider phenomenon. To name but a few examples, anatomical bequeathal appears to have developed along the same lines in France, Britain and the United States, and the legal, social, scientific and material distinctions between dissection and autopsy were not confined to Belgium either. While much of historiographical focus has been on how the poor were coerced into relinquishing their bodies and on maintaining networks of cadaver supply against popular protests, historians, so this book suggests, should start looking at the other end: the gradual move to anatomical donations from around the turn of the twentieth century. The decades around 1900 might prove to have been a pivotal moment in the history of anatomy. As the cultural significance of the individual became more important than the scientific power of the fragment, the corpse in anatomy transformed from a nobody into a somebody.

NOTES

1. Samuel J.M.M. Alberti, *Morbid Curiosities: Medical Museums in Nineteenth-Century Britain* (Oxford: Oxford University Press, 2011), 71.
2. Ibid., 72.
3. Michel Foucault, *The Birth of the Clinic: An Archeology of Medical Perception*, trans. A.M. Sheridan (New York: Routledge, 1973), 124–48; Nicholas D. Jewson, 'The Disappearance of the Sick-Man from Medical Cosmology, 1770–1870', *Sociology* 10, no. 2 (1976): 225–44.
4. Ruth Richardson, *Death, Dissection and the Destitute*, 2nd ed. (Chicago and London: The University of Chicago Press, 2000), 30–51.
5. Quotes from: Léon Marcq, *Essai sur l'histoire de la médecine belge contemporaine* (Brussels: Manceaux, 1866), 130–3.
6. Helen Lambert and Maryon McDonald, 'Introduction', in *Social Bodies*, eds. Helen Lambert and Maryon McDonald (Oxford and New York: Berghahn, 2011), 1–15.

7. Alberti, *Morbid Curiosities*, 164.
8. Tatjana Buklijas, 'Cultures of Death and Politics of Corpse Supply: Anatomy in Vienna, 1848–1914', *Bulletin of the History of Medicine* 82, no. 3 (2008): 570–607.
9. Christopher Clark and Wolfram Kaiser, eds., *Culture Wars: Secular–Catholic Conflict in Nineteenth-Century Europe* (Cambridge: Cambridge University Press, 2003).
10. Alfons Labisch, 'From Traditional Individualism to Collective Professionalism: State, Patient, Compulsory Health Insurance and the Panel Doctor Question in Germany 1883–1931', in *Medicine and Modernity: Public Health and Medical Care in Nineteenth- and Twentieth-Century Germany*, eds. Manberg Berg and Geoffrey Cocks (Cambridge: Cambridge University Press, 1997), 18–34; Keir Waddington, 'Unsuitable Cases: The Debate over Outpatient Admissions, the Medical Profession and Late-Victorian London Hospitals', *Medical History* 42, no. 1 (1998): 26–46.
11. Christian Bonah, *Histoire de l'expérimentation humaine en France* (Paris: Les Belles Lettres, 2007), 114–25; Barbara Elkeles, 'The German Debate on Human Experimentation between 1880–1914', in *Twentieth Century Ethics of Human Subject Research*, eds. Volker Roelcke and Giovanni Maio (Stuttgart: Franz Steiner Verlag, 2004), 18–33; Andreas Holger Maehle, *Doctors, Honour and the Law. Medical Ethics in Imperial Germany* (Houndmills: Palgrave Macmillan, 2009), 69–94; Ruth R. Faden and Tom L. Beauchamp, *A History and Theory of Informed Consent* (Oxford: Oxford University Press, 1986), 114–50.
12. Most recently: Thomas W. Laqueur, *The Work of the Dead: A Cultural History of Mortal Remains* (Princeton: Princeton University Press, 2015).
13. Elizabeth T. Hurren, 'World Without Welfare. Pauper Perspectives on Medical Care under the Late Victorian Poor Law 1870–1900', *Obligation, Entitlement and Dispute under the English Poor Laws*, eds. Peter Jones and Steven King (Newcastle upon Tyne: Cambridge Scholars, 2015), 292–320.
14. Karin Stukenbrock, *'Der zerstückte Cörper': Zur Sozialgeschichte der anatomischen Sektionen in der frühen Neuzeit 1650–1800* (Stuttgart: Franz Steiner Verlag, 2001) and Buklijas, 'Cultures of Death', 570–607.
15. Fiona Hutton, *The Study of Anatomy in Britain, 1700–1900* (London: Pickering and Chatto, 2013), 12.

BIBLIOGRAPHY

PRIMARY RESEARCH

ARCHIVAL RESEARCH

CITY ARCHIVES OF BRUSSELS (ARCHIEF VAN DE STAD BRUSSEL/ARCHIVES DE VILLE DE BRUXELLES)

ASB Bienfaisance S107/2: Construction de dépôts mortuaires, 1885–1887.
ASB Cultes 1866: Règlement—tarif sur les concessions de sépultures, 1877–1880; Règlement sur les transports funèbres et inhumations. Minutes, expéditions, rapports.
ASB Cultes 2699: Anciens règlements sur les transports funèbres, inhumations, concessions de sépultures, 1880–1965.
ASB Cultes 2843: Corps conservés dans des collections scientifiques, 1878–1986.

ARCHIVES OF THE SOCIAL SERVICES OF BRUSSELS (ARCHIEF OCMW BRUSSEL/ARCHIVES CPAS BRUXELLES)

Affaires générales 8: Minutes—Services hospitaliers, 1824–1932.
Affaires générales 34: Cultes, 1811–1895.
Affaires générales 35: Cultes, 1896–1920.

© The Author(s) 2019
T. Claes, *Corpses in Belgian Anatomy, 1860–1914*, Medicine and
Biomedical Sciences in Modern History,
https://doi.org/10.1007/978-3-030-20115-9

Affaires générales 38: Inhumations, 1848–1894.
Affaires générales 39: Inhumations, 1895–1925.
Affaires générales 66: Université, 1834–1922.
Affaires générales 69: Collections scientifiques, 1834–1878.
Affaires générales 76: Instructions concernant les hôpitaux, 1864–1895.
Affaires générales 105: Autopsies—règlement, 1861–1899 & 1869–1913.
Affaires générales 106: Autopsies, 1876–1908.
Affaires générales 107: Autopsies, 1873–1924.
Affaires générales 113: Hôpital St. Jean, généralités, 1864–1887.
Affaires générales 114: Hôpital St. Jean, généralités, 1888–1890.
Affaires générales 116: Hôpital St. Jean, généralités, 1896–1900.
Affaires générales 123: Hôpital St. Jean, généralités, 1906–1910.
Affaires générales 130: Hôpital St. Pierre, divers, 1870–1886.
Affaires générales 131: Hôpital St. Pierre, divers, 1887–1891.
Affaires générales 133: Hôpital St. Pierre, divers, 1896–1902.
Affaires générales 135: Hôpital St. Pierre, divers, 1910–1924.
Affaires générales 248: Maternité, section universitaire, 1886–1922.
Affaires générales 261: Sana. Roger de Grimberghe, Matériel médical, s.d.
Cartes et plans 12: Projet de transformation et d'agrandissements, 1917.
Fonds du directeur 292: Décès—autopsies, 1852–1922.
Travaux, établissements hospitaliers 21: Construction d'un nouveau dépôt mortuaire et amphithéâtre, 1904–1909.

Archives of the Free University of Brussels (Archives de l'Université Libre de Bruxelles)

Procès-verbaux des séances du conseil d'administration, 1881–1930.
1FD 701–702: Anatomie humaine (démonstrations anatomiques): Répertoires des macchabées remis à l'Institut d'anatomie, 1899–1907 & 1899–1911.
1FD 703: Corps légués ou mis à la disposition du service d'anatomie, 1945–1968.

University Archives of Ghent (Universiteitsarchief Gent)

46 (100): Autopsies, 1899.
4A 25: 1873.
4A 255 (71): lijkentekort voor het anatomisch instituut (1935/36).
4A 27: 1875–1876.
4A28: 1876–1877.
4A29: 1877–1878.

4A34: 1880.
4A40: 1885.
4A43: 1886–1887.
4A44: 1888.
4A45: 1889.

ARCHIVES OF THE SOCIAL SERVICES OF GHENT (ARCHIEF OCMW GENT)

BG12: Hospitaal De Bijloke, 1702–1925.

UNIVERSITY ARCHIVES OF LIÈGE (ARCHIVES DE L'UNIVERSITÉ DE LIÈGE)

Fonds du secrétariat central 254: Faculté de médecine, 1882–1912.
Fonds du secrétariat central 255: Faculté de médecine, 1891–1914.
Fonds du secrétariat central 293: Faculté de médecine, 1920.
Fonds du secrétariat central 327: Faculté de médecine, 1910–1921.

CITY ARCHIVES OF LEUVEN (STADSARCHIEF LEUVEN)

Modern Archief 50002 (641/8): lijkenvervoer van en naar het anatomisch instituut van de universiteit, 1844–1940.

ARCHIVES OF THE SOCIAL SERVICES OF LEUVEN (ARCHIEF OCMW LEUVEN)

1: Autopsie, 1839–1901.
22: Anatomisch Amfitheater, s.d.

PRIVATE ARCHIVES

Register of bodies brought to the Institut d'anatomie, 1920–1968, Private Archives of Stéphane Louryan, Brussels.

STATE ARCHIVES BRUSSELS (ARCHIVES DE L'ÉTAT À BRUXELLES/RIJKSARCHIEF IN BRUSSEL)

T015/583: Dossier concernant la collection phrénologique à la Maison de Force à Gand, 1844.

PERIODICALS AND OTHER SERIAL SOURCES

Annales des Universités de Belgique.
Annuaire de l'Université Catholique de Louvain.
Bulletin communal de Bruxelles.
Bulletin de l'Académie royale de médecine de Belgique.
Bulletin de la société anatomo-pathologique de Bruxelles.
Bulletin de la société d'anthropologie de Bruxelles.
Dalloz: recueil périodique et critique de jurisprudence.
Het Handelsblad.
Journal de Bruxelles.
La Belgique judiciaire.
La libre Belgique.
La Meuse.
L'Indépendance belge.
Le Courrier de l'Escaut.
Le Journal de Paris.
Le Peuple.
Pasicrisie belge. Recueil général de la jurisprudence des cours de Belgique.
Vingtième Siècle.
Vooruit.

DATABASES

Belgica Press Database. Koninklijke Bibliotheek van België/Bibliothèque royale de Belgique. Accessed December 29, 2018. http://opac.kbr.be/belgicapress. php.
Plenum Database: Proceedings of the Plenary Sessions of the Belgian Chamber of Representatives. Google Sites. Accessed December 29, 2018. https://sites. google.com/site/bplenum/.

OTHER PUBLISHED SOURCES

1837. Règlement pour l'amphithéâtre d'anatomie et les salles de dissection, January 15, 1836. *Annuaire de l'Université catholique de Louvain* 1: 37–41.
1849. *Loi organique de l'enseignement supérieur du 27 septembre 1835, modifiée par la loi du 15 juillet 1849.* Brussels: G. Stapleaux.
1862. Prix quinquennaux décernés à MM les professeurs Van Kempen et Van Beneden. *Annuaire de l'Université Catholique de Louvain* 27: 405–406.
1864. *Assemblée générale des catholiques en Belgique, première session à Malines, 18–22 Août 1863.* Brussels: Goemaere.

1868. *Assemblée générale des catholiques en Belgique, troisième session à Malines, 2–7 septembre 1867*. Brussels: Victor Devaux.

1883. *Recueil de documents concernant la révision de la loi du 20 Mai 1876 sur la collation des grades académiques et les programmes des examens universitaires*. Brussels: Adolphe Mertens.

1893. *Actes du troisième congrès international d'anthropologie criminelle tenu à Bruxelles*. Brussels: Hayez.

1897. *Règlement sur les inhumations et les transports funèbres: règlement sur le prix des transports funèbres*. Brussels: Baertsoen.

1907. *Manifestation en l'honneur du Dr. Daniel Van Duyse*. Ghent: Vander Haeghen.

1911. *Règlement pour la collation des grades institués par les lois du 10 avril 1890 et du 3 juillet 1891*. Leuven: Van Linthout.

1925. *Manifestation en l'honneur du Dr. Charles Firket*. Liège: Vaillant-Carmanne.

Administration du service de santé et de l'hygiène. 1910. *Recueil des dispositions légales et réglementaires concernant l'hygiène et la salubrité publiques*. Brussels: Weissenbruch.

Baunard, Louis. 1926. *Les deux frères: Cinquante années de l'Action Catholique dans le nord: Philibert Vrau, Camille Feron-Vrau 1829–1908*. Paris: Maison de la bonne presse.

Berger, André. 1893. *Guide de l'étudiant à l'hôpital: examen clinique—autopsies*. Paris: Masson.

Bizzozero, Giolio, and Charles Firket. 1883. *Manuel de microscopie clinique*. Brussels: Manceaux.

Boddaert, Richard. 1870. *De l'importance des études pratiques en médecine*. Ghent: Hebbelynck.

Boitard, Pierre. 1890. *Nouveau manuel complet du naturaliste préparateur. Deuxième partie. Taxidermie, préparation des pièces anatomiques*. Paris: Roret.

Bonjean, R.J., J.B. Bivort, J.J. Cloes, and E.A. Dubois. 1837. *Revue de l'administration et du droit administratif de la Belgique*. Liège: Imprimerie de Veuve Verhoven-Debeur.

Bormans, Théodore. 1870. De la violation des cadavres. *La Belgique judiciaire*, 19 April.

Bouchut, Eugène, and Armand Després. 1883. *Dictionnaire de médecine et de thérapeutique médicale et chirurgicale*. Paris: Librairie Germer Ballière.

Bourneville, Désiré-Magloire, and Paul Bricon. 1887. *Manuel de technique des autopsies*. Paris: Librairie du progrès médical.

Brocas, Roger. 1938. *Le droit d'autopsie: étude historique et juridique*. Paris: Faculté de droit.

Burggraeve, Adolphe. 1840a. *Cours théorique et pratique de l'anatomie*. Ghent: Impens.

———. 1840b. *Discours sur le médecin P. E. Wauters, prononcé le jour de son enterrrement.* Ghent: Gyselynck.

———. 1841. *Etudes sur André Vésale, précédées d'une notice historique sur sa vie et ses écrits.* Ghent: Annoot-Braeckman.

Carnoy, Jean-Baptiste. 1889. *Révision de la loi de 1876. Les programmes des examens de sciences naturelles et de médecine.* Leuven: Fonteyn.

Colard, Armand. 1952. *Souvenirs du vieux Saint-Pierre.* Brussels: Arcsia.

Cunningham, Daniel. 1890. *Guide de dissection, résumé d'anatomie topographique.* Translated by Pierre Kuborn. Liège: Nierstrasz.

Dallemagne, Jules. 1894. *Les stigmates anatomiques de la criminalité.* Paris: Masson.

Debierre, Charles Marie. 1895. *Le crâne des criminels.* Lyon: Masson.

De Brouckère, Charles, and Franciscus Thielemans. 1838. Cadavre. In *Répertoire de l'administration et du droit l'administratif de la Belgique*, vol. 4, 59–61. Brussels: Weissenbruch.

De Gronckel, Charles. 1884. *Hospices civils et bureaux de bienfaisance. Précis du régime légal de l'assistance public.* Brussels: L. Bourlard et V. Havaux.

De Laet, Maurice. 1927. *Les responsabilités du médecin.* Brussels: Imprimerie médicale et scientifique.

———. 1948. Le droit à l'autopsie et à prélèvements post-mortem. *Académie royale de médecine de Belgique* 107: 84–92.

Delcour, Charles. 1876. *Situation de l'enseignement supérieur donné aux frais de l'état. Rapport triennal 1871–1872 et 1873.* Brussels: Gobbaerts.

Delcourt, Albert. 1899. La rachitisme. Sa pathogénie. PhD diss., Université Libre de Bruxelles.

Demogue, René. 1909. La notion de sujet de droit: caractère et conséquences. *Revue trimestrielle de droit civil* 3: 611–655.

Deroubaix, Louis. 1870. *Traité des fistules uro-génitales de la femme: comprenant les fistules vésico-vaginales, vésicales cervico-vaginales, uréthro-vaginales, vésicales cervico-utérines, vesico-utérines, urétéro-vaginales et urétérales cervico-utérines.* Brussels: Manceaux.

De Spoelberch de Lovenjoul, Molly. 1893. *Belgique charitable. Bruxelles: charité, bienfaisance, philanthropie, etc.* Brussels: Larcier.

Firket, Charles. 1883. *Du but et de l'organisation des services d'autopsies.* Liège: Vaillant-Carmanne.

———. 1893. *L'éducation médicale en Angleterre, en Ecosse et en Irlande.* Paris: Armand Colin.

Francotte, Polydore. 1885. *Manuel de technique microscopique: applicable à l'histologie, l'anatomie comparée, l'embryologie et la botanique.* Brussels: Lebègue.

———. 1887. *Résumé d'une conférence sur la microphotographie appliquée à l'histologie, l'anatomie comparée et l'embryologie.* Brussels: Manceaux.

Frédericq, Léon. 1881. L'enseignement de physiologie à l'Université de Berlin. *Revue de Belgique* 3: 118–137.

———. 1884. *Le corps humain: anatomie et physiologie populaires*. Brussels: Lebègue.

———. 1892. *Manipulations de physiologie: guide de l'étudiant au labouratoire pour les travaux pratiques et les démonstrations de physiologie*. Paris: J.B. Baillière.

Fritz, André. 1879. *Du danger des inhumations précipitées*. Brussels: Jamin et Coosemans.

Gluge, Gottlieb. 1849. Sur les progrès que l'anatomie et la physiologie humaine ont faits dans les derniers temps en Belgique. *Bulletin et mémoires de l'Académie royale de médecine de Belgique* 8: 684–699.

Goubert, Emile. 1867. *Manuel de l'art des autopsies cadavériques, surtout dans ses applications à l'anatomie pathologique*. Paris: Germer Ballière.

Guislain, Joseph. 1840. *Lettres médicales sur l'Italie, avec quelques renseignements sur la Suisse*. Ghent: Gyselynck.

Harris, Thomas. 1888. *Manuel d'autopsies ou méthode de pratiquer les examens cadavériques au point de vue clinique et médico-légal*. Translated by H. Surmont. Brussels: Manceaux.

Héger, Paul. 1873. *Expériences sur la circulation du sang dans des organes isolés*. Brussels: Manceaux.

———. 1888–1889. Création d'un musée anthropologique à Bruxelles. *Bulletin de la société d'anthropologie de Bruxelles* 7: 284–286.

———. 1919. Notice sur la vie et les œuvres de Willem Rommelaere. In *Rapport sur l'Année académique, 1914–1918*, 41–55. Brussels: Université Libre de Bruxelles.

Héger-Gilbert, Fernand. 1908. *La radiographie foetale envisagée au point de vue médico-légal*. Brussels: Piette.

———. 1928. *Manuel de déontologie médicale: résumé du cours professé à l'Université Libre de Bruxelles*. Brussels: Imprimerie médicale et scientifique.

Icard, Séverine. 1910. *La constatation des décès dans les hôpitaux en France et à l'étranger et la nécessité de la pratique hâtive des autopsies*. Paris: Maloine.

Laurent, François. 1880–1881. *Droit Civil International*. 4 vols. Brussels: Bruylant.

Leboucq, Hector. 1884a. *Le musée anatomique de l'Université de Gand*. Ghent: Vanderhaeghen.

———. 1884b. *Un mot sur la technique des coupes en séries*. Ghent: Annoot-Braeckman.

———. 1907. *L'anatomie humaine et les tendances modernes de la morphologie*. Ghent: Annoot-Braeckman.

Ledresseur, Charles. 1882. *Résumé du cours d'anatomie des régions professé à l'université catholique*. Leuven: Peeters-Ruelens.

Le Roy, Alphonse. 1869. *Liber memorialis. L'Université de Liège depuis sa fondation*. Liège: Vaillant Carmanne.

Letulle, Maurice. 1903. *La pratique des autopsies*. Paris: Naud.

Marcq, Léon. 1866. *Essai sur l'histoire de la médecine belge contemporaine*. Brussels: Manceaux.

Marcq, Philippe-Antoine. 1821. *De l'état actuel de l'enseignement médical en Belgique*. Brussels: Demanet.

Mareska, Daniel, and J. Heyman. 1845. *Enquête sur le travail et la condition physique et morale des ouvriers employés dans les manufactures de coton à Gand*. Ghent: Gyselinck.

Marinus, J.R. 1837. Nécrologie de Philippe Antoine Marcq, mort en 1837, à l'age de 40 ans. *Bulletin médical belge* 19: 153–156.

Maygrier, Jacques-Pierre. 1818. *Manuel de l'anatomiste*. 4th ed. Paris: Gabon.

Moreau, Alfred. 1891. *La responsabilité médicale*. Brussels: Bruylant-Christophe.

Picard, Edmond. 1913. *Théodore Hauben, médecin: une vie belge au XIXe siècle*. Brussels: Larcier.

Picard, Edmond, and Napoléon d'Hoffschmidt. 1885. Cadavre. In *Pandectes belges. Répertoire général de législation, de doctrine et de jurisprudence belges donnant pour toutes les matières du droit belge*, vol. 15, 244–265. Brussels: Larcier.

Picard, Eugène. 1839. *Travaux pratiques d'anatomie pathologique: autopsie et microscopie*. 3rd ed. Leuven: S.D. Dewallens.

Poelman, Charles. 1839. *Notice sur les embaumements: procédé Gannal*. Ghent: Rousseau.

———. 1868. *Catalogue des collections d'anatomie comparée y compris les ossements fossiles de l'université de Gand*. Ghent: Vanderhaeghen.

Responsabilité médicale: notes de jurisprudence. Brussels: Ch. Miguet.

Rommelaere, Guillaume. 1866. *Des institutions médicales et hospitalières en Angleterre*. Brussels: Bols-Wittouck.

Saint-Vincent, Ludovic, and Charles Vloeberghs. 1904. *Belgique charitable*. Brussels: Dewit.

Salsmans, Jozef. 1919. *Geneeskundige plichtenleer*. Leuven: Vlaamsche Boekenhalle.

Spring, Joseph Antoine. 1863. De l'esprit scientifique à notre époque et dans nos universités. *Annales des Universités de Belgique* 2: 386–396.

Stocquart, Alfred. 1892. La théorie d'Albrecht concernant la signification morphologique de bec-de-lièvre compliqué de fissure palatine. *Bulletin de la société d'anthropologie de Bruxelles* 11: 185–193.

Suau de Varennes, Edouard. 1846. *Die Mysterien von Brüssel*. Stuttgart: Franck Verlag.

Tillaux, Paul Jules. 1879. *Traité d'anatomie topographique avec applications à la chirurgie*. Paris: Asselin.

Timbal, Gabriel. 1903. *La condition juridique des morts*. Toulouse: Edouard Privat.

Tissandier, Gaston. 1882. *Les martyrs de la science*. 2nd ed. Paris: Maurice Dreyfous.

Tricot-Royer, Jean-Joseph. 1936. *L'Eglise et la mutilation du cadavre humain. Décarnisation, dissection pour enseignement, embaumement, autopsie.* Leuven: Nova et Vetera.

Urban, Louis Joseph. 1805. *Établissement d'une école de médecine à Bruxelles, avec l'approbation du ministre de l'intérieur.* Brussels: Imprimerie Urban.

Uytterhoeven, André. 1852. *Notice sur l'hôpital Saint-Jean de Bruxelles ou l'étude sur la meilleure manière de construire et d'organiser un hôpital de malades.* Brussels: N.J. Gregoir.

Van Duyse, Daniel. 1896. *Technique des autopsies.* Ghent: Annoot-Braeckman.

———. 1904. *Premiers éléments de diagnostic anatomo-pathologique au cours des autopsies.* Ghent: Van Goethem.

Van Kempen, Etienne. 1844. Rapport adressé à M. le Ministre de l'Intérieur, sur l'état de physiologie en Allemagne. *Annales des Universités de Belgique* 3: 1054–1057.

———. 1854. *Traité d'anatomie descriptive et d'histologie spéciale.* Leuven: Vanlinthout.

———. 1859. *Expériences physiologiques sur la transmission de la sensibilité et du mouvement dans la moelle épinière.* Brussels: Mortier.

———. 1860. *Manuel d'anatomie générale.* Leuven: Vanlinthout.

Van Overbergh, Cyrille. 1900. *Réforme de la bienfaisance en Belgique.* Brussels: Lesigne.

Van Schoor, Joseph. 1878. *Discours d'ouverture prononcés en séance publique le 14 octobre 1878.* Brussels: Mayolez.

Verriest, Gustave, and Jean-Baptiste Carnoy. 1883. De l'organisation des études médicales. *Revue médicale* 2: 44–53.

Ville de Gand. 1865. *Règlement sur les inhumations. Rapport de la commission du contentieux.* Ghent: C. Annoot.

Ville de Liège. 1889. *Réorganisation du service des sépultures. Rapport présenté par M. Gustave Kleyer.* Liège: Thiriart.

Virchow, Rudolf. 1847. Ueber die Standpunkte in der wissenschaftlichen Medicin. *Archiv für pathologische Anatomie und Physiologie und für klinische Medicin* 1: 1–19.

———. 1893. *Die Sektions-Technik im Leichenhause des Charité-Krankenhauses mit besonderer Rücksicht auf gerichtsärztliche Praxis.* 4th ed. Berlin: August Hirschwald.

SECONDARY RESEARCH

Ackerknecht, Erwin. 1967. *Medicine at the Paris Hospital 1794–1848.* Baltimore: Johns Hopkins University Press.

Adamson, Glenn. 2009. The Case of the Missing Footstool. Reading the Absent Object. In *History and Material Culture: A Student's Guide to Approaching Alternative Sources,* ed. Karen Harvey, 192–207. London: Routledge.

Aerts, Erik, Claude Beaud, and Jean Stengers, eds. 1990. *Liberalism and Paternalism in the 19th Century*. Leuven: Leuven University Press.

Ahrén, Eva. 2009. *Death, Modernity and the Body: Sweden 1870–1940*. Rochester: University of Rochester Press.

Alberti, Samuel J.M.M. 2005. Objects and the Museum. *Isis* 96: 559–571.

———. 2011. *Morbid Curiosities: Medical Museums in Nineteenth-Century Britain*. Oxford: Oxford University Press.

Aleksiun, Natalia. 2012. Jewish Students and Christian Corpses in Interwar Poland: Playing with the Language of Blood Libel. *Jewish History* 26: 327–342.

Al-Gailani, Salim. 2016. The "Ice Age" of Anatomy and Obstetrics: Hand and Eye in the Promotion of Frozen Sections around 1900. *Bulletin of the History of Medicine* 90: 611–642.

Altman, Lawrence. 1986. *Who Goes First? The Story of Self-Experimentation in Medicine*. Berkeley, Los Angeles and London: University of California Press.

Andrews, Jonathan. 2012. Death and the Dead-House in Victorian Asylums. Necroscopy versus Mourning at the Royal Edinburgh Asylum c.1832–1901. *History of Psychiatry* 23: 6–26.

Annas, George J., and Michael A. Grodin, eds. 1992. *The Nazi Doctors and the Nuremberg Code*. Oxford: Oxford University Press.

Ariès, Philippe. 1977. *L'homme devant la mort*. Paris: Baron.

Armstrong, David. 1998. Bodies of Knowledge/Knowledge of Bodies. In *Reassessing Foucault: Power, Medicine and the Body*, ed. Colin Jones and Roy Porter, 17–27. London and New York: Routledge.

Arnold, Ken, and Danielle Olsen, eds. 2003. *Medicine Man: The Forgotten Museum of Henry Wellcome*. London: British Museum Press.

Arnout, Anneleen. 2019. *Streets of Splendour: Shopping Culture and Spaces in a European Capital City (Brussels, 1830–1914)*. London and New York: Routledge.

Baker, Robert. 1993. Deciphering Percival's Code. In *The Codification of Medical Morality: Historical and Philosophical Studies of the Formalisation of Western Medical Morality in the Eighteenth and Nineteenth Centuries*, eds. Robert Baker, Dorothy Parker, and Roy Porter, vol. 2, 179–212. Dordrecht: Springer.

Baker, Robert, Dorothy Parker, and Roy Porter, eds. 1993. *The Codification of Medical Morality: Historical and Philosophical Studies of the Formalisation of Western Medical Morality in the Eighteenth and Nineteenth Centuries*. 2 vols. Dordrecht: Springer.

Bardez, Renaud. 2016. La Faculté de médecine de l'Université Libre de Bruxelles: entre création, circulation et enseignement des savoirs, 1795–1914. PhD diss., Université Libre de Bruxelles.

Bauer, Axel. 1990. Die Institutionalisierung der Pathologischen Anatomie im 19. Jahrhundert an den Universitäten Deutschlands, der deutschen Schweiz und Österreichs. *Gesnerus* 47: 303–328.

Belgisch raadgevend comité voor bio-ethiek. 2011. Advies betreffende bepaalde ethische aspecten van de wijzigingen door de wet van 25 februari 2007 aangebracht aan de wet van 13 juni 1986 betreffende het wegnemen en transplanteren van organen. Accessed July 17, 2017. https://www.health.belgium.be/sites/default/files/uploads/fields/fpshealth_theme_file/19104555/Advies%2050.pdf.

Benedek, Thomas. 2014. Case Neisser. Experimental Design, the Beginnings of Immunology and Informed Consent. *Perspectives in Biology and Medicine* 57: 249–267.

Bennett, Tony. 1995. *The Birth of the Museum: History, Theory, Politics*. London: Routledge.

Berg, Manfred, and Geoffrey Cocks, eds. 1997. *Medicine and Modernity: Public Health and Medical Care in Nineteenth- and Twentieth-Century Germany*. Cambridge: Cambridge University Press.

Berkowitz, Carin. 2012. Systems of Display: The Making of Anatomical Knowledge in Enlightenment Britain. *The British Journal for the History of Science* 46: 359–387.

Bertrand, Régis. 2011. *Mort et mémoire. Provence, XVIIIe–XXe siècles, Une approche d'historien*. Marseille: La Thune.

Betta, Emmanuel. 2015. Between Law and Profession: The Origins of Informed Consent (1840–1900). In *Doctors and Patients. History, Representation, Communication from Antiquity to the Present*, ed. Maria Malatesta, 108–133. San Francisco: University of California Medical Humanities Press.

Beyers, Leen. 2003. Rasdenken tussen geneeskunde en natuurwetenschap: Emile Houzé en de Société d'Anthropologie de Bruxelles 1882–1921. In *Degeneratie in België 1860–1940: een geschiedenis van ideeën en praktijken*, eds. Jo Tollebeek and Geert Vanpaemel en Kaat Wils, 43–78. Leuven: Leuven University Press.

Biesbrouck, Maurits, Luc Missotten, and Omer Steeno. 2014. De Vesalius-schilderijen van E.J.C. Hamman (1819–1888). In *Heel-meesters. Befaamde artsen en figuren uit de geschiedenis van de geneeskunde*, eds. Bob Van Hee and Cornelis Van Tilburg, 19–39. Antwerp: Garant.

Block, Jane, ed. 1997. *Belgium, the Golden Decades (1880–1914)*. New York: Peter Lang.

Bonah, Christian. 2007. *Histoire de l'expérimentation humaine en France*. Paris: Les Belles Lettres.

Bowler, Peter J., and John V. Pickstone, eds. 2009. *The Cambridge History of Science: The Modern Biological and Earth Sciences*. Vol. 6. Cambridge: Cambridge University Press.

Bracegirdle, Brian. 1978. *A History of Microtechnique: The Evolution of the Microtome and the Development of Tissue Preparation*. London: Heinemann.

Brandes, Inga, and Katrin Marx-Jaskulski, eds. 2008. *Armenfürsorge und Wohltätigkeit. Ländliche Gesellschaften in Europa, 1850–1930/Poor Relief and Charity. Rural Societies in Europe, 1850–1930*. Frankfurt am Main: Peter Lang.

Brandt, Allan M., and David C. Sloane. 1999. Of Beds and Benches: Building the Modern American Hospital. In *The Architecture of Science*, eds. Peter Galison and Emily Thompson, 281–308. Cambridge, MA: MIT Press.

Brockliss, Laurence. 2000. The New Paris Medical School and the Invention of the Clinic. In *The French Experience from Republic to Monarchy, 1792–1824: New Dawns in Politics, Knowledge and Culture*, eds. Mairé F. Cross and David Williams, 120–139. Basingstoke: Palgrave Macmillan.

Brooks-Gordon, Belinda, et al., eds. 2007. *Death Rites and Rights*. Oxford: Hart Publishing.

Buklijas, Tatjana. 2008. Cultures of Death and Politics of Corpse Supply: Anatomy in Vienna, 1848–1914. *Bulletin of the History of Medicine* 82: 570–607.

Buls, Charles, and Marcel Bots. 1987. *Het dagboek van C. Buls*. Ghent: Liberaal Archief.

Burney, Ian. 2000. *Bodies of Evidence: Medicine and the Politics of the English Inquest, 1830–1926*. Baltimore and London: Johns Hopkins University Press.

Burrell, Sean, and Geoffrey Gill. 2005. The Liverpool Cholera Epidemic of 1832 and Anatomical Dissection: Medical Mistrust and Civil Unrest. *Journal of the History of Medicine and Allied Sciences* 60: 478–498.

Bynum, William. 1988. Reflections on the History of Human Experimentation. In *The Use of Human Beings in Research. With Special Reference to Clinical Trials*, ed. Stuart Spickert et al., 29–46. Dordrecht: Springer.

———. 1994. *Science and the Practice of Medicine in the Nineteenth Century*. Cambridge: Cambridge University Press.

Cahan, David. 2003. Looking at Nineteenth-Century Science: An Introduction. In *From Natural Philosophy to the Sciences: Writing the History of Nineteenth-Century Science*, ed. David Cahan, 3–15. Chicago: University of Chicago Press.

Carol, Anne. 2004. *Les médecins et la mort XIXe—XXe siècle*. Paris: Aubier.

———. 2016. *Embaumement: une passion romantique, France XIXe siècle*. Lyon: Champ Vallon.

Chamayou, Grégoire. 2011. *Les corps vils: expérimenter sur les êtres humains au XVIIIe et XIXe siècle*. Paris: La découverte.

Chaplin, Simon. 2014. Anatomy or an Ottamy? Bodies on Show in Georgian London. In *The Morbid Anatomy Anthology*, eds. Joanna Ebenstein and Colin Dickey, 254–270. New York: Morbid Anatomy Press.

Claes, Tinne. 2018. By What Right does the Scalpel Enter the Pauper's Corpse? Dissections and Consent in Late Nineteenth-Century Belgium. *Social History of Medicine* 31: 258–277.

Claes, Tinne, and Veronique Deblon. 2015. Van panoramisch naar preventief. Populariserende anatomische musea in de Lage Landen. *De Negentiende Eeuw* 39: 289–306.

———. 2018. When Nothing Remains: Anatomical Collections, the Ethics of Stewardship and the Meanings of Absence. *Journal of the History of Collections* 30: 351–362.

Claes, Tinne, and Pieter Huistra. 2016. Il importe d'établir une distinction entre la dissection et l'autopsie. Lijken en medische disciplinevorming n laatnegentiende-eeuws België. *BMGN Low Countries Historical Review* 131: 26–53.

Clark, Christopher, and Wolfram Kaiser, eds. 1993. *Culture Wars: Secular-Catholic Conflict in Nineteenth-Century Europe*. Cambridge: Cambridge University Press.

Close-Koenig, Tricia. 2011. Betwixt and Between: Production and Commodification of Knowledge in a Medical School Pathological Anatomy Laboratory, Strasbourg (Mid-19th Century to 1939). PhD diss., University of Strasbourg.

———. 2015. Cataloguing Collections: The Importance of Paper Records of Strasbourg's Medical School Pathological Anatomy Collection. In *The Fate of Anatomical Collections*, eds. Rina Knoeff and Rob Zwijnenberg, 211–227. New York: Ashgate.

Conn, Steven. 1998. *Museums and American Intellectual Life, 1876–1926*. Chicago: University of Chicago Press.

Couttenier, Maarten. 2005. *Congo tentoongesteld: een geschiedenis van de Belgische antropologie en het museum van Tervuren (1882–1925)*. Leuven: University Press.

Cullen, L.T. 2017. Post-mortem in the Victorian Asylum: Practice, Purpose and Findings at the Littlemore County Lunatic Asylum, 1886–1887. *History of Psychiatry* 28: 280–296.

Cunningham, Andrew. 2002–2003. The Pen and the Sword: Recovering the Disciplinary Identity of Physiology and Anatomy before 1800: I&II. *Studies in History and Philosophy of Biological and Biomedical Sciences* 33 & 34: 631–665 and 51–76.

———. 2010. *The Anatomist Anatomis'd: An Experimental Discipline in Enlightenment Europe*. Farnham: Ashgate.

Cunningham, Andrew, and Perry Williams, eds. 1992. *The Laboratory Revolution in Medicine*. Cambridge: Cambridge University Press.

Daston, Lorraine, and Peter Galison. 2007. *Objectivity*. New York: Zone Books.

Deblon, Veronique. 2014. Een nieuw beroep en een onverstoorbare ziel: de rol van de anatomie in de gezondheidszorg rond 1800. In *Vesalius: het lichaam in beeld*, ed. Geert Vanpaemel, 86–93. Leuven: Davidsfonds.

———. 2015. Het zachte gedruis van het leven. De preparaten van Adolphe Burggraeve. In *Post Mortem. Vesalius tussen kunst en wetenschap*, ed. Marjan Doom, 63–65. Ghent: Universiteit Gent.

———. 2017. Imitating Anatomy: Recycling Anatomical Illustrations in Nineteenth-Century Atlases. In *Bodies Beyond Borders: Moving Anatomies, 1750–1950*, eds. Kaat Wils, Raf de Bont, and Au Sokhieng, 115–137. Leuven: Leuven University Press.

———. 2020. Anatomie gereanimeerd. PhD diss., KU Leuven.

Deblon, Veronique, and Pieter Huistra. 2016. Het geheim van de anatoom: de ontwikkeling van de Belgische anatomie in de negentiende eeuw. *Studium: Tijdschrift voor Wetenschaps-en Universiteitsgeschiedenis/Revue de l'Histoire des Sciences et des Universités* 9: 202–216.

Deblon, Veronique, and Kaat Wils. 2017. Overcoming Death: Conserving the Dead Body in the Nineteenth Century. In *When is Death? Interdisciplinary Perspectives on Death and Its Timings*, ed. Shane McCorristine, 49–67. London: Palgrave Macmillan.

De Bont, Raf. 2003. Onbeschaamde geleerden hebben zijn naaktheid betast. Genialiteit, waanzin en degeneratie omstreeks 1900. In *Degeneratie in België 1860–1940: Een geschiedenis van ideeën en praktijken*, eds. Jo Tollebeek, Geert Vanpaemel, and Kaat Wils, 121–154. Leuven: Leuven University Press.

———. 2008. *Darwins kleinkinderen: de evolutietheorie in België 1865–1945*. Nijmegen: Uitgeverij Vantilt.

Deferme, Jo. 2007. *Uit de ketens van de vrijheid. Het debat over de sociale politiek in België, 1886–1914*. Leuven: Leuven University Press.

De Ganck, Julie. 2014. De verzorging van het vrouwelijk geslacht: een maatschappelijke kwestie? *Historica* 16: 9–16.

———. 2016. Cultiver la différence: Histoire du développement de la gynécologie à Bruxelles, 1870–1935. PhD diss., Université Libre de Bruxelles.

De Maeyer, Jan, and Jo Deferme. 2008. Vrouwelijke religieuzen in de openbare en private gezondheidszorg in het België van de negentiende en twintigste eeuw: tussen traditie en moderniteit. In *Bezielde zorg. Verpleging door katholieke religieuzen in Nederland en Vlaanderen (negentiende—twintigste eeuw)*, eds. Liesbeth Labbeke, Vefie Poels, and Rob Wolf, 10–28. Hilversum: Uitgeverij Verloren.

Deneckere, Gita. 1998. *Geuzengeweld. Antiklerikaal straatrumoer in de politieke geschiedenis van België 1831–1914*. Brussels: VUB Press.

———. 2006. *1900: België op het breukvlak van twee eeuwen*. Tielt: Lannoo.

De Raes, Wouter. 2002. Roger Soenen: een Vlaams-nationalistische rassentheoreticus (1902–1977). *Wetenschappelijke Tijdingen* 61: 79–86.

De Rooy, Laurens. 2011. *Snijburcht: Lodewijk Bolk en de bloei van de Nederlandse anatomie*. Amsterdam: Amsterdam University Press.

De Spiegeleer, Christoph. 2010. Tussen banketten en begrafenissen: de radicaal-liberale burgerlijke cultuur rond Charles Potvin in Brussel tijdens de tweede helft van de 19e eeuw. *De Negentiende Eeuw* 34: 289–308.

———. 2015. Secularisering van stedelijke begraafplaatsen in de tweede helft van de negentiende eeuw in België. In *R.I.P. Aspecten van 200 jaar begrafeniscultuur in Vlaanderen*, ed. Tamara Ingels, 5–23. Ghent: Academia Press.

———. 2016. Sterven, begraven en herdenken van koninklijke en politieke elites in België tussen 1830 en 1940: een culturele en politieke geschiedenis. PhD diss., Vrije Universiteit Brussel.

De Spiegeleer, Christoph, and Jeffrey Tyssens. 2017. Secularising Funerary Culture in Nineteenth-Century Belgium: A Product of Political and Religious Controversy. *Death Studies* 41: 14–21.

Despy-Meyer, Andrée, Robert Halleux, Jan Vandersmissen, and Geert Vanpaemel, eds. 2001. *Geschiedenis van de wetenschappen in België, 1815–2000*. Brussels: Dexia/La Renaissance du livre.

Dhondt, Pieter. 2004. La situation précaire de l'enseignement supérieur dans les départements belges entre 1797 et 1815. *Revue belge de philologie et d'histoire/ Belgisch tijdschrift voor philologie en geschiedenis* 82: 935–967.

———. 2011. *Un double compromis: enjeux et débats relatifs à l'enseignement universitaire en Belgique au XIXe siècle*. Ghent: Academia Press.

Dickstein-Bernard, Claire. 1985–1986. Naissance des services spéciaux dans les hôpitaux belges au XIXe siècle: Réflexions sur le cas bruxellois. *Annales de la Société Belge d'Histoire des Hôpitaux et de la Santé publique/Annalen van de Belgische Vereniging voor de Geschiedenis van de Hospitalen en de Volksgezondheid* 23–24: 49–66.

Dierig, Sven, Jens Lachmund, and Andrew J. Mendelsohn. 2003. Introduction: Toward an Urban History of Science. *Osiris* 18: 1–19.

Digby, Anne. 1982. *The Poor Law in Nineteenth-Century England and Wales*. London: Historical Association.

Dittmar, Jenna M., and Mitchell Piers. 2016. From Cradle to Grave via the Dissection Room: The Role of Foetal and Infant Bodies in Anatomical Education from the Late 1700s to Early 1900s. *Journal of Anatomy* 229: 713–722.

Douglas, Mary. 1984. *Purity and Danger*. 2nd ed. London: Routledge.

Dracobly, Alex. 2003. Ethics and Experimentation on Human Subjects in Mid-Nineteenth-Century France: The Story of the 1859 Syphilis Experiments. *Bulletin of the History of Medicine* 77: 332–366.

Durant, Guy, et al. 2000. *Histoire de l'éthique médicale et infirmière: context socio-culturel et scientifique*. Montréal: Les presses de l'université de Montréal.

Dutton, Paul. 2002. *Origins of the French Welfare State: The Struggle for Social Reform in France, 1914–1947*. Cambridge: Cambridge University Press.

———. 2007. *Differential Diagnoses: A Comparative History of Health Care Problems and Solutions in the United States and France*. Ithaca and London: Cornell University Press.

Ebenstein, Joanna, and Colin Dickey, eds. 2014. *The Morbid Anatomy Anthology*. New York: Morbid Anatomy Press.

Elkeles, Barbara. 2004. The German Debate on Human Experimentation between 1880–1914. In *Twentieth Century Ethics of Human Subject Research*, eds. Volker Roelcke and Giovanni Maio, 18–33. Stuttgart: Franz Steiner Verlag.

Engstrom, Eric. 2003. *Clinical Psychiatry in Imperial Germany. A History of Psychiatric Practice*. Ithaca and London: Cornell University Press.

Eulner, Hans-Heinz. 1970. *Die Entwicklung der medizinischen Spezialfächer an den Universitäten des deutschen Sprachgebietes*. Stuttgart: Ferdinand Enke Verlag.

Everaert, Guido, et al. 1997. Het Anatomisch Instituut in het Bijlokehospitaal te Gent. *Stadsarcheologie* 11: 4–21.

Faden, Ruth R., and Tom L. Beauchamp. 1986. *A History and Theory of Informed Consent*. Oxford: Oxford University Press.

Fahlander, Fredrik, and Anna Kjellström, eds. 2010. *Making Sense of Things: Archaeologies of Sensory Perception*. Stockholm: Stockholm University Press.

Fennell, Phil. 1996. *Treatment without Consent. Law, Psychiatry and the Treatment of Mentally Disordered People since 1845*. London: Routledge.

Ferber, Sarah, and Sally Wilde, eds. 2012. *The Body Divided: Human Beings and Human 'Material' in Modern Medical History*. Farnham: Ashgate.

Foucault, Michel. 1973. *The Birth of the Clinic. An Archeology of Medical Perception*. Translated by A.M. Sheridan. New York: Routledge.

Fowler, Louise, and Natasha Powers. 2012. *Doctors, Dissection and Resurrection Men: Excavations in the 19th-Century Burial Ground of the London Hospital, 2006*. London: Museum of London Archaeology Service.

Frewer, Andreas, and Ulf Schmidt, eds. 2007. *Standards der Forschung. Historische Entwicklung und ethische Grundlagen klinischer Studien*. Frankfurt am Main: Peter Lang Verlag.

Galison, Peter, and Emily Thompson, eds. 1999. *The Architecture of Science*. Cambridge, MA: MIT Press.

Garment, Ann, Susan Lederer, Naomi Rogers, and Lisa Boult. 2007. Let the Dead Teach the Living: The Rise of Body Bequeathal in 20th-Century America. *Academic Medicine* 82: 1000–1005.

Geison, Gerald. 1978. *Michael Foster and the Cambridge School of Physiology: The Scientific Enterprise in Late Victorian Society*. Princeton: Princeton University Press.

Gelfand, Toby. 1972. The "Paris Manner" of Dissection: Student Anatomical Dissection in Early Eighteenth-Century Paris. *Bulletin of the History of Medicine* 46: 99–130.

Gieryn, Thomas. 1999a. *Cultural Boundaries of Science. Credibility on the Line*. Chicago: University of Chicago Press.

———. 1999b. Two Faces on Science: Building Identities for Molecular Biology and Biotechnology. In *The Architecture of Science*, ed. Peter Galison and Emily Thompson, 423–455. Cambridge, MA: MIT Press.

Gijbels, Jolien. 2018. Reassessing the Pauper Burial: The Disposal of Corpses in Nineteenth-Century Brussels. *Mortality: Promoting the Interdisciplinary Study of Death and Dying* 23: 184–198.

Godeau, Emmanuelle. 2007. *L'esprit de corps: Sexe et mort dans la formation des internes en médecine*. Paris: Maison des sciences de l'homme.

Goldberg, Daniel. 2011. Suffering and Death among Early American Roentgenologists: The Power of Remotely Anatomizing the Living Body. *Bulletin of the History of Medicine* 85: 1–28.

Greenhalgh, Trisha, and Brian Hurwitz, eds. 1998. *Narrative Based Medicine: Dialogue and Discourse in Clinical Practice*. London: Wiley.

Grodin, Michael. 1992. Historical Origins of the Nuremberg Code. In *The Nazi Doctors and the Nuremberg Code*, ed. George J. Annas and Michael A. Grodin, 121–144. Oxford: Oxford University Press.

Guerrini, Anita. 2004. Anatomists and Entrepreneurs in Early Eighteenth-Century London. *Journal of the History of Medicine and Allied Sciences* 59: 219–239.

Gunst, Petra, and Fien Danniau. 2015. De Vriese, Bertha (1877–1958). UGentMemorie. Accessed December 26, 2018. www.ugentmemorie.be/personen/de-vriese-bertha-1877-1958.

Haakonssen, Lisbeth. 1997. *Medicine and Morals in the Enlightenment: John Gregory, Thomas Percival and Benjamin Rush*. Amsterdam and Atlanta: Rodopi.

Hagner, Michael. 2003. Scientific Medicine. In *From Natural Philosophy to the Sciences: Writing the History of Nineteenth-Century Science*, ed. David Cahan, 49–87. Chicago: University of Chicago Press.

Hagner, Michael, and Hans-Jorg Rheinberger, eds. 1993. *Die Experimentalisierung des Lebens: Experimentalsysteme in den biologischen Wissenschaften 1850–1930*. Berlin: Akademie.

Hallam, Elizabeth. 2007. Anatomical Bodies and Materials of Memory. In *Death Rites and Rights*, ed. Belinda Brooks-Gordon et al., 279–298. Oxford: Hart Publishing.

———. 2012. Articulating Bones: An Epilogue. *Journal of Material Culture* 17: 465–492.

———. 2016. *Anatomy Museum: Death and the Body Displayed*. London: Reaktion Books.

Halleux, Robert. 2001. Naar de kern van het leven: de biologie. In *Geschiedenis van de wetenschappen in België, 1815–2000*, eds. Andrée Despy-Meyer, Robert Halleux, Jan Vandersmissen, and Geert Vanpaemel, vol. 1, 289–304. Brussels: Dexia/La Renaissance du livre.

Hamilakis, Yannis, Marc Pluciennik, and Sarah Tarlow, eds. 2002. *Thinking Through the Body: Archeologies of Corporeality*. New York: Kluwer Academic and Plenum Publishers.

Hammerborg, Morten. 2011. The Laboratory and the Clinic Revisited: The Introduction of Laboratory Medicine into the Bergen General Hospital, Norway. *Social History of Medicine* 24: 758–775.

Hardcastle, Rohan. 2007. *Law and the Human Body: Property Rights, Ownership and Control*. Oxford and Portland: Hart Publishing.

Harvey, Karen, ed. 2009. *History and Material Culture: A Student's Guide to Approaching Alternative Sources*. London: Routledge.

Havelange, Carl. 1987. L'hôpital à la croisée des chemins: la question des malades payants. *Annales belges d'Histoire des Hôpitaux et de la Santé Publique/Annalen van de Belgische Vereniging voor de Geschiedenis van de Hospitalen en de Volksgezondheid* 25: 83–94.

———. 1992. Rupture ou Continuité? La Création de la Faculté de Médecine de l'Université de Liège en 1817. In *Regards sur 175 ans de Science à l'Université de Liège, 1817–1992*, ed. Anne-Catherine Bernes, 42–52. Liège: Université de Liège.

Hecht, Jennifer. 1997. French Scientific Materialism and the Liturgy of Death: The Invention of a Secular Version of Catholic Last Rites (1876–1914). *French Historical Studies* 20: 703–735.

Hendriksen, Marieke. 2014. *Elegant Anatomy: The Eighteenth Century Leiden Anatomical Collections*. Leiden: Brill.

Hennock, Peter. 2007. *The Origin of the Welfare State in England and Germany, 1850–1914: Social Policies Compared*. Cambridge: Cambridge University Press.

Hertz, Robert. 2004. *Death and the Right Hand*. 2nd ed. Oxford: Routledge.

Herzig, Rebecca. 2005. *Suffering for Science: Reason and Sacrifice in Modern America*. New Brunswick: Rutgers University Press.

Hess, Volker. 2000. Raum und Disziplin: Klinische Wissenschaft im Krankenhaus. *Berichte zur Wissenschafts-Geschichte* 23: 317–329.

Hess, Volker, and Andrew Mendelsohn. 2010. Case and Series: Medical Knowledge and Paper Technology, 1600–1900. *History of Science* 48: 287–314.

Hess, Volker, and Sophie Ledebur. 2001. Taking and Keeping: A Note on the Emergence and Function of Hospital Patient Records. *Journal of the Society of Archivists* 32: 21–32.

Hetherington, Kevin. 2004. Secondhandness: Consumption, Disposal and Absent Presence. *Environment and Planning D: Society and Space* 22: 157–173.

Hildebrandt, Sabine. 2016. Insights into the Freiburg Anatomical Institute during National Socialism, 1933–1945. *Annals of Anatomy* 205: 90–102.

Hopwood, Nick. 1999. Giving Body to Embryos: Modeling, Mechanism and the Microtome in Late Nineteenth-Century Anatomy. *Isis* 90: 462–496.

———. 2000. Producing Development: The Anatomy of Human Embryos and the Norms of Wilhelm His. *Bulletin of the History of Medicine* 74: 29–79.

———. 2007. Artist versus Anatomist, Models against Dissection: Paul Zeiller of Munich and the Revolution of 1848. *Medical History* 51: 279–308.

———. 2009. Embryology. In *The Cambridge History of Science: The Modern Biological and Earth Sciences*, ed. Peter J. Bowler and John V. Pickstone, vol. 6, 285–315. Cambridge: Cambridge University Press.

Howell, Joel. 1996. *Technology in the Hospital: Transforming Patient Care in the Early Twentieth Century*. Baltimore and London: Johns Hopkins University Press.

Huisman, Frank, and John H. Warner, eds. 2004. *Locating Medical History: The Stories and their Meanings*. London and Baltimore: Johns Hopkins University Press.

Huistra, Hieke. 2010. Weg met pottenkijkers! Hoe het publiek verdween uit het Leids anatomisch kabinet. *De Negentiende Eeuw* 34: 193–208.

———. 2019. *The Afterlife of the Leiden Anatomical Collections: Hands On, Hands Off*. London and New York: Routledge.

Hurcombe, Linda. 2014. *Perishable Material Culture in Prehistory: Investigating the Missing Majority*. London: Routledge.

Hurren, Elizabeth T. 2007. *Protesting about Pauperism: Poverty, Politics and Poor Relief in Late-Victorian England, 1870–1900*. Woodbridge: Boydell Press.

———. 2008. Whose Body Is It Anyway? Trading the Dead Poor, Coroner's Disputes, and the Business of Anatomy at Oxford University, 1885–1929. *Bulletin of the History of Medicine* 82: 775–818.

———. 2012a. Abnormalities and Deformities: The Dissection and Interment of the Insane Poor, 1832–1929. *Journal of the History of Psychiatry* 23: 65–77.

———. 2012b. *Dying for Victorian Medicine: English Anatomy and its Trade in the Dead Poor c. 1834–1929*. Houndmills: Palgrave Macmillan.

———. 2015. World Without Welfare. Pauper Perspectives on Medical Care under the Late Victorian Poor Law 1870–1900. In *Obligation, Entitlement and Dispute under the English Poor Laws*, ed. Peter Jones and Steven King, 292–320. Newcastle upon Tyne: Cambridge Scholars.

———. 2016. *Dissecting the Criminal Corpse: Staging Post-execution Punishment in Early Modern England*. Basingstoke: Palgrave Macmillan.

———. 2018. Other Spaces for the Dangerous Dead of Provincial England, c. 1752–1832. *History* 103: 27–59.

Hurren, Elizabeth T., and Steve King. 2005. Begging for a Burial: Form, Function and Conflict in Nineteenth-Century Pauper Burial. *Social History* 30: 321–341.

Hutton, Fiona. 2013. *The Study of Anatomy in Britain, 1700–1900*. London: Pickering & Chatto.

Ingels, Tamara, ed. 2015. *R.I.P. Aspecten van 200 jaar begrafeniscultuur in Vlaanderen*. Ghent: Academia Press.

Jacyna, Stephen. 1988. The Laboratory and the Clinic: The Impact of Pathology on Surgical Diagnosis in the Glasgow Western Infirmary, 1875–1910. *Bulletin of the History of Medicine* 62: 384–406.

Jenkins, Tiffany. 2011. *Contesting Human Remains in Museum Collections: The Crisis of Cultural Authority*. New York: Routledge.

Jewson, Nicholas. 1976. The Disappearance of the Sick-Man from Medical Cosmology, 1770–1870. *Sociology* 10: 225–244.

Jones, Peter, and Steven King, eds. 2015. *Obligation, Entitlement and Dispute under the English Poor Laws*. Newcastle upon Tyne: Cambridge Scholars.

Jones, Colin, and Roy Porter, eds. 1998. *Reassessing Foucault: Power, Medicine and the Body*. London and New York: Routledge.

Jones, Gareth D., and Maja Whitaker. 2013. The Contested Realm of Displaying Dead Bodies. *Journal of Medical Ethics* 39: 652–653.

Katz, Jay. 1984. *The Silent World of Doctor and Patient*. New York: Macmillan.

Keel, Othmar. 2002. *L'avènement de la médecine clinique moderne en Europe, 1750–1815: politiques, institutions et saviors*. Montréal: Géorg éditeur et presses de l'université de Montréal.

Kelly, Laura. 2011. Anatomical Dissections and Student Experience at Irish Universities c.1900s-1960s. *Studies in History and Philosophy of Biological and Biomedical Sciences* 42: 467–474.

Keymolen, Denise. 1975. Feminisme in België. De eerste vrouwelijke artsen (1873–1914). *BMGN Low Countries Historical Review* 90: 38–58.

King, Peter. 2006. *Crime and Law in England, 1750–1840: Remaking Justice from the Margins*. Cambridge: Cambridge University Press.

King, Steven. 2018. *Sickness, Medical Welfare and the English Poor, 1750–1834*. Manchester: Manchester University Press.

Klestinec, Cynthia. 2011. *Theatres of Anatomy: Students, Teachers, and Traditions of Dissection in Renaissance Venice*. Baltimore: Johns Hopkins University Press.

Knoeff, Rina. 2015. Touching Anatomy: On the Handling of Preparations in the Anatomical Cabinets of Frederik Ruysch 1638–1731. *Studies in History and Philosophy of Biological and Biomedical Sciences* 49: 32–44.

Knoeff, Rina, and Rob Zwijnenberg, eds. 2015. *The Fate of Anatomical Collections*. New York: Ashgate.

Kraft, Alison, and Samuel J.M.M. Alberti. 2003. Equal though Different. Laboratories, Museums and the Institutional Development of Biology in Late-Victorian Northern England. *Studies in History and Philosophy of Biological and Biomedical Sciences* 34: 203–236.

Kremer, Richard. 2009. Physiology. In *The Cambridge History of Science: The Modern Biological and Earth Sciences*, eds. Peter J. Bowler and John V. Pickstone, vol. 6, 342–366. Cambridge: Cambridge University Press.

Krmpotich, Cara, Joos Fontein, and John Harries. 2010. The Substance of Bones: The Emotive Materiality and Affective Presence of Human Remains. *Journal of Material Culture* 15: 371–384.

Kselman, Thomas. 1988. Funeral Conflicts in Nineteenth-Century France. *Comparative Studies in Society and History* 30: 312–332.

———. 1993. *Death and Afterlife in Modern France*. Princeton: Princeton University Press.

Labbeke, Liesbeth, Veffie Poels, and Rob Wolf, eds. 2008. *Bezielde zorg. Verpleging door katholieke religieuzen in Nederland en Vlaanderen (negentiende—twintigste eeuw)*. Hilversum: Uitgeverij Verloren.

Labisch, Alfons. 1997. From Traditional Individualism to Collective Professionalism: State, Patient, Compulsory Health Insurance and the Panel Doctor Question in Germany 1883–1931. In *Medicine and Modernity: Public Health and Medical Care in Nineteenth- and Twentieth-Century Germany*, eds. Manfred Berg and Geoffrey Cocks, 18–34. Cambridge: Cambridge University Press.

Lambert, Helen, and Maryon McDonald, eds. 2011. *Social Bodies*. Oxford and New York: Berghahn.

Langendries, Elienne, and Anne-Marie Van der Meersch. 1999. *Het Rommelaere complex: onderdeel van het gebouwenmasterplan voor de Gentse universiteit op het einde van de 19de eeuw*. Ghent: RUG Archief.

Laqueur, Thomas. 1983. Bodies, Death, and Pauper Funerals. *Representations* 1: 109–131.

———. 2015. *The Work of the Dead: A Cultural History of Mortal Remains*. Princeton and Oxford: Princeton University Press.

Lawrence, Christopher. 1985. Incommunicable Knowledge: Science, Technology and the Clinical Art in Britain 1850–1914. *Journal of Contemporary History* 20: 503–520.

Lawrence, Susan. 1998. Beyond the Grave. The Use and Meaning of Human Body Parts: A Historical Introduction. In *Stored Tissue Samples: Ethical, Legal and Public Policy Implications*, ed. Robert F. Weir, 111–142. Iowa: University of Iowa Press.

———. 2009. Anatomy, Histology and Cytology. In *The Cambridge History of Science: The Modern Biological and Earth Sciences*, ed. Peter J. Bowler and John V. Pickstone, vol. 6, 265–284. Cambridge: Cambridge University Press.

Leclerq, Valérie. 2017. Guérir, travailler, désobéir. Une histoire des interactions hospitalières avant l'ère du 'patient autonome (Bruxelles 1870–1930). PhD diss., Université Libre de Bruxelles.

Lederer, Susan. 1995. *Subjected to Science: Human Experimentation in America Before the Second World War*. Baltimore and London: Johns Hopkins University Press.

———. 2015. Bodies for Science. Paper presented at the *Bodies Beyond Borders Conference*, Leuven.

Lesch, John. 1984. *Science and Medicine in France: The Emergence of Experimental Physiology, 1790–1855*. Cambridge, MA: Harvard University Press.

Leusen, Isidoor. 2000. *125 jaar fysiologie in de faculteit geneeskunde van de universiteit te Gent, 1817–1942*. Ghent: Archief RUG.

Lindskoug, Henrik B., and Anne Gustavsson. 2015. Stories from Below: Human Remains at the Gothenburg Museum of Natural History and the Museum of World Culture. *Journal of the History of Collections* 27: 97–109.

Lis, Catharina, Hugo Soly, and Dirk Van Damme. 1985. *Op vrije voeten? Sociale politiek in West-Europa, 1450–1914*. Leuven: Kritak Uitgeverij.

Louryan, Stéphane. 2008. Un portrait des enseignants d'anatomie humaine à l'Université libre de Bruxelles entre 1834 et 1905. *Revue médicale de Bruxelles* 29: 63–69.

———. 2012. L'encadrement des travaux pratiques d'anatomie entre 1834 et 1940. *Revue médicale de Bruxelles* 33: 117–124.

Louryan, Stéphane, and Paul Kinnaert, eds. 2009. *Le Pôle Santé de l'ULB: Histoire de lieux, de personnages, de découvertes*. Brussels: Memogrames.

Louryan, Stéphane, and Nathalie Vanmuylder. 2009. L'Institut d'Anatomie Raoul Warocqué de l'ULB (1893–1928). In *Le Pôle Santé de l'ULB: Histoire de lieux, de personnages, de découvertes*, eds. Stéphane Louryan and Paul Kinnaert, 107–124. Brussels: Memogrames.

Lucey, Donnacha. 2015. *The End of the Irish Poor Law? Welfare and Healthcare Reform in Revolutionary and Independent Ireland*. Manchester: Manchester University Press.

MacDonald, Helen. 2006. *Human Remains: Dissection and Its Histories*. 2nd ed. New Haven and London: Yale University Press.

———. 2007. A Scandalous Act: Regulating Anatomy in a British Settler Colony, Tasmania 1869. *Social History of Medicine* 20: 39–56.

———. 2010. *Possessing the Dead. The Artful Science of Anatomy*. Melbourne: Melbourne University Press.

———. 2012. A Body Buried is a Body Wasted: The Spoils of Human Dissection. In *The Body Divided: Human Beings and Human 'Material' in Modern Medical History*, eds. Sarah Ferber and Sally Wilde, 9–27. Farnham: Ashgate.

MacKinnon, Dolley. 2011. Bodies of Evidence: Dissecting Madness in Colonial Victoria (Australia). In *The Body Divided: Human Beings and Human 'Material' in Modern Medical History*, eds. Sarah Ferber and Sally Wilde, 75–107. Farnham: Ashgate.

MacKinnon, Mary. 1987. English Poor Law Policy and the Crusade Against Outrelief. *The Journal of Economic History* 47: 603–625.

Maehle, Andreas-Holger. 2009. *Doctors, Honour and the Law. Medical Ethics in Imperial Germany*. Houndmills: Palgrave Macmillan.

Maehle, Andreas-Holger, and Johanna Geyer-Kordesch, eds. 2002. *Historical and Philosophical Perspectives on Biomedical Ethics: From Paternalism to Autonomy?* Farnham: Ashgate.

Maerker, Anna. 2006. The Anatomical Models of La Specola: Production, Uses and Reception. *Nuncius: Istituto e museo di storia della scienza* 21: 295–321.

———. 2011. *Model Experts: Wax Anatomies and Enlightenment in Florence and Vienna, 1775–1815*. Manchester: Manchester University Press.

Malatesta, Maria, ed. 2015. *Doctors and Patients. History, Representation, Communication from Antiquity to the Present*. San Francisco: University of California Medical Humanities Press.

Mandressi, Rafael. 2016. Affected Doctors: Dead Bodies and Affective and Professional Cultures in Early Modern European Anatomy. *Osiris* 31: 119–136.

Margócsy, Daniel. 2014. *Commercial Visions: Science, Trade and Visual Culture in the Dutch Golden Age*. Chicago: University of Chicago Press.

Martin, Susan. 2012. Dissection, Anatomy Acts and the Appropriation of Bodies in 19th Century Australia: "The Government's Brains" and the Benevolent Asylum. In *The Body Divided: Human Beings and Human 'Material' in Modern Medical History*, ed. Sarah Ferber and Sally Wilde, 53–74. Farnham: Ashgate.

Matossian, Chaké, ed. 2007. *Art, Anatomie: trois siècles d'évolution des répresentations du corps*. Brussels: Part de l'œil.

Maulitz, Russell. 2002. *Morbid Appearances: The Anatomy of Pathology in the Early Nineteenth Century*. Cambridge: Cambridge University Press.

———. 2009. Pathology. In *The Cambridge History of Science: The Modern Biological and Earth Sciences*, ed. Peter J. Bowler and John V. Pickstone, vol. 6, 367–381. Cambridge: Cambridge University Press.

McLeary, Erin. 2001. Science in a Bottle: The Medical Museum in North-America, 1860–1940. PhD diss., University of Pennsylvania.

Menenteau, Sandra. 2009. *L'autopsie judiciaire. Histoire d'une pratique ordinaire au 19e siècle*. Rennes: Presses universitaires de Rennes.

———. 2011. Stigmata of the Autopsy. Operative Liberties and Protocol in Forensic Examination of the Dead Body in Nineteenth-Century France. *Intertexts* 15: 20–38.

Meskell, Lynn. 2002. Negative Heritage and Past Mastering in Archaeology. *Anthropological Quarterly* 75: 557–574.

Meyer, Morgan. 2012. Placing and Tracing Absence: A Material Culture of the Immaterial. *Journal of Material Culture* 17: 103–110.

Michael, Pamela, and David Hirst. 2011. Recording the Many Faces of Death at the Denbigh Asylum, 1848–1938. *History of Psychiatry* 23: 40–51.

Morantz-Sanchez, Regina. 2000. Negotiating Power at the Bedside: Historical Perspectives on Nineteenth-Century Patients and their Gynaecologists. *Feminist Studies* 26: 287–309.

Morgan, Lynn. 2004. A Social Biography of Carnegie Embryo No. 836. *The Anatomical Record* 276: 3–7.

———. 2009. *Icons of Life: A Cultural History of Human Embryos*. Berkeley: University of California Press.

———. 2010. Properly Disposed Of. A History of Embryo Disposal and the Changing Claims on Foetal Remains. *Medical Anthropology* 21: 247–274.

Mulnard, Jacques. 1992. The Brussels School of Embryology. *International Journal of Developmental Biology* 36: 17–24.

————. 2009. Albert Dalcq. In *Le Pôle Santé de l'ULB: Histoire de lieux, de personnages, de découvertes*, eds. Stéphane Louryan and Paul Kinnaert, 149–158. Brussels: Memorages.

Nandrin, Jean-Pierre. 1990. Les libéraux et la genèse du droit social en Belgique. Peut-on parler d'un modèle paternaliste libéral? In *Liberalism and Paternalism in the 19th Century*, eds. Erik Aerts, Claude Beaud, and Jean Stengers, 94–103. Leuven: Leuven University Press.

Neville Bonner, Thomas. 1995. *Becoming a Physician: Medical Education in Great Britain, France, Germany and the United States, 1750–1945*. New York and Oxford: Oxford University Press.

Nuard, Pierre. 1970. *Sciences, médecine, pharmacie, de la révolution à l'empire*. Paris: Dacosta.

Nussbaum, Martha. 1995. Objectification. *Philosophy and Public Affairs* 24: 249–291.

Nyberg, Jenny. 2010. A Peaceful Sleep and Heavenly Celebration for the Pure and Innocent: The Sensory Experience of Death during the Long Eighteenth Century. In *Making Sense of Things: Archaeologies of Sensory Perception*, eds. Fredrik Fahlander and Anna Kjellström, 15–33. Stockholm: Stockholm University Press.

Nye, Robert. 2006. Médecins, éthique médicale et État en France 1789–1947. *Le Mouvement Social* 214: 19–36.

Nyhart, Lynn. 1995. *Biology Takes Form: Animal Morphology and the German Universities, 1800–1900*. Chicago: University of Chicago Press.

Nys, Liesbet. 2003. De ruiters van de apocalyps: alcoholisme, tuberculosis, syfilis en degeneratie in Belgische medische kringen, 1870–1940. In *Degeneratie in België 1860–1940: een geschiedenis van ideeën en praktijken*, eds. Jo Tollebeek, Geert Vanpaemel, and Kaat Wils, 11–42. Leuven: Leuven University Press.

————. 2012. *De intrede van het publiek. Museumbezoek in België, 1830–1914*. Leuven: Leuven University Press.

Olejaz, Maria. 2013. Anatomical Dissection Revisited: Exchanging the Dead Body in 18th Century Denmark. In *(aus)tauschen: Erkundungen Einer Praxisform*, eds. Sebastian Mohr, Lydia Maria Quart, and Vetter Andrea, 96–106. Berlin: Berliner Blätter.

Olesko, Kathryn. 1988. Commentary: On Institutes, Investigations and Scientific Training. In *The Investigative Enterprise: Experimental Physiology in Nineteenth-Century Medicine*, eds. William Coleman and Frederic L. Holmes, 295–332. Berkeley: University of California Press.

Onghena, Sofie. 2013. Spektakelstukken. De mise-en-scène van de wetenschap in de Belgische stad, 1860–1914. In *Tussen beleving en verbeelding. De stad in de negentiende-eeuwse literatuur*, ed. Inge Bertels et al., 43–69. Leuven: Leuven University Press.

Onuigbo, Wilson. 1962. The Paradox of Virchow's Views on Cancer Metastasis. *Bulletin of the History of Medicine* 36: 444–449.

O'Sullivan, Lisa, and Ross Jones. 2015. Two Australian Fetuses: Frederic Wood Jones and the Work of an Anatomical Specimen. *Bulletin of the History of Medicine* 89: 243–266.

Park, Katherine. 1995. The Life of the Corpse: Division and Dissection in Late Medieval Europe. *The Journal of the History of Medicine and Allied Sciences* 50: 111–132.

Peeters, Yvo. 1982. *Karel Buls (1837–1914), burgemeester op de raaklijn van twee kulturen.* Antwerp: De Nederlanden.

Pellizoni, Luigi. 2014. Construction, Co-production, and Beyond. Academic Disputes and Public Concerns in the Recent Debate on Nature and Society. *Sociology Compass* 8: 851–864.

Pernick, Martin. 1982. The Patient's Role in Medical Decision-Making: A Social History of Informed Consent in Medical Therapy. In *President's Commission for the Study of Ethical Problems in Medicine and Biomedical and Behavioural Research, Making Healthcare Decisions,* vol. 3, 1–35. Washington: The Commission.

Pickstone, John. 1985. *Medicine and Industrial Society: A History of Hospital Development in Manchester and Its Region 1752–1946.* Manchester: Manchester University Press.

———. 2001. *Ways of Knowing: A New History of Science, Technology and Medicine.* Manchester: University of Chicago Press.

Piechocki, Werner. 1965. Zur Leichenversorgung der halleschen Anatomie im 18. und 19. Jahrhundert. *Acta Historica Leopoldina* 2: 67–105.

Pirson, Chloé. 2009. *Corps à corps. Les modèles anatomiques entre art et médecine.* Paris: Mare et Martin.

Popu, Hélène. 2009. *La dépouille mortelle, chose sacrée. A la découverte d'une catégorie juridique oubliée.* Paris: L'Harmattan.

Powderly, Kathleen. 2000. Patient Consent and Negotiation in the Brooklyn Gynaecological Practice of Alexander J.C. Skene: 1863–1900. *Journal of Medicine and Philosophy* 25: 12–27.

Prüll, Cay-Rüdiger, ed. 1998. *Pathology in the 19th and 20th Centuries: The Relationship between Theory and Practice.* Sheffield: EAHMH Publications.

———. 2000. No Law, No Rights? Autopsy in Germany since 1800. In *Coping with Sickness: Medicine, Law and Human Rights—Historical Perspectives,* eds. John Woodward and Robert Jütte, 30–53. Sheffield: EAHMH Publications.

———. 2004. *Medizin am Toten oder am Lebenden? Pathologie in Berlin und in London, 1900–1945.* Basel: Schwabe Verlag.

Prüll, Cay-Rüdiger, and Marianne Sinn. 2002. Problems of Consent to Surgical Procedures and Autopsies in Twentieth-Century Germany. In *Historical and Philosophical Perspectives on Biomedical Ethics: From Paternalism to Autonomy?* ed. Andreas-Holger Maehle and Johanna Geyer-Kordesch, 73–93. Farnham: Ashgate.

Reinarz, Jonathan. 2005. The Age of Museum Medicine: The Rise and Fall of the Medical Museum at Birmingham's School of Medicine. *Social History of Medicine* 18: 419–437.

Richardson, Ruth. 1998. Organ Music. In *Narrative Based Medicine: Dialogue and Discourse in Clinical Practice*, eds. Trisha Greenhalgh and Brian Hurwitz, 266–272. London: Wiley.

———. 2000. A Necessary Inhumanity? *Medical Humanities* 26: 104–106.

———. 2001. *Death, Dissection and the Destitute*. 2nd ed. Chicago and London: University of Chicago Press.

———. 2003. Human Remains. In *Medicine Man: The Forgotten Museum of Henry Wellcome*, eds. Ken Arnold and Danielle Olsen, 319–345. London: British Museum Press.

Roelcke, Volker, and Giovanni Maio, eds. 2004. *Twentieth Century Ethics of Human Subject Research*. Stuttgart: Franz Steiner Verlag.

Rugg, Julie. 2013. Constructing the Grave: Competing Burial Ideals in Nineteenth-Century England. *Social History* 38: 328–345.

Ruisinger, Marion Maria. 2007. Geschichte des Humanexperiments. Zur Entwicklung der Forschung am Menschen. In *Standards der Forschung. Historische Entwicklung und ethische Grundlagen klinischer Studien*, ed. Andreas Frewer and Ulf Schmidt, 19–35. Frankfurt am Main: Peter Lang Verlag.

Sahmland, Irmtraut. 2008. Verordnete Körperspende: Das Hospital Haina als Bezugsquelle für Anatomieleichen (1786–1855). In *An der Wende zur Moderne. Die hessischen Hohen Hospitäler im 18. und 19. Jahrhundert*, ed. Friedrich Von Arnd, Christina Vanja, and Irmtraut Sahmland, 65–105. Petersberg: Michael Imhof Verlag.

Sappol, Michael. 2002. *A Traffic of Dead Bodies: Anatomy and Embodied Social Identity in Nineteenth-Century America*. Princeton and Oxford: Princeton University Press.

Schmiedebach, Heinz-Peter. 1993. Pathologie bei Virchow und Traube. Experimentalstrategien in unterschiedlichem Kontext. In *Die Experimentalisierung des Lebens: Experimentalsysteme in den biologischen Wissenschaften 1850–1930*, eds. Michael Hagner and Hans-Jorg Rheinberger, 116–134. Berlin: Akademie.

Schütz, Mathias. 2019. Erzwungener Wandel: Die Transformation der anatomischen Leichenbeschaffung in Bayern nach 1945. *Medizinhistorisches Journal* 54: 70–92.

Schwartz, Margaret. 2013. An Iconography of the Flesh: How Corpses Mean as Matter. *Communication +1* 2: 1–16.

Secord, James. 2004. Knowledge in Transit. *Isis* 95: 654–672.

Shapin, Steven. 2008. *The Scientific Life. A Moral History of a Late Modern Vocation*. Chicago and London: University of Chicago Press.

Simões, Ana, Maria Paula Diogo, and Kostas Gavrogly, eds. 2015. *Sciences in the Universities of Europe, Nineteenth and Twentieth Century: Academic Landscapes.* Dordrecht: Springer.

Smets, Marcel. 1995. *Charles Buls: Les principes de l'art urbain.* Liège: Editions Mardaga.

Smith, Laurajane. 2003. The Repatriation of Human Remains: Problem or Opportunity? *Antiquity* 78: 404–413.

Sofaer, Joanna. 2006. *The Body as Material Culture: A Theoretical Osteoarcheology.* Cambridge: Cambridge University Press.

Spickert, Stuart, et al., eds. 1988. *The Use of Human Beings in Research. With Special Reference to Clinical Trials.* Dordrecht: Springer.

Stockman, René. 2008. *Pro Deo. De geschiedenis van de christelijke gezondheidszorg.* Leuven: Davidsfonds.

Strange, Julie-Marie. 2000. Death and Dying: Old Themes and New Directions. *Journal of Contemporary History* 35: 491–499.

———. 2003. Only a Pauper Whom Nobody Owns: Reassessing the Pauper Grave c.1880–1914. *Past & Present* 178: 148–175.

———. 2005. *Death, Grief and Poverty in Britain 1870–1914.* Cambridge: Cambridge University Press.

Stukenbrock, Karin. 2001. *'Der zerstückte Cörper': Zur Sozialgeschichte der anatomischen Sektionen in der frühen Neuzeit 1650–1800.* Stuttgart: Franz Steiner Verlag.

Sturdy, Steve. 1992. The Political Economy of Scientific Medicine: Science, Education and the Transformation of Medical Practice in Sheffield, 1890–1922. *Medical History* 36: 125–159.

———. 2011. Looking for Trouble: Medical Science and Clinical Practice in the Historiography of Modern Medicine. *Social History of Medicine* 24: 739–757.

Tarlow, Sarah. 2002. The Aesthetic Corpse in Nineteenth-Century Britain. In *Thinking Through the Body: Archeologies of Corporeality*, ed. Yannis Hamilakis, Mark Pluciennik, and Sarah Tarlow, 85–95. New York: Kluwer Academic/Plenum Publishers.

Theriot, Nancy. 2001. Negotiating Illness: Doctors, Patients and Families in the Nineteenth Century. *Journal of the History of the Behavorial Sciences* 37: 349–368.

Tihon, André. 1976. Les religieuses en Belgique du XVIIIe siècle Approche statistique. *Belgisch Tijdschrift voor Nieuwste Geschiedenis* 7: 1–54.

Tollebeek, Jo, Geert Vanpaemel, and Kaat Wils, eds. 2003. *Degeneratie in België 1860–1940: Een geschiedenis van ideeën en praktijken.* Leuven: Leuven University Press.

Topp, Leslie. 2012. Complexity and Coherence: The Challenge of the Asylum Mortuary in Europe, 1898–1908. *Journal of the Society of Architectural Historians* 71: 8–41.

Troyer, John. 2007. Embalmed Vision. *Mortality: Promoting the Interdisciplinary Study of Death and Dying* 12: 22–47.

Van Damme, Dirk. 1985. Divergerende wegen van sociale beheersing. In *Op vrije voeten? Sociale politiek in West-Europa, 1450–1914*, eds. Catharina Lis, Hugo Soly, and Dirk Van Damme, 171–178. Leuven: Kritak Uitgeverij.

Vandendriessche, Joris. 2017. Anatomy and Sociability in Nineteenth-Century Belgium. In *Bodies Beyond Borders: Moving Anatomies, 1750–1950*, eds. Kaat Wils, Raf de Bont, and Au Sokhieng, 51–72. Leuven: Leuven University Press.

———. 2019. *Medical Societies and Scientific Culture in Nineteenth-Century Belgium*. Manchester: Manchester University Press.

Van de Perre, Stijn. 2008. Public Charity and Private Assistance in Nineteenth-Century Belgium. In *Armenfürsorge und Wohltätigkeit. Ländliche Gesallschaften in Europa, 1850–1930/Poor Relief and Charity. Rural Societies in Europe, 1850–1930*, ed. Inga Brandes and Katrin Marx-Jaskulski, 93–123. Frankfurt am Main: Peter Lang.

Van Hee, Bob, and Cornelis Van Tilburg, eds. 2014. *Heel-meesters. Befaamde artsen en figuren uit de geschiedenis van de geneeskunde*. Antwerp and Apeldoorn: Garant.

Van Meulder, Griet. 1997. Mutualiteiten en ziekteverzekering in België (1886–1914). *Belgisch Tijdschrift voor Nieuwste Geschiedenis* 27: 83–134.

Van Molle, Leen, ed. 2017. *Charity and Social Welfare. The Dynamics of Religious Reform in Northern Europe, 1780–1920*. Leuven: Leuven University Press.

Vanpaemel, Geert, ed. 2014. *Vesalius: het lichaam in beeld*. Leuven: Davidsfonds.

———. 2015. The German Model of Laboratory Science and the European Periphery (1860–1914). In *Sciences in the Universities of Europe, Nineteenth and Twentieth century: Academic Landscapes*, ed. Ana Simões, Maria Paula Diogo, and Kostas Gavrogly, 211–225. Dordrecht: Springer.

———. 2017. *Wetenschap als roeping: een geschiedenis van de Leuvense faculteit voor wetenschappen*. Leuven: Leuven University Press.

Vanpaemel, Geert, Marc Derez, and Jo Tollebeek, eds. 2012. *Album van een wetenschappelijke wereld/Album of a Scientific World*. Leuven: Leuven University Press.

Velle, Karel. 1992. *Begraven of cremeren: de crematiekwestie in België*. Ghent: Stichting mens en kultuur.

Vervliet, Arianne. 2013. Het juridische statuut van een lijk. MA diss., University of Antwerp.

Viré, Liliane. 1974. La "Cité scientifique" du Parc Léopold à Bruxelles, 1890–1920. *Cahiers Bruxellois: Revue de l'histoire urbaine* 19: 86–179.

Von Arnd, Friedrich, Christina Vanja, and Irmtraut Sahmland, eds. 2008. *An der Wende zur Moderne. Die hessischen Hohen Hospitäler im 18. und 19. Jahrhundert*. Petersberg: Michael Imhof Verlag.

Vovelle, Michelle. 1983. *La Mort et l'Occident. De 1300 à nos jours*. Paris: Gallimard.

Waddington, Ivan. 1973. The Role of the Hospital in the Development of Modern Medicine: A Sociological Analysis. *Sociology* 7: 211–224.

Waddington, Keir. 1998. Unsuitable Cases: The Debate over Outpatient Admissions, the Medical Profession and Late-Victorian London Hospitals. *Medical History* 42: 26–46.

———. 2002. Mayhem and Medical Students. Image, Conduct and Control in the Victorian and Edwardian London Teaching Hospital. *Social History of Medicine* 15: 45–64.

Wagener, Silke. 1995. …wenigstens im Tode der Welt noch nützlich und brauchbar. Die Göttinger Anatomie und ihre Leichen. *Göttinger Jahrbuch* 43: 63–90.

Warner, John Harley. 1998. *Against the Spirit of the System: The French Impulse in Nineteenth-Century American Medicine*. Baltimore and London: Princeton University Press.

Warner, John Harley, and James M. Edmonson. 2009. *Dissection: Photographs of a Rite of Passage in American Medicine, 1880–1930*. New York: Blast Books.

Warner, John Harley, and Lawrence J. Rizzolo. 2006. Anatomical Instruction and Training for Professionalism from the 19th to the 21st Centuries. *Clinical Anatomy* 19: 403–414.

Wartelle, Jean-Claude. 2004. La Société d'Anthropologie de Paris de 1859 à 1920. *Revue d'histoire des sciences humaines* 10: 125–171.

Weiner, Dora. 1993. *The Citizen-Patient in Revolutionary and Imperial Paris*. Baltimore and London: Johns Hopkins University Press.

Weiner, Dora, and Michael J. Sauter. 2003. The City of Paris and the Rise of Clinical Medicine. *Osiris* 18: 23–42.

Weir, Robert F., ed. 1998. *Stored Tissue Samples: Ethical, Legal and Public Policy Implications*. Iowa: University of Iowa Press.

Weisz, George. 2003. The Emergence of Medical Specialisation in the Nineteenth Century. *Bulletin of the History of Medicine* 77: 536–575.

———. 2006. *Divide and Conquer: A Comparative History of Medical Specialisation*. Oxford: Oxford University Press.

Wilde, Sally. 2009. Truth, Trust, and Confidence in Surgery, 1890–1910: Patient Autonomy, Communication and Consent. *Bulletin of the History of Medicine* 83: 302–330.

Williams, Karel. 1981. *From Pauperism to Poverty*. London: Routledge.

Wils, Kaat. 2012. Institut Vésale. In *Album van een wetenschappelijke wereld/ Album of a Scientific World*, eds. Geert Vanpaemel, Marc Derez, and Jo Tollebeek, 162–169. Leuven: Leuven University Press.

Wils, Kaat, Raf de Bont, and Au Sokhieng, eds. 2017. *Bodies Beyond Borders: Moving Anatomies, 1750–1950*. Leuven: Leuven University Press.

Withycombe, Shannon. 2015. From Women's Expectations to Scientific Specimens: The Fate of Miscarriage Materials in Nineteenth-Century America. *Social History of Medicine* 28: 245–262.

Witte, Els. 1993. The Battle for Monasteries, Cemeteries and Schools: Belgium. In *Culture Wars: Secular-Catholic Conflict in Nineteenth-Century Europe*, eds. Christopher Clark and Wolfram Kaiser, 102–128. Cambridge: Cambridge University Press.

Woodward, John, and Robbert Jütte, eds. 2000. *Coping with Sickness: Medicine, Law and Human Rights—Historical Perspectives*. Sheffield: EAHMH Publications.

Wright, David, Laurie Jacklin, and Tom Themeles. 2013. Dying to Get Out of the Asylum: Mortality and Madness in Four Mental Hospitals in Victorian Canada, c. 1841–1891. *Bulletin of the History of Medicine* 87: 591–621.

Index

© The Author(s) 2019

T. Claes, *Corpses in Belgian Anatomy, 1860–1914*, Medicine and Biomedical Sciences in Modern History,
https://doi.org/10.1007/978-3-030-20115-9

The manufacturer's authorised representative in the EU is Springer
Nature Customer Service Centre GmbH, Europaplatz 3, 69115 Heidelberg,
Germany. If you have any concerns regarding our products, please
contact ProductSafety@springernature.com

Printed and bound by CPI Group (UK) Ltd, Croydon, CR0 4YY
29/04/2026
02099450-0008